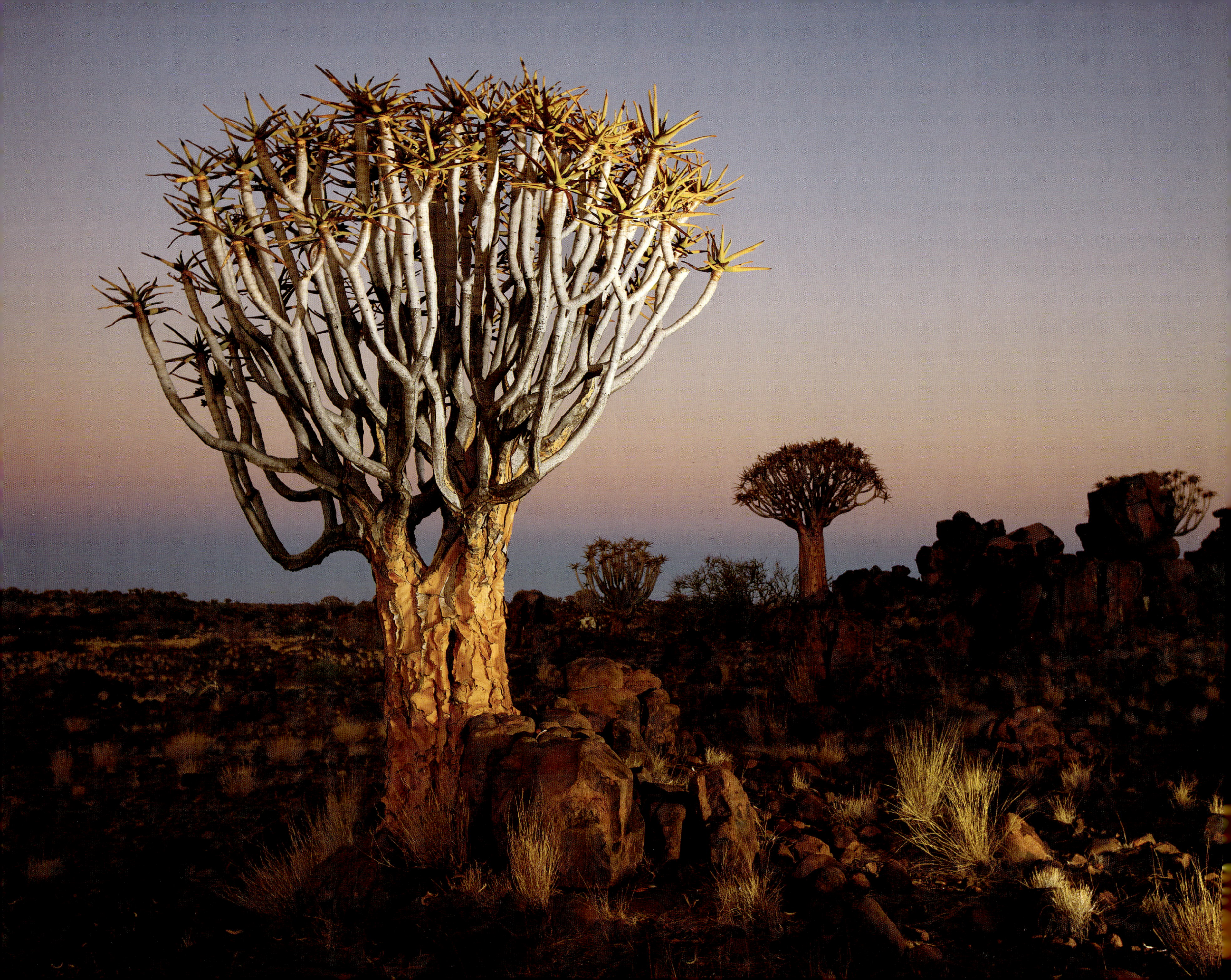

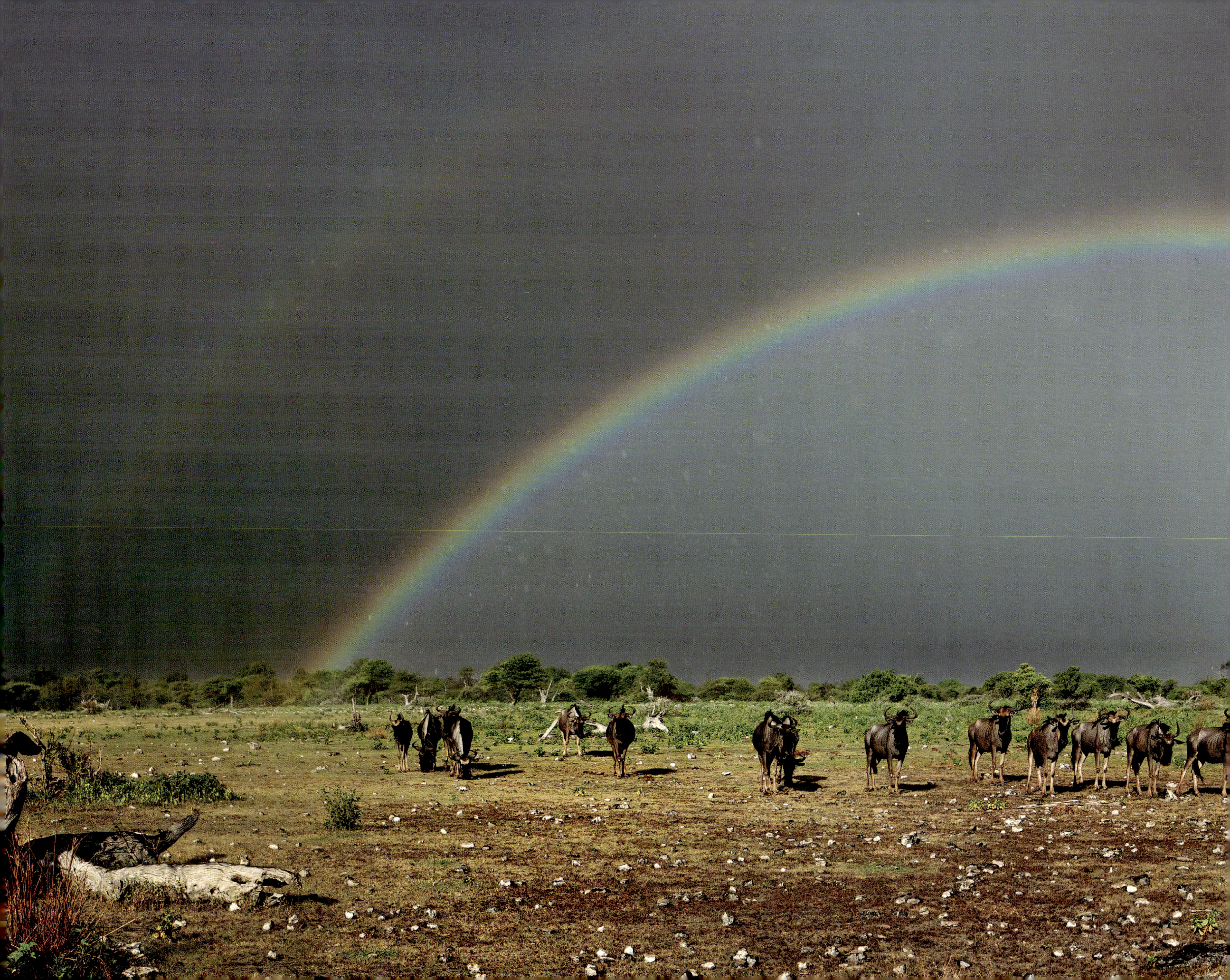

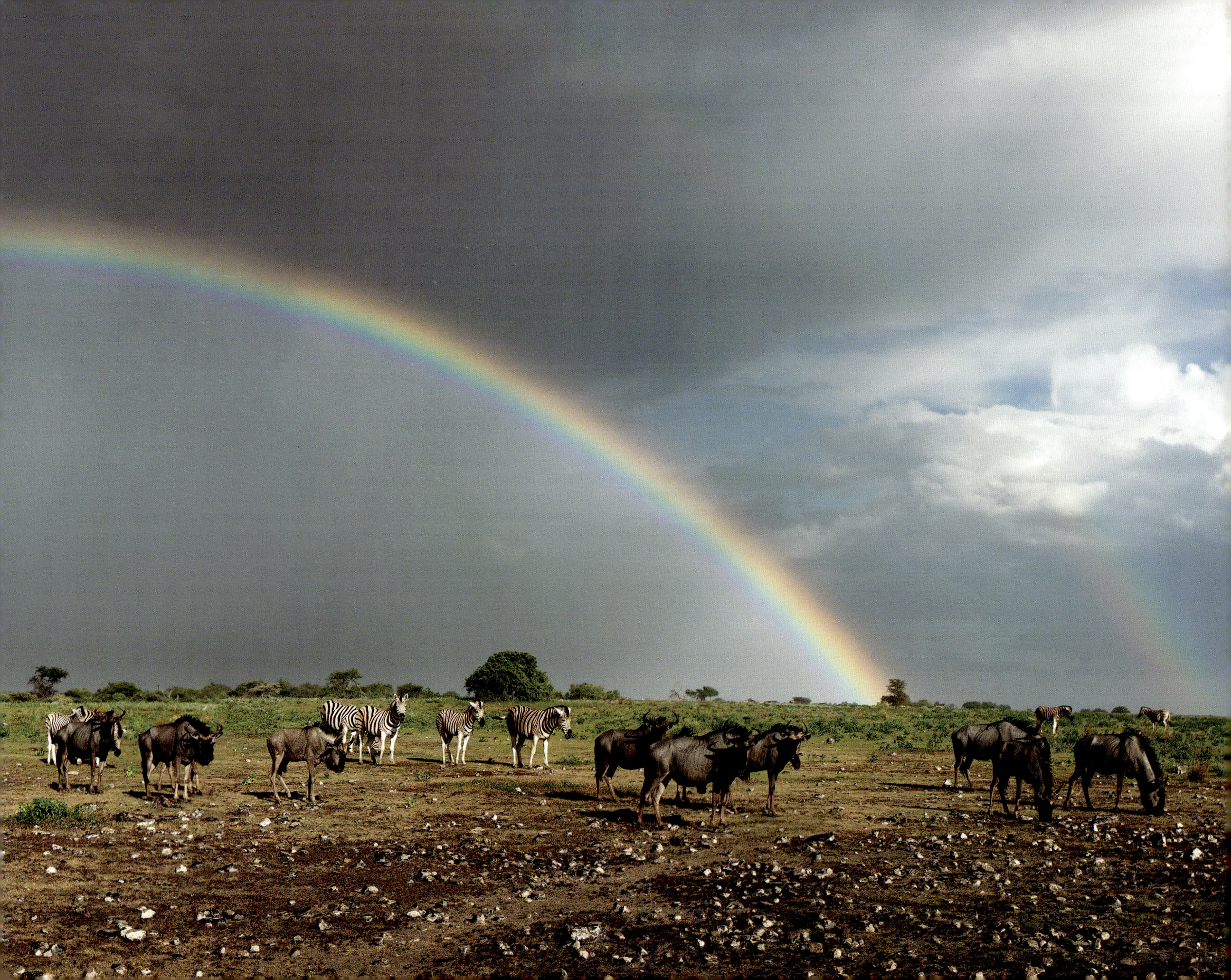

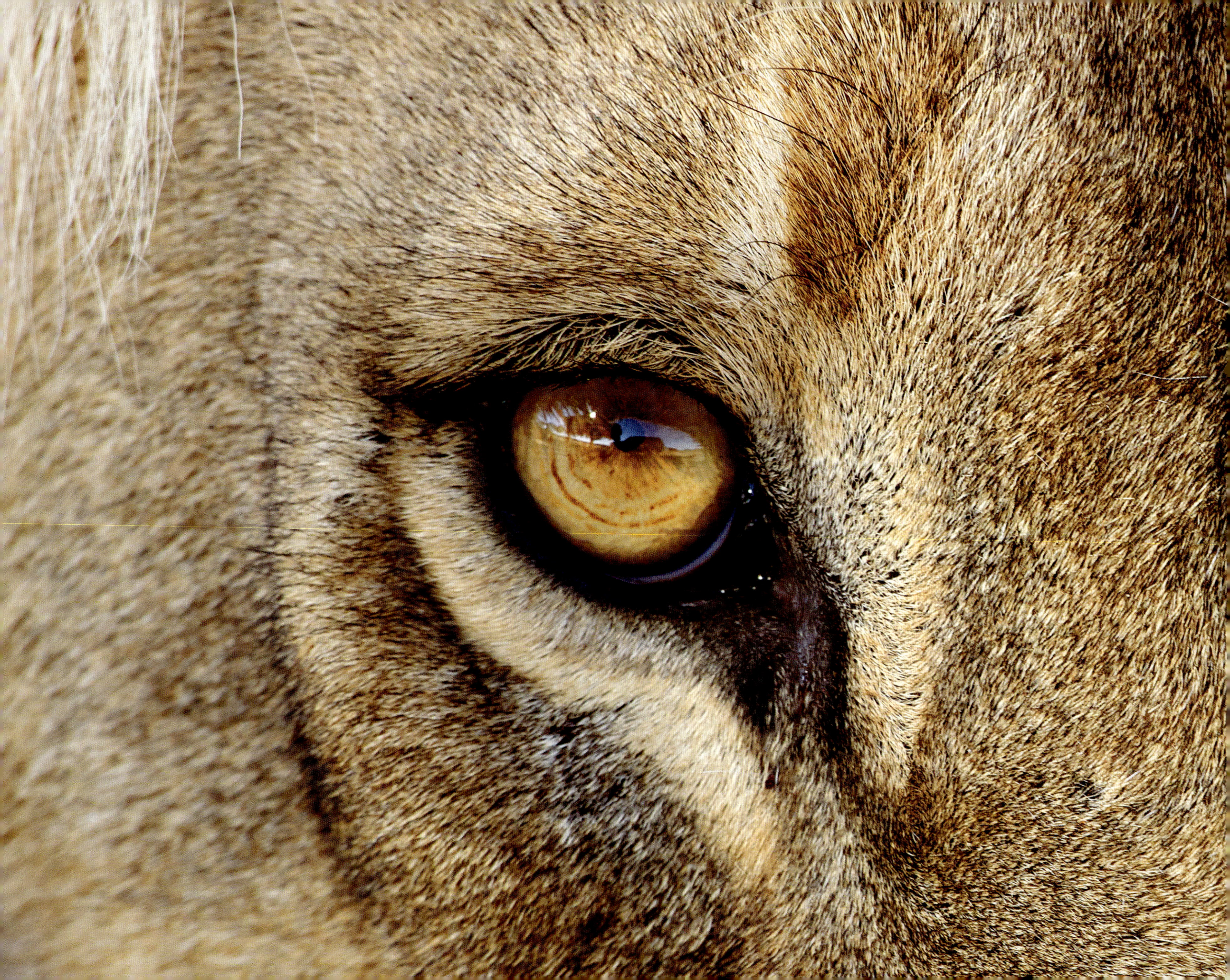

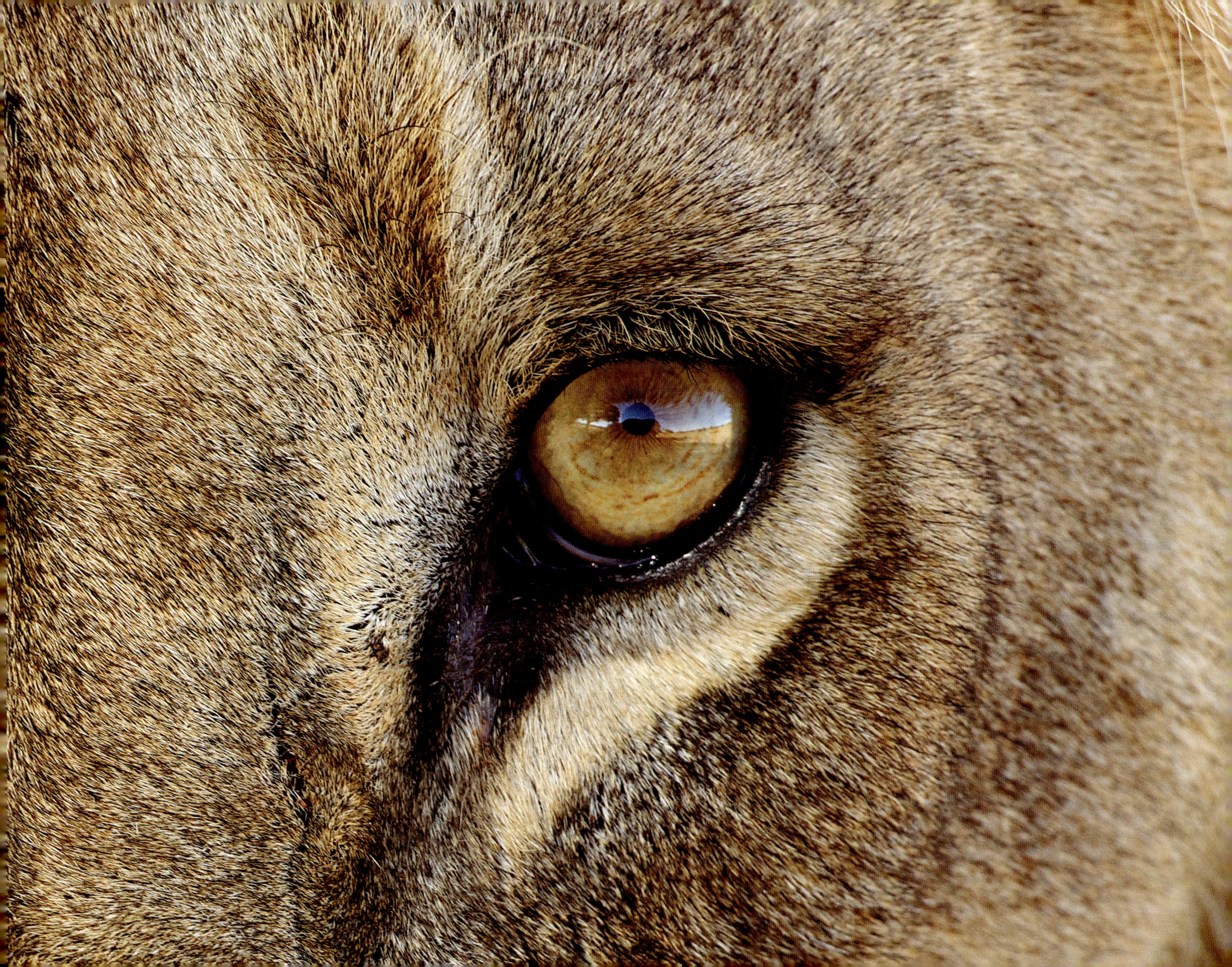

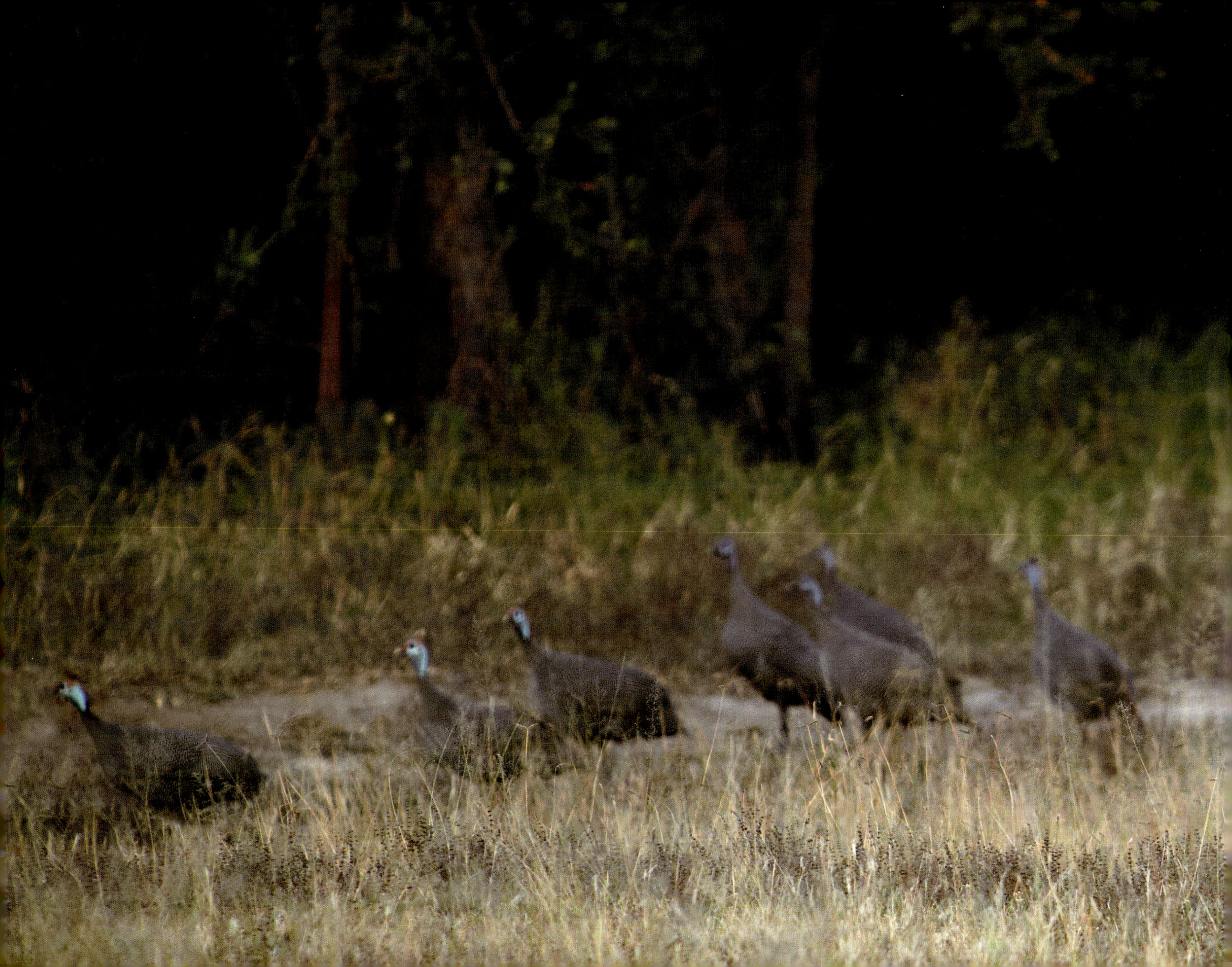

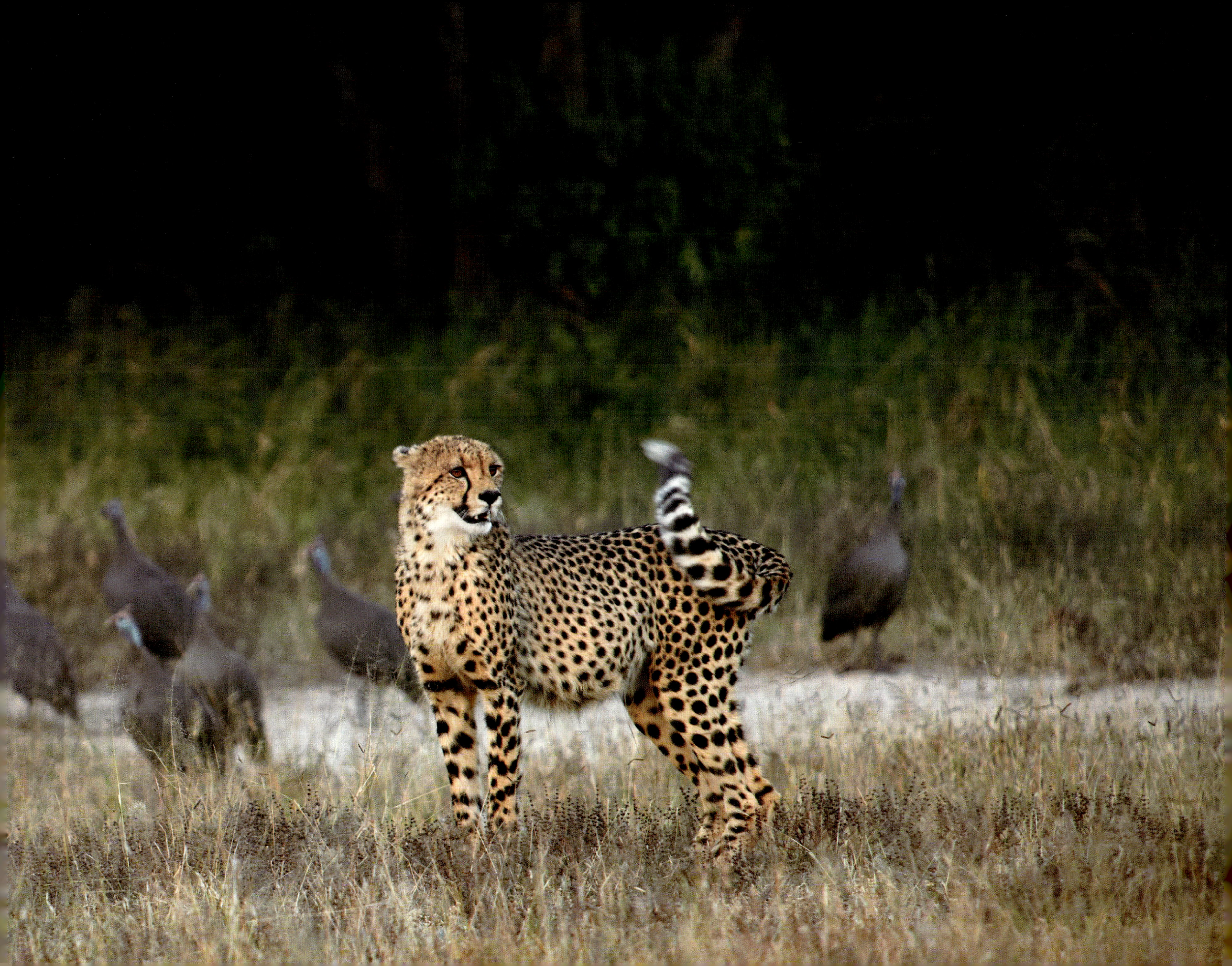

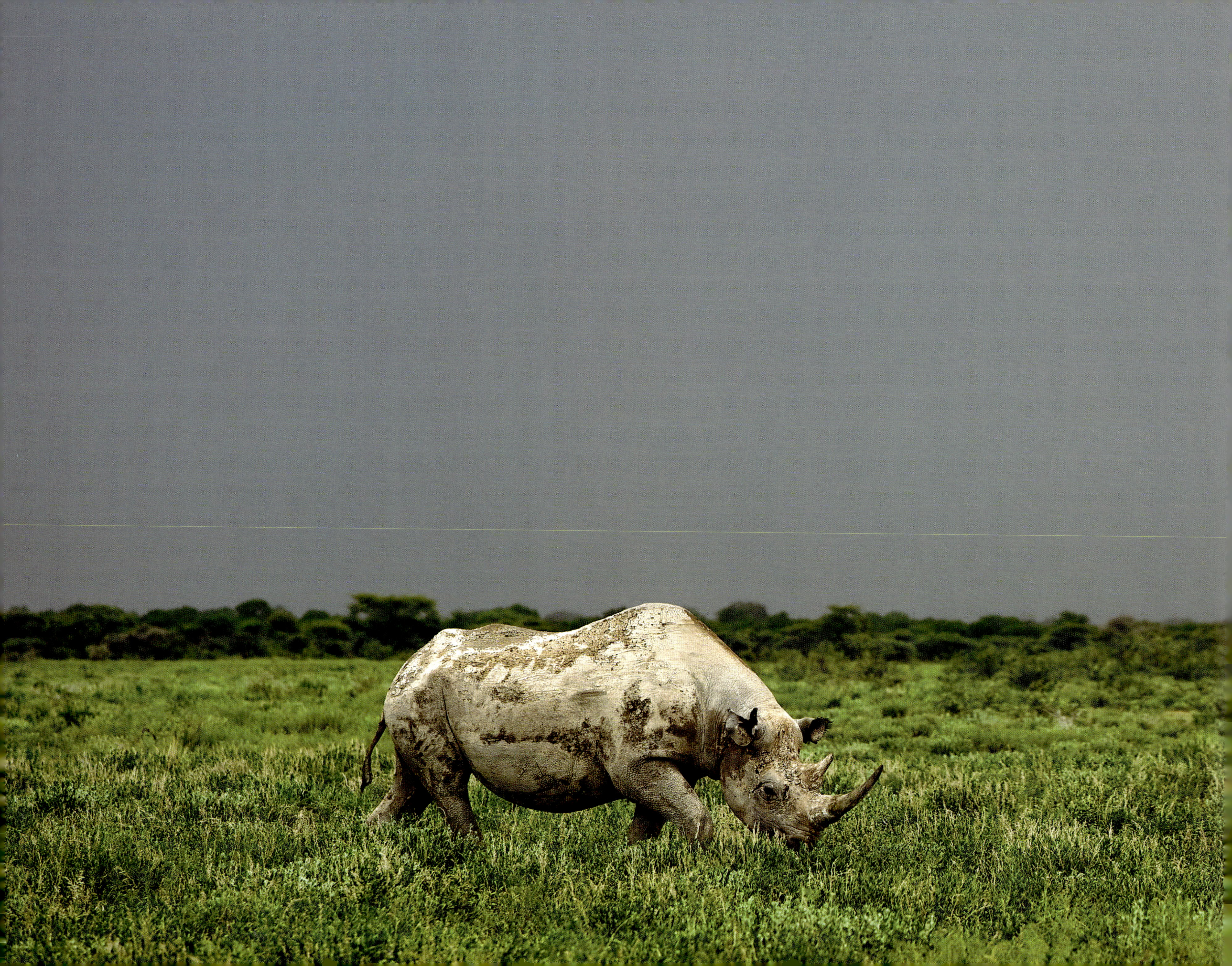

land *of* contrast

southern africa's diverse biomes

Photography by:

Heinrich & Dana van den Berg

Philip & Ingrid van den Berg

Text by:

Ingrid van den Berg

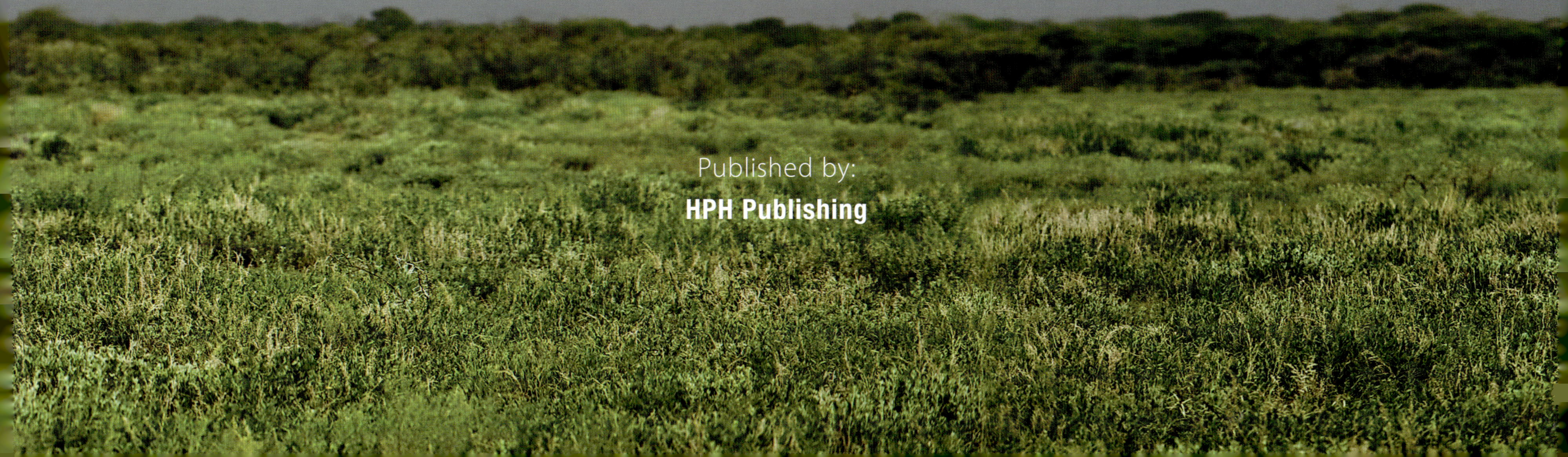

Published by:

HPH Publishing

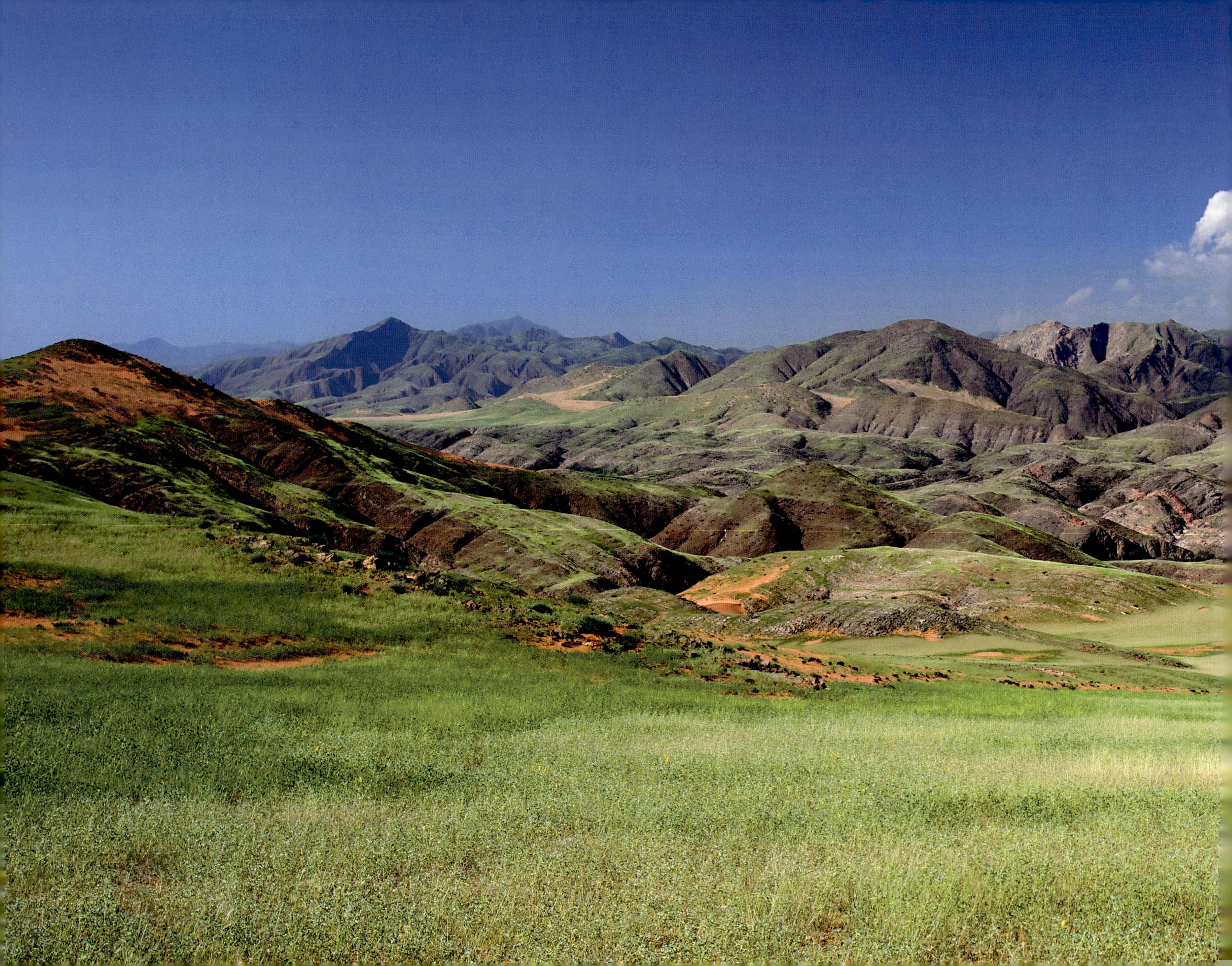

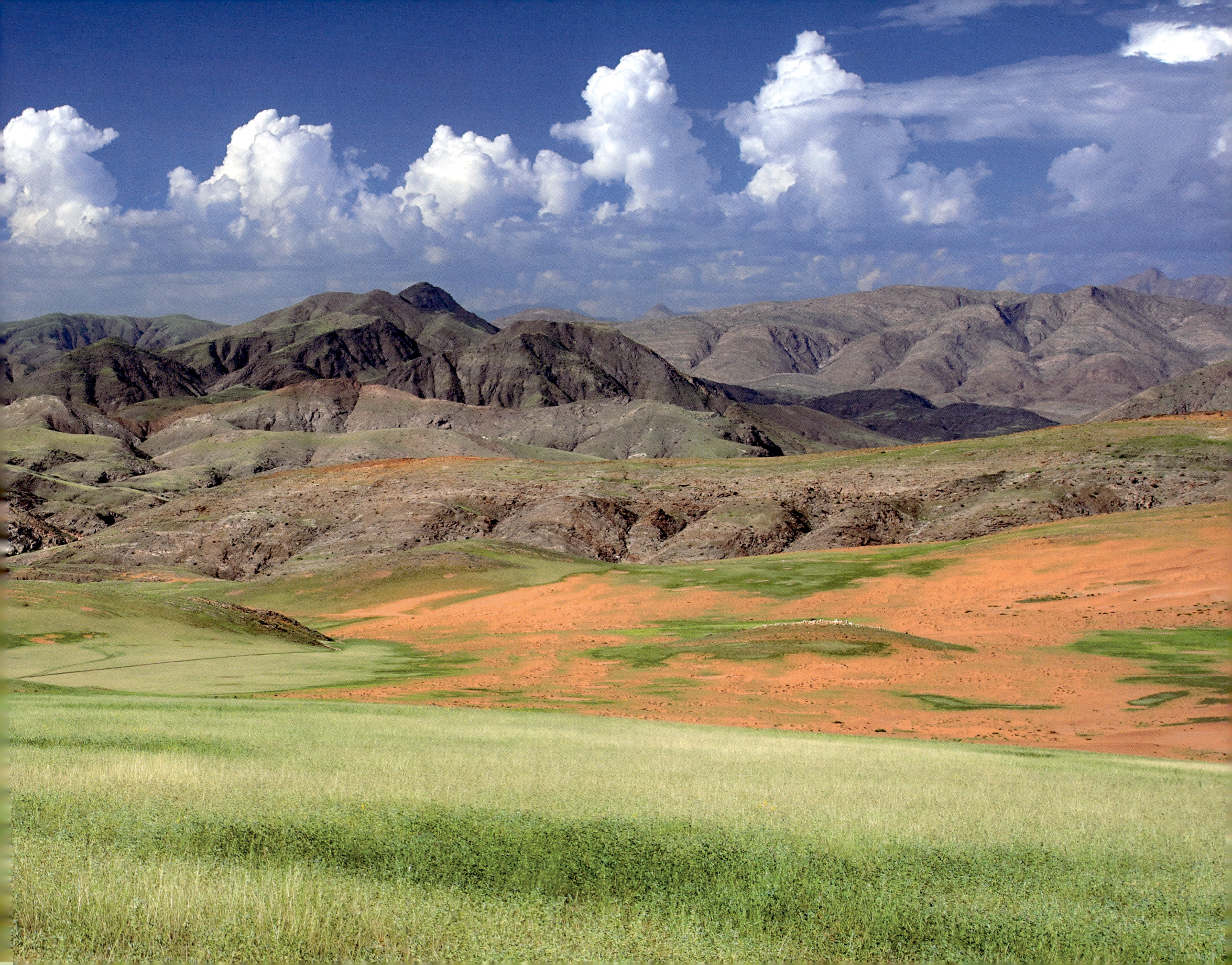

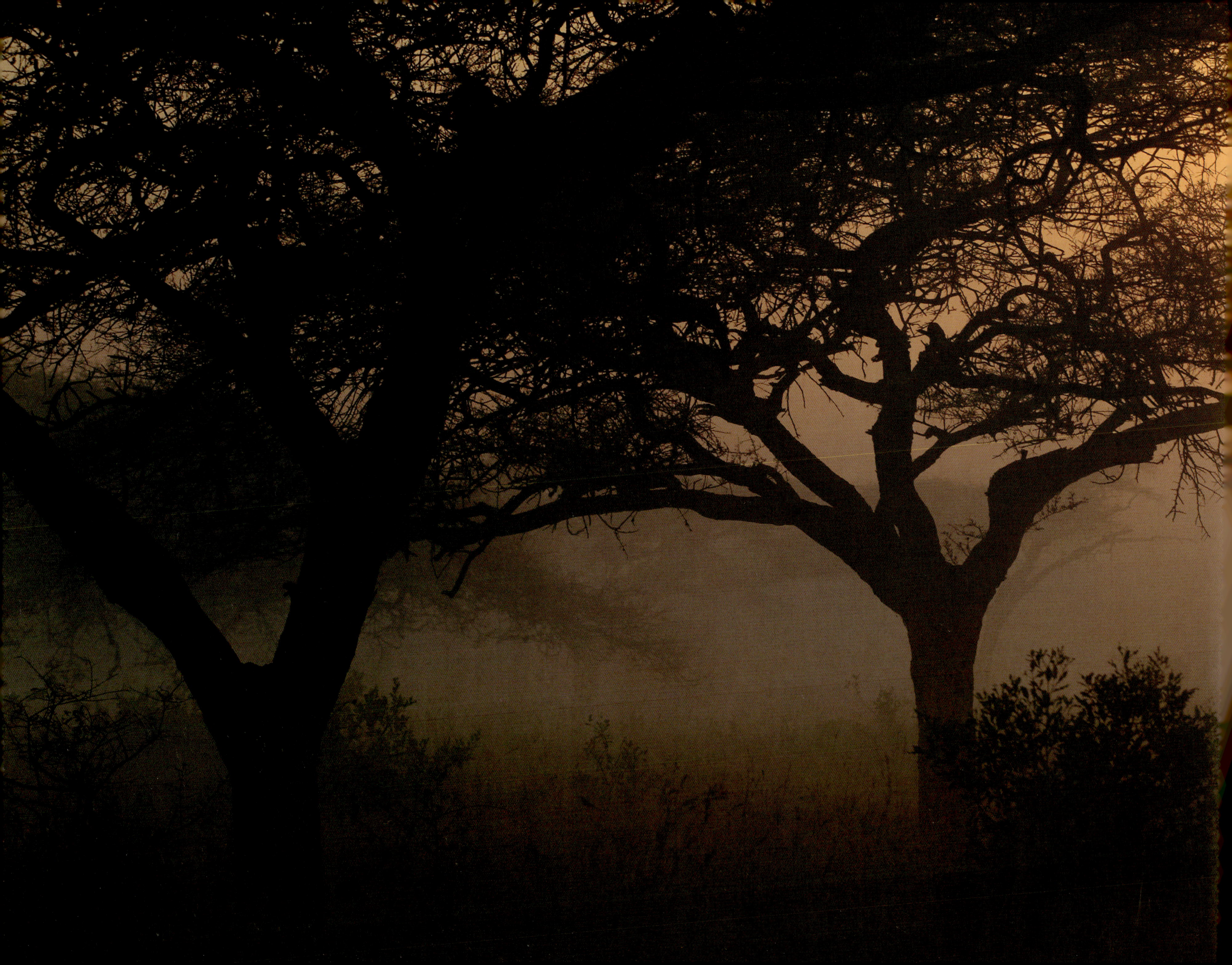

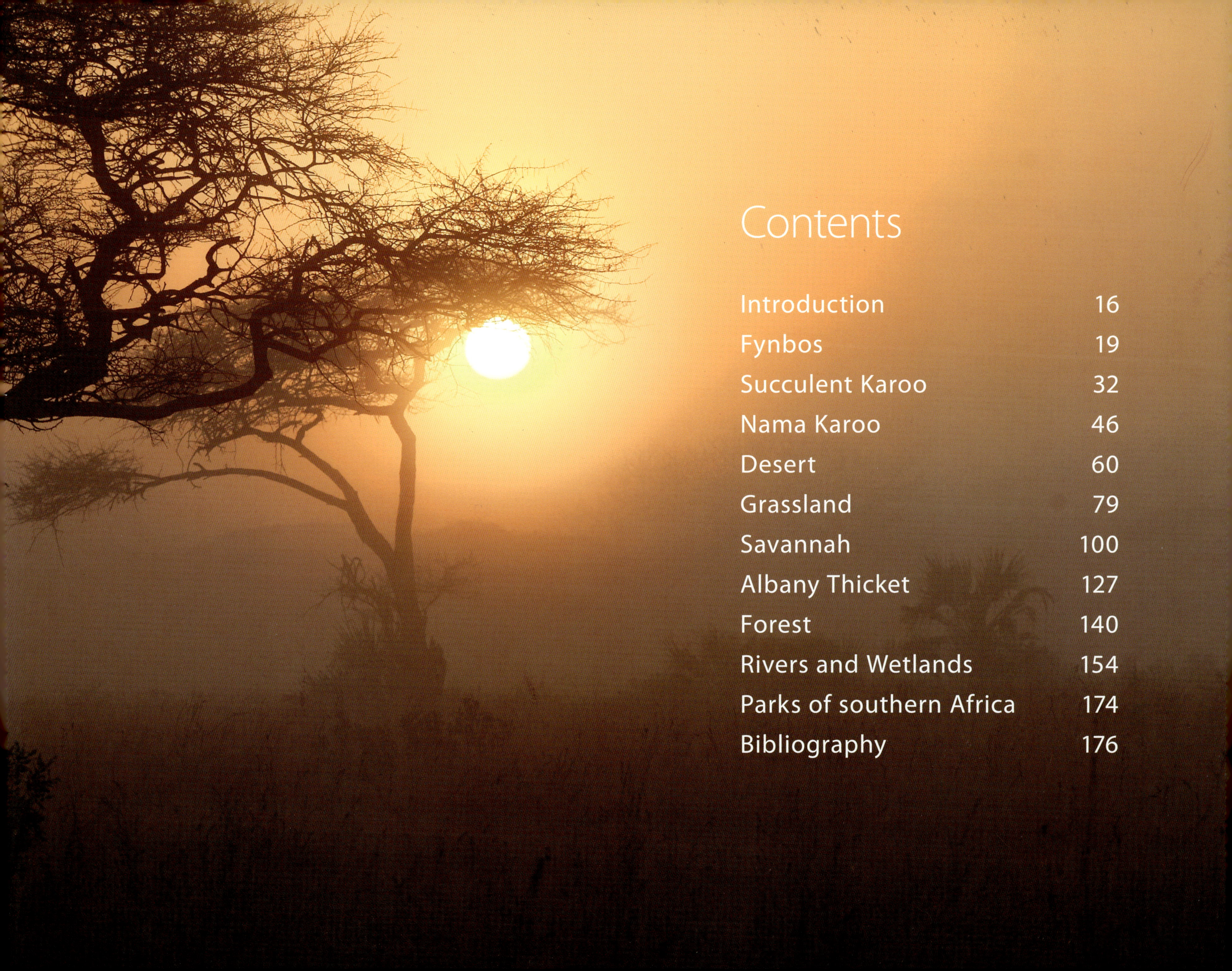

Contents

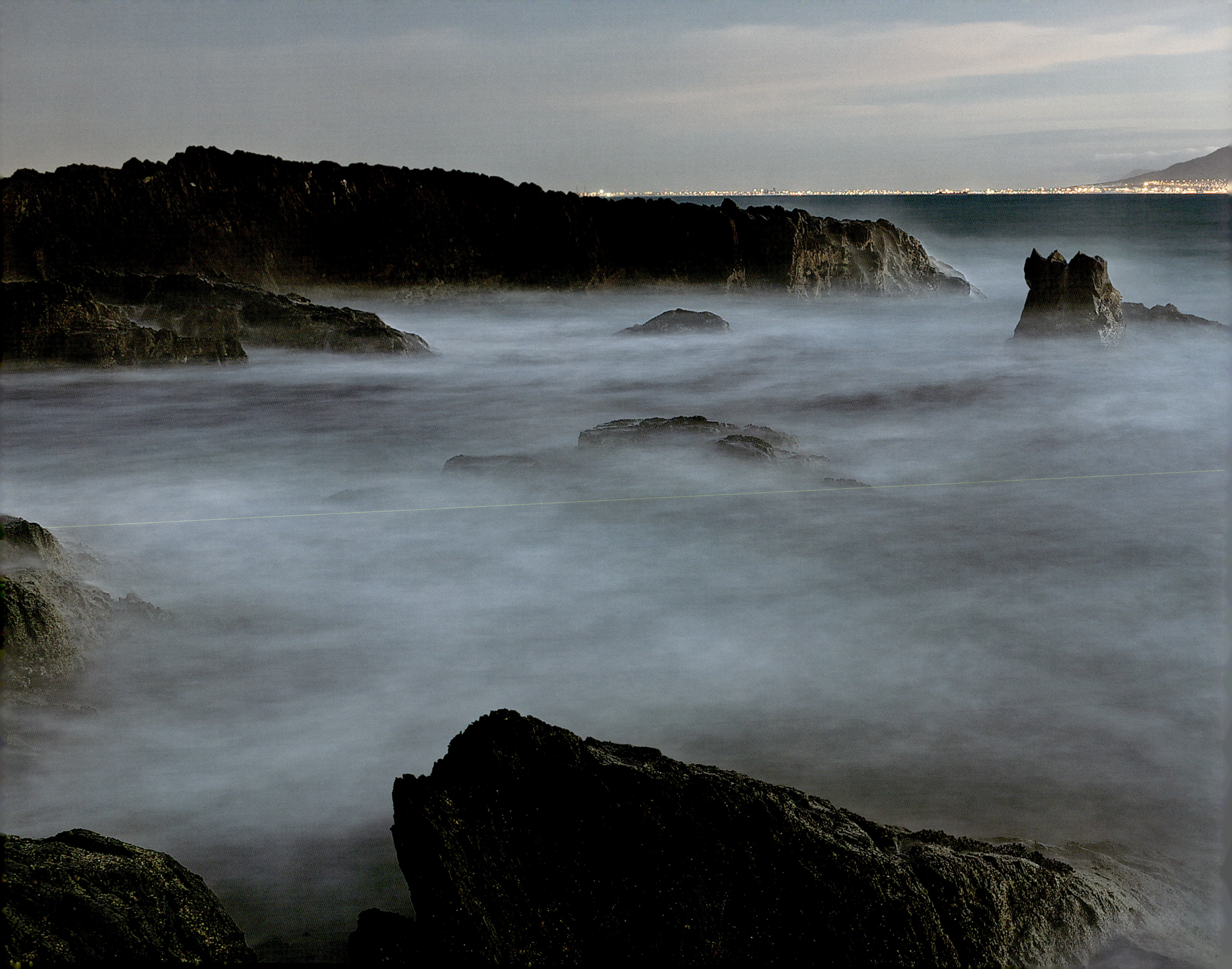

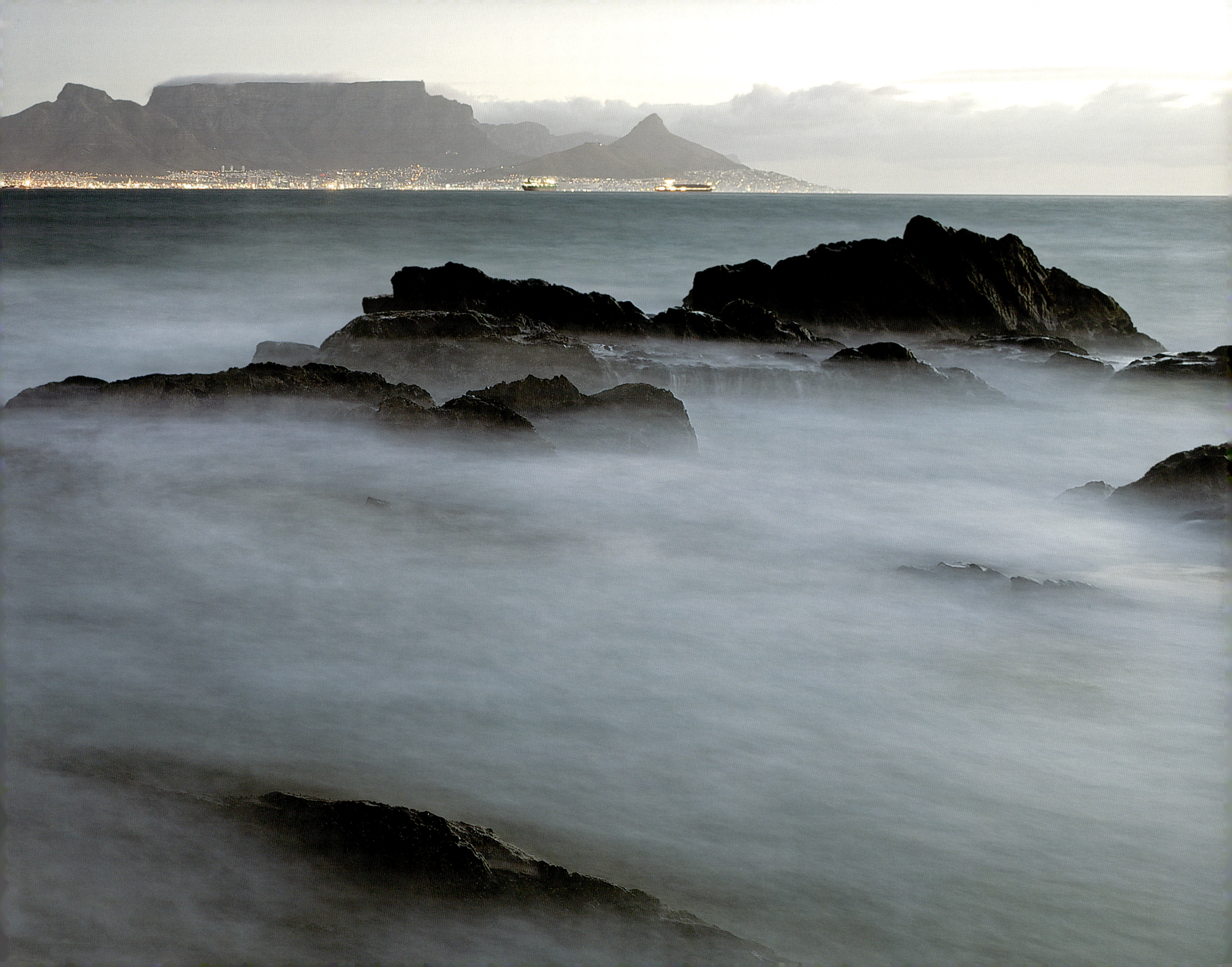

Southern Africa – land of captivating diversity

Looking down from outer space, astronauts have seen that planet Earth appears as a mystical light blueish ball wrapped in wisps of white cloud. A closer look reveals vast blue areas of ocean and the shapes of the different continents in shades of brown, yellow and green. We, here on Earth, use our magic machines and Google Earth to see what astronauts see. We try to experience their enchantment as we zoom in on our world, the most dynamic of all planets in our solar system.

We see Africa, the vast continent, in all shades from dark to khaki green stretching over the middle and down the eastern areas. These are sandwiched by a broad band of beige, brown and yellow on the heavy northern half and another smaller but similarly shaded band towards the western parts of the tip of Africa.

As we zoom in on the south of this magic continent our enchantment intensifies.

At the most southerly tip, Cape Agulhas, the ocean currents meet – the warm Mozambique Current of the Indian Ocean, flowing down the eastern coast of southern Africa from the equatorial regions, meets the cold waters of the Atlantic's Benguela Current , which often causes fog out at sea. The Cape Peninsula is a narrow finger of land with many valleys, bays and beaches, bounded on all sides by the cold Atlantic .

The landforms are fascinating and make up an incredibly varied terrain.

Zooming in further, we see the topography more clearly. There is a large interior plateau rimmed by a marginal zone. The escarpment is the edge that separates the highland plateau from the narrow coastal strip. On the eastern side, the escarpment lies predominantly within South Africa and Lesotho but extends into eastern Zimbabwe where it separates much of that country from Mozambique. On the western side, the escarpment extends into Namibia and Angola, roughly dividing the central plateaus of those countries from their western coastal plains – which in Namibia's case is a very arid area.

Relief and ocean currents determine the distinctive patterns of the occurring climatic regions. The western parts of southern Africa are significantly drier than the eastern parts. Only the low-lying areas of Mozambique and those along the border of Zimbabwe have a tropical climate. The southern tip is sufficiently far south to be influenced by weather systems associated with the belt of westerly ocean winds south of the continent in winter. Because of this the southern coastal regions get their rains in winter while the eastern coastlands and the interior get most of theirs in summer. However, most of southern Africa has generally fine weather, averaging eight to ten hours of daily sunshine.

Zooming in even further and moving the cursor over the land shows that the countryside is not only geologically diverse but also covered by a wide variety of vegetation types. The different climatic zones and the geologically varying landscapes give rise to distinct kinds of vegetation, which in turn form the basis of animal diversity and distribution. Such distinct communities of plants and animals are called biomes.

Major community types or biomes found on our globe are grasslands, deserts and forests (all on land) and aquatic communities, both in fresh and sea water. Within each of these major types, various subdivisions can be recognised.

A biome is a region with a similar geology, where climate and soil types are much the same. This brings about a grouping, or a community, of plant types that can tolerate the specific climatic conditions and thrive within the given soil types. Plants support animals, and while some animal species can tolerate a wide range of environmental conditions, others only occur in specific biomes.

The biomes evident in southern Africa are Fynbos, Succulent Karoo, Nama Karoo, Desert, Grassland, Savannah and Forest. Life-supporting water forms part of all biomes, even the very small quantities found in deserts. The extraordinary variety of biomes in southern Africa contributes to the irresistible attraction of its wild places.

Not all scientists agree on this particular classification of biomes and may add others, depending on how the biomes (or subdivisions) are defined. Biomes are characterised by a similar association of species, comparable climates and consistent soil types. Within each biome there are ecosystems, which are communities of organisms that inhabit specific physical environments.

This book is about the charm of the fascinating landscapes and the beauty and diversity of the southern African environments. It is also about captivating encounters with wildlife in several of these subdivisions or biomes and their ecosystems.

Southern Africa has a great number of well-managed conservation areas, which are relatively untouched by the burgeoning human population and the resultant need for increased food production. Significant numbers of various wildlife species are still flourishing in natural habitats where it is possible to visit, watch and be enthralled by the theatre of the bush.

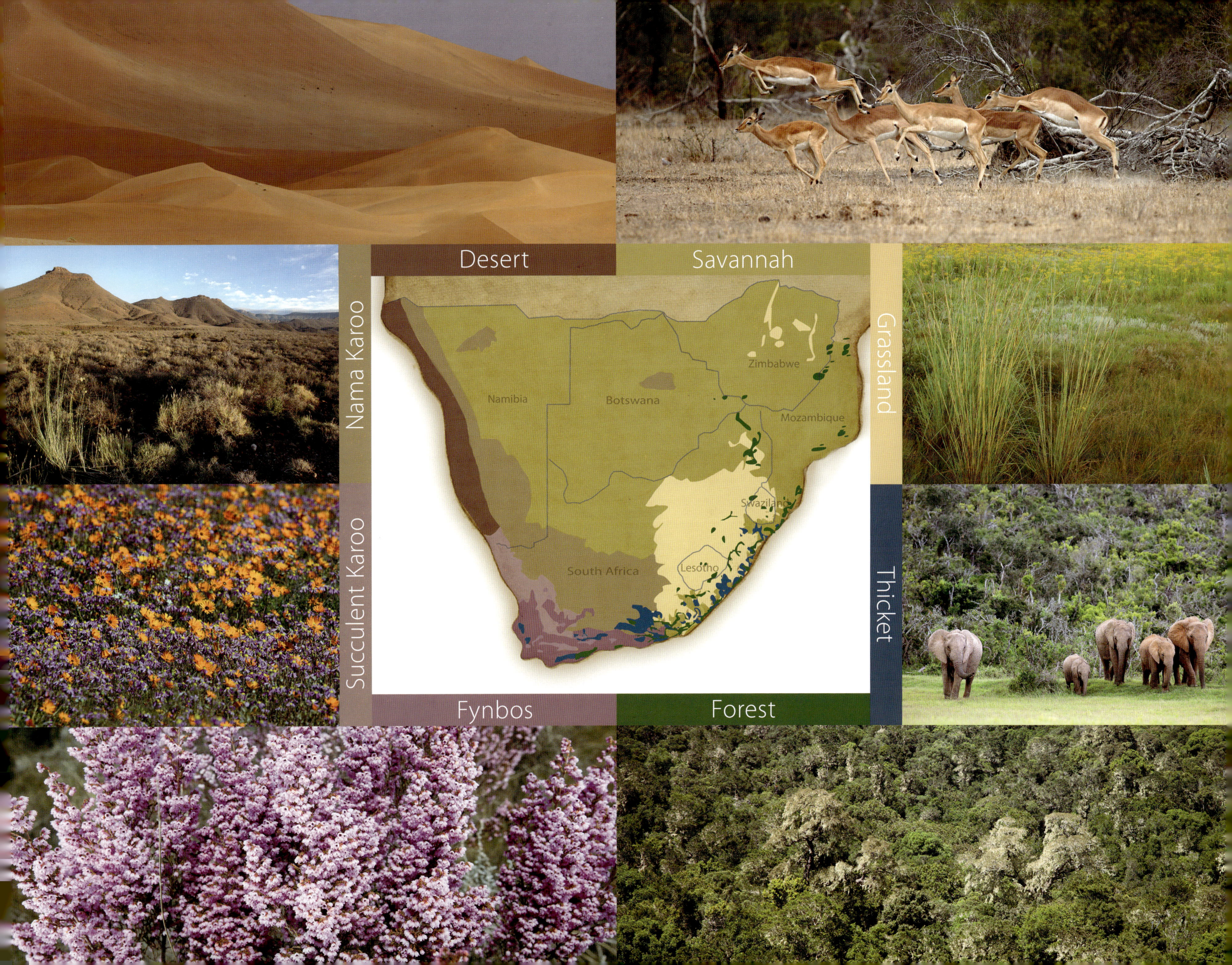
Desert
Savannah
Nama Karoo
Grassland
Succulent Karoo
Thicket
Fynbos
Forest
Namibia
Botswana
Zimbabwe
Mozambique
South Africa
Swaziland
Lesotho

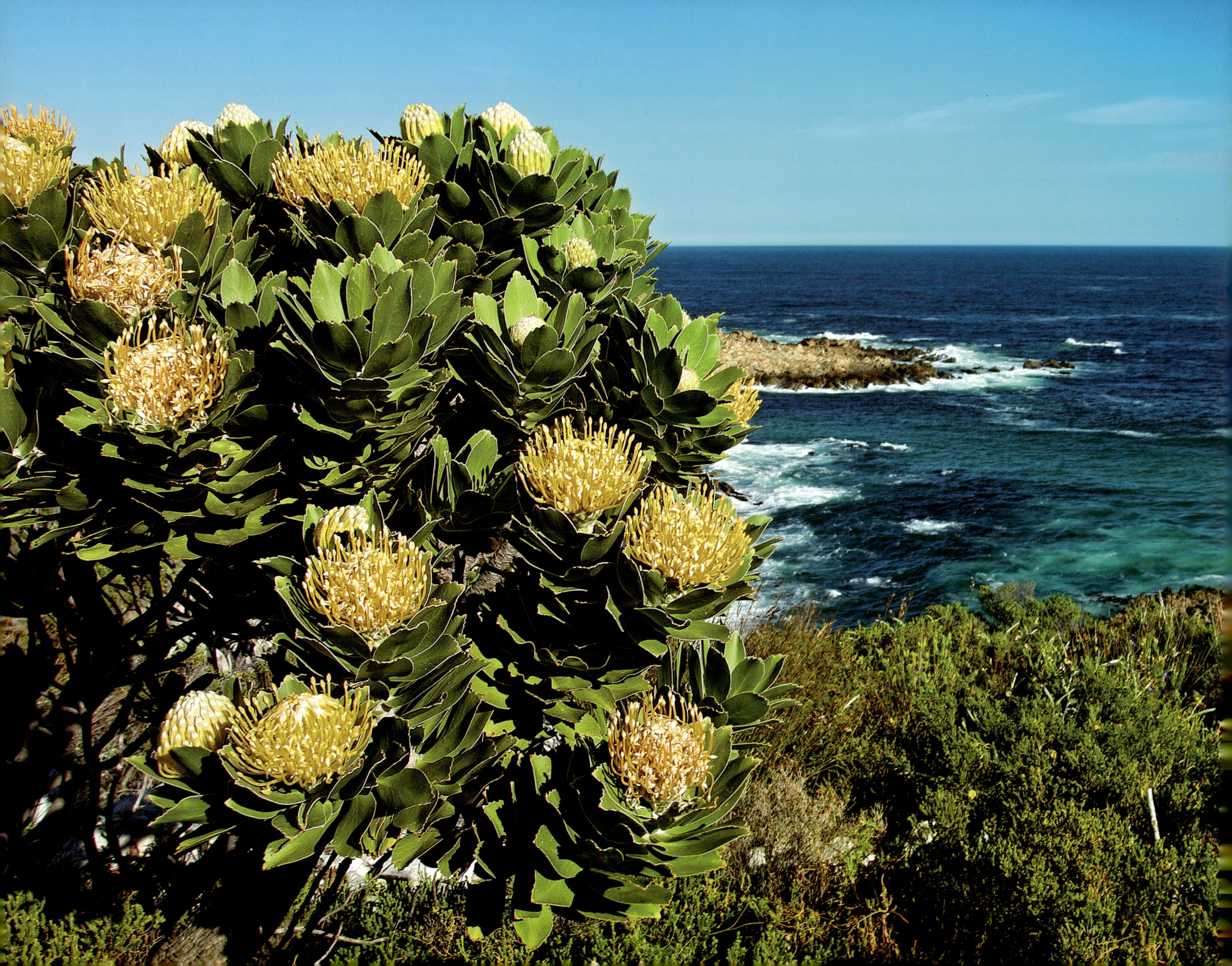

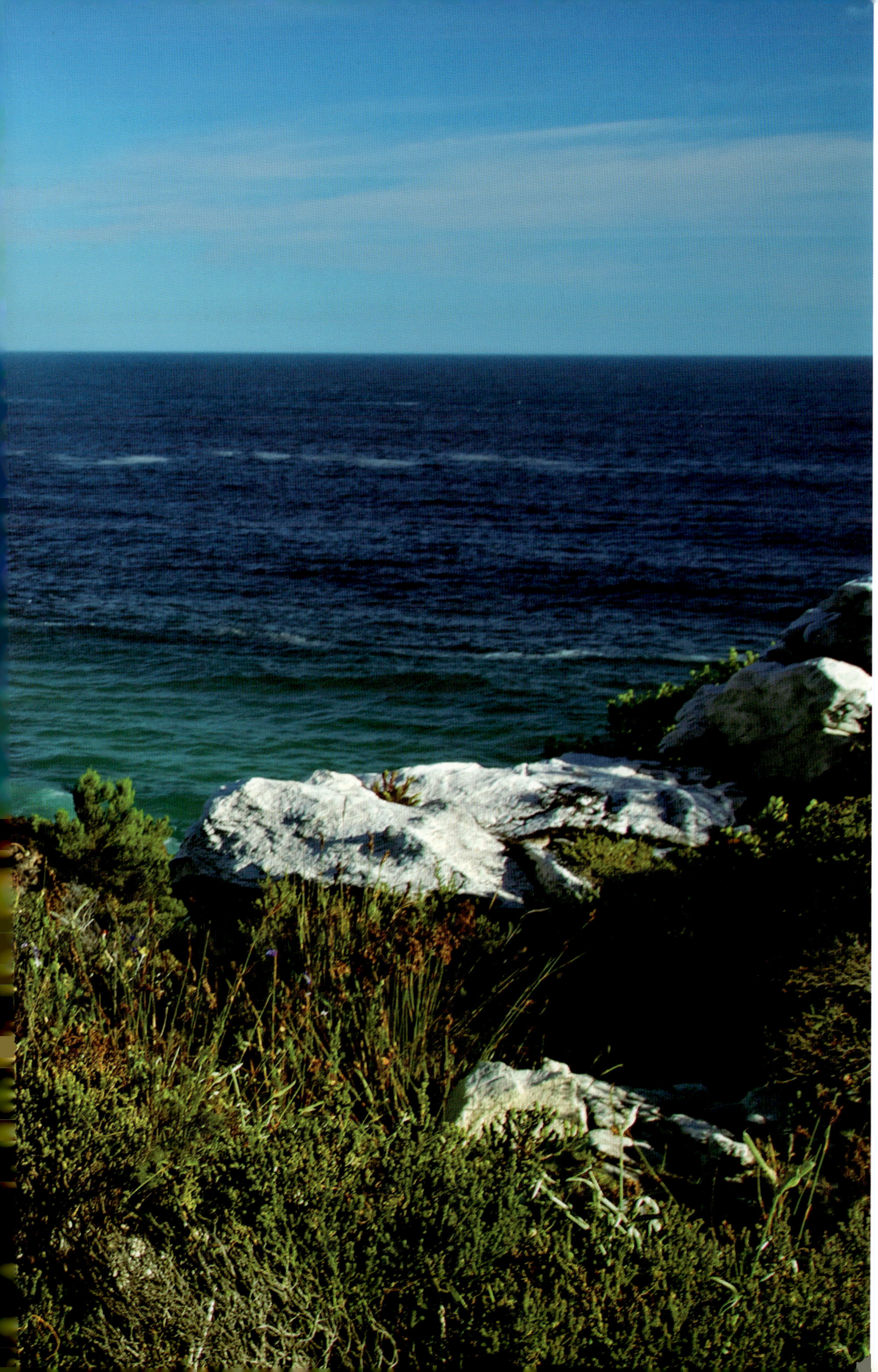

Fynbos

Jewel of the tip of Africa

The fynbos region (or biome) is unique in the world and typical of the South Western Cape. It covers a triangular area from Vanrhynsdorp north of the Cedarberg, to the Cape Peninsula and Boland mountains, and then extends in an easterly direction to Port Elizabeth and Grahamstown. The summers are not only warm but often very hot with long sunny days. A strong south-easterly wind, generally known as the Cape Doctor, often occurs during summer and in extreme conditions can reach speeds of up to 160km/h. The winters, however, are mild but often rainy – a typical Mediterranean climate.

In and around Cape Town, most of the annual rainfall occurs during the winter months of June, July and August. To the east of Cape Town, the Fynbos Biome merges into a region where some rain occurs throughout the year but where temperature conditions are similar.

The soil types where fynbos occurs are usually comparatively infertile, acid and coarse-grained. A typical terrain would be sandstone mountain ranges with valleys, shale flats and limestone outcrops. This is one of the explanations for the richness and diversity of the flora found in these areas.

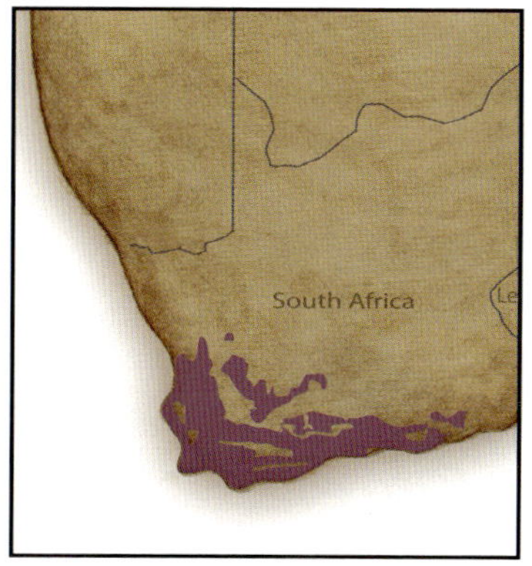

The name 'fynbos' is derived from the Dutch words 'fijn bosch', translating as delicate bush, which is the collective name for a myriad of evergreen shrub-like plants with small, often rolled, firm leaves and woody plants with hard, broad and leathery leaves. This unique flora is of particular botanical interest because of the great number of species, many of which are endemic (occurring nowhere else).

The typical Cape Fynbos is also called heath and includes fynbos on the mountains, in coastal areas and some plants resembling grass. But there is also another kind of vegetation in this biome called Renosterveld. This is a transitional kind of vegetation and highly endangered. It occurs where soils are mainly clays and silts and therefore more fertile. Unfortunately, very little undisturbed Renosterveld is left since most of it has now been converted to agricultural land.

Places to experience fynbos

Almost the entire Western Cape, including the West Coast National Park and the Cedarberg area, still has huge natural areas covered with fynbos. Ericas, proteas and daisies grow in abundance along road verges and delight travellers. Special places to go and enjoy the diversity are the Kirstenbosh National Botanical Gardens in Cape Town, the Table Bay National Park comprising large areas of the Cape Peninsula, the Harold Porter National Botanical Garden at Betty's Bay, and various smaller nature reserves such as Fernkloof, Walker Bay, Salmonsdam, Agulhas, De Mond, Waenhuiskraal and De Hoop. Inland are Vrolijkheid, Marloth, Bontebok National Park, Anysberg, Swartberg and parts of the Garden Route National Park, which includes places such as the Tsitsikamma area and the Outeniqua mountains.

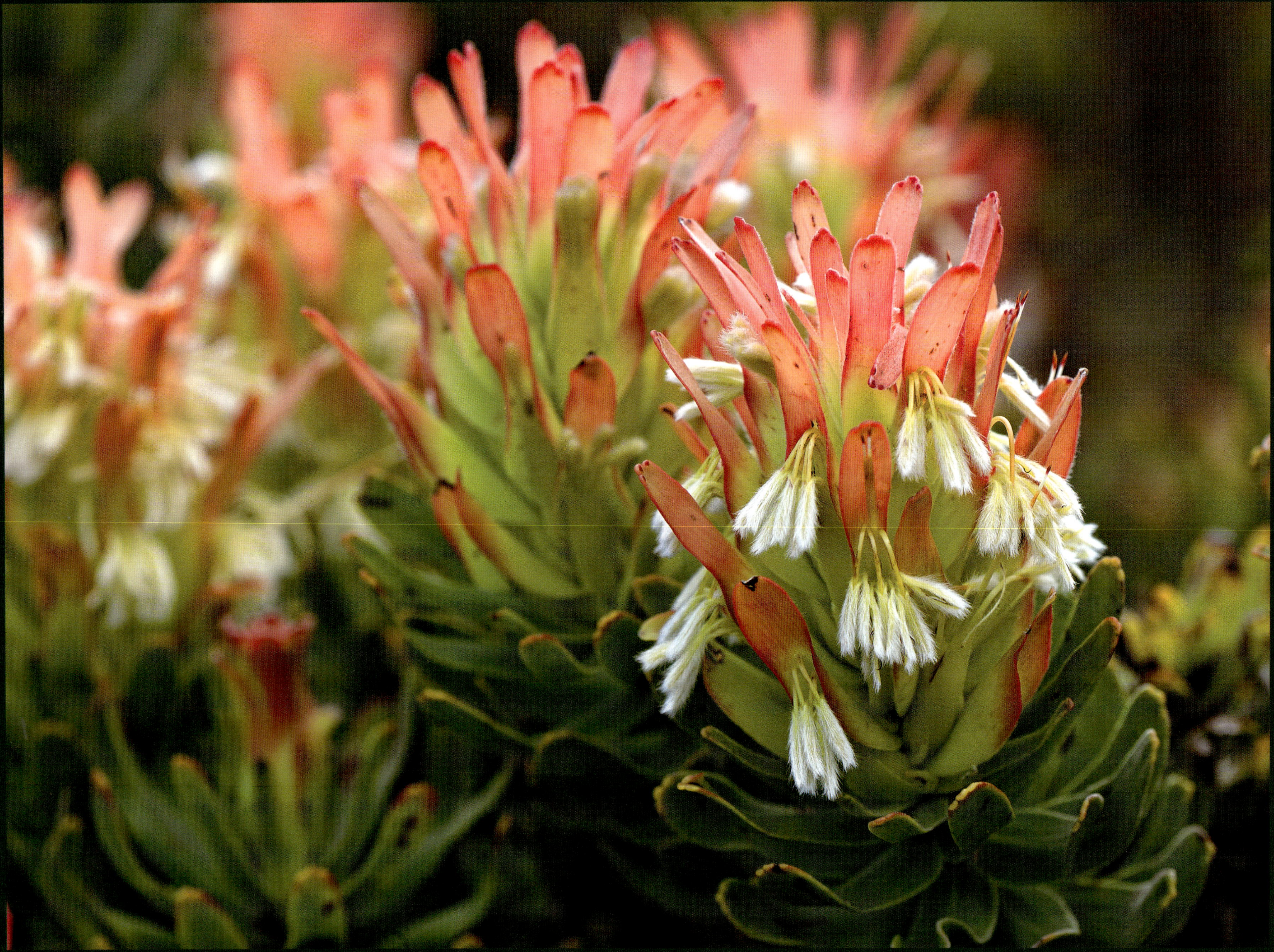

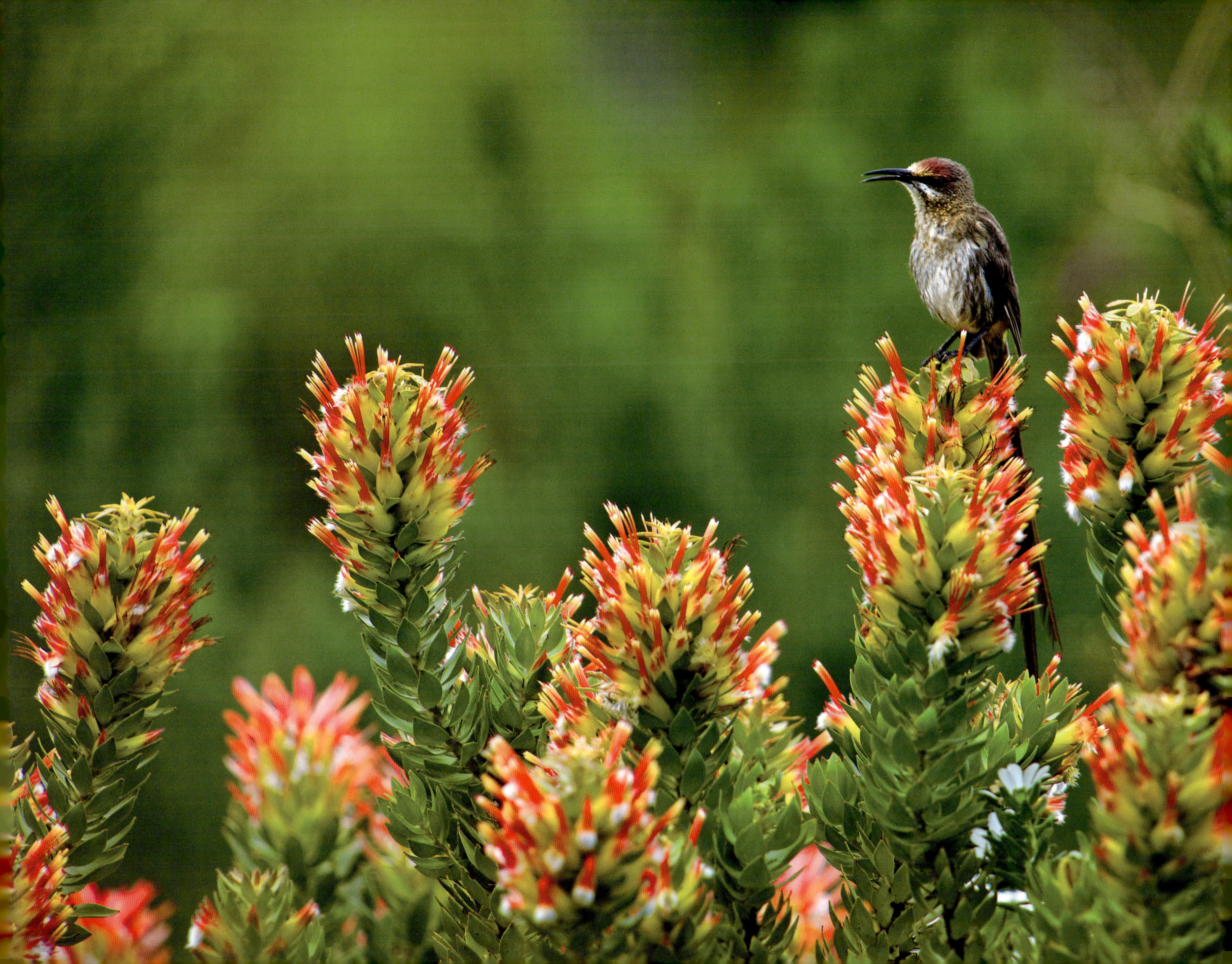

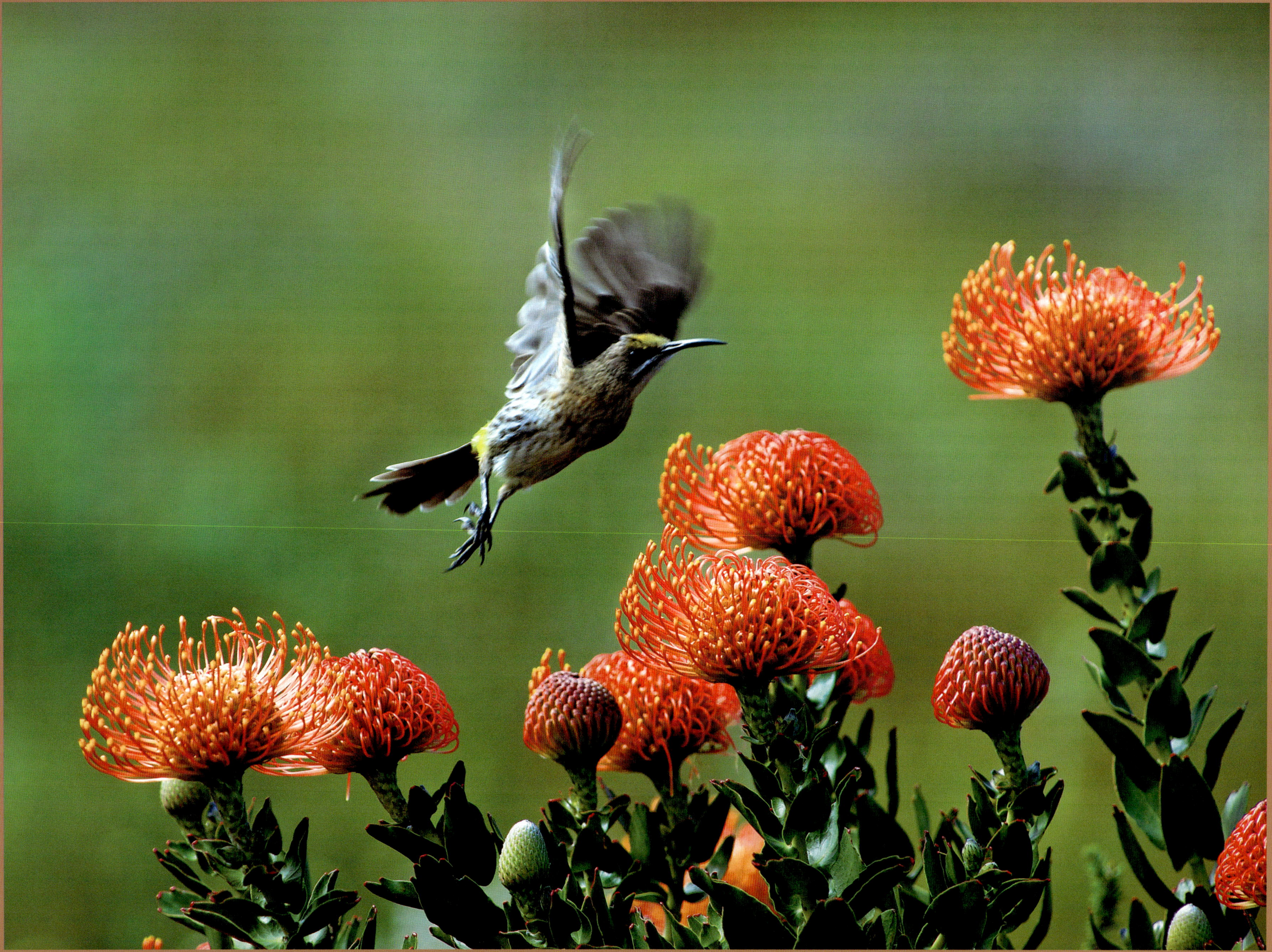

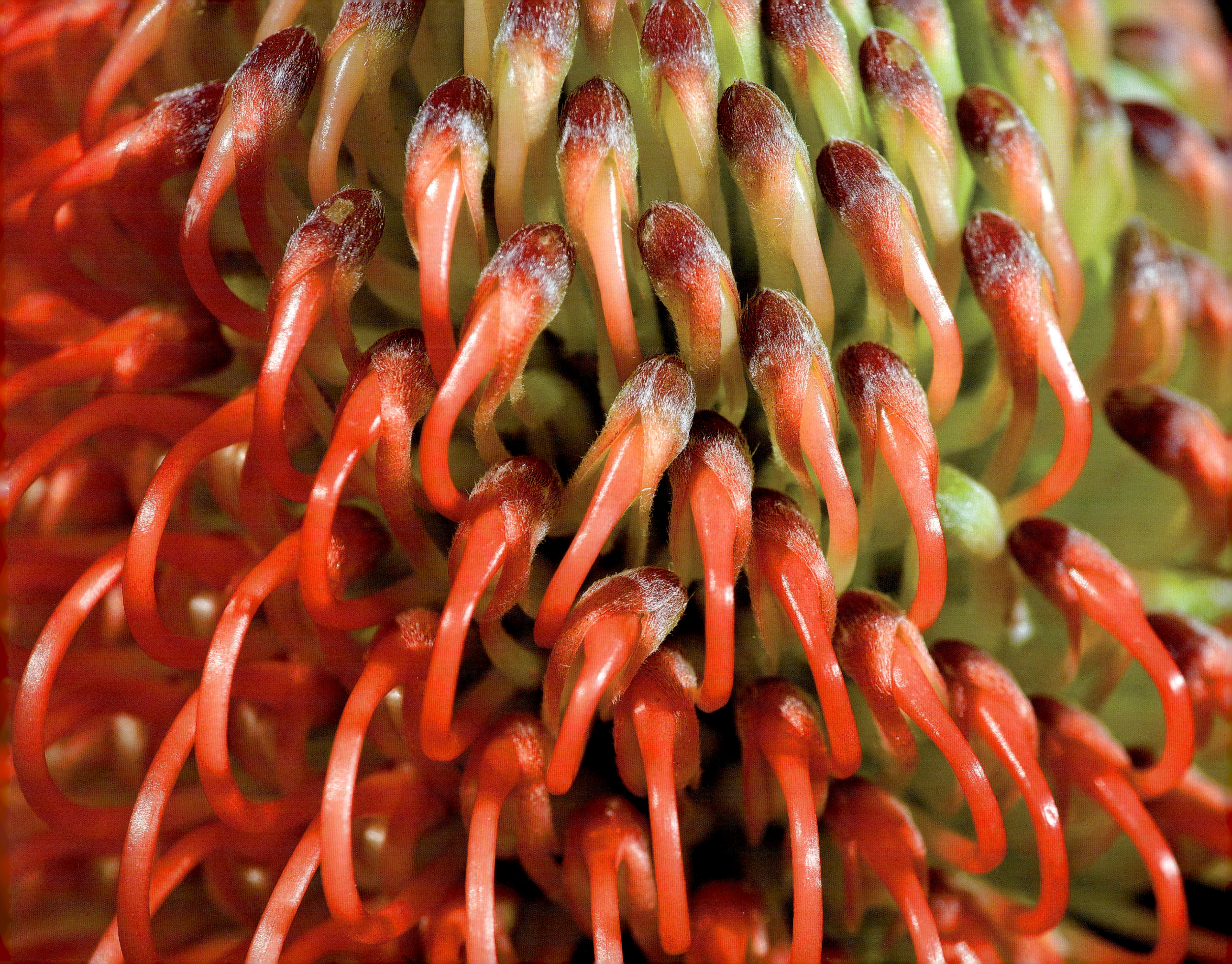

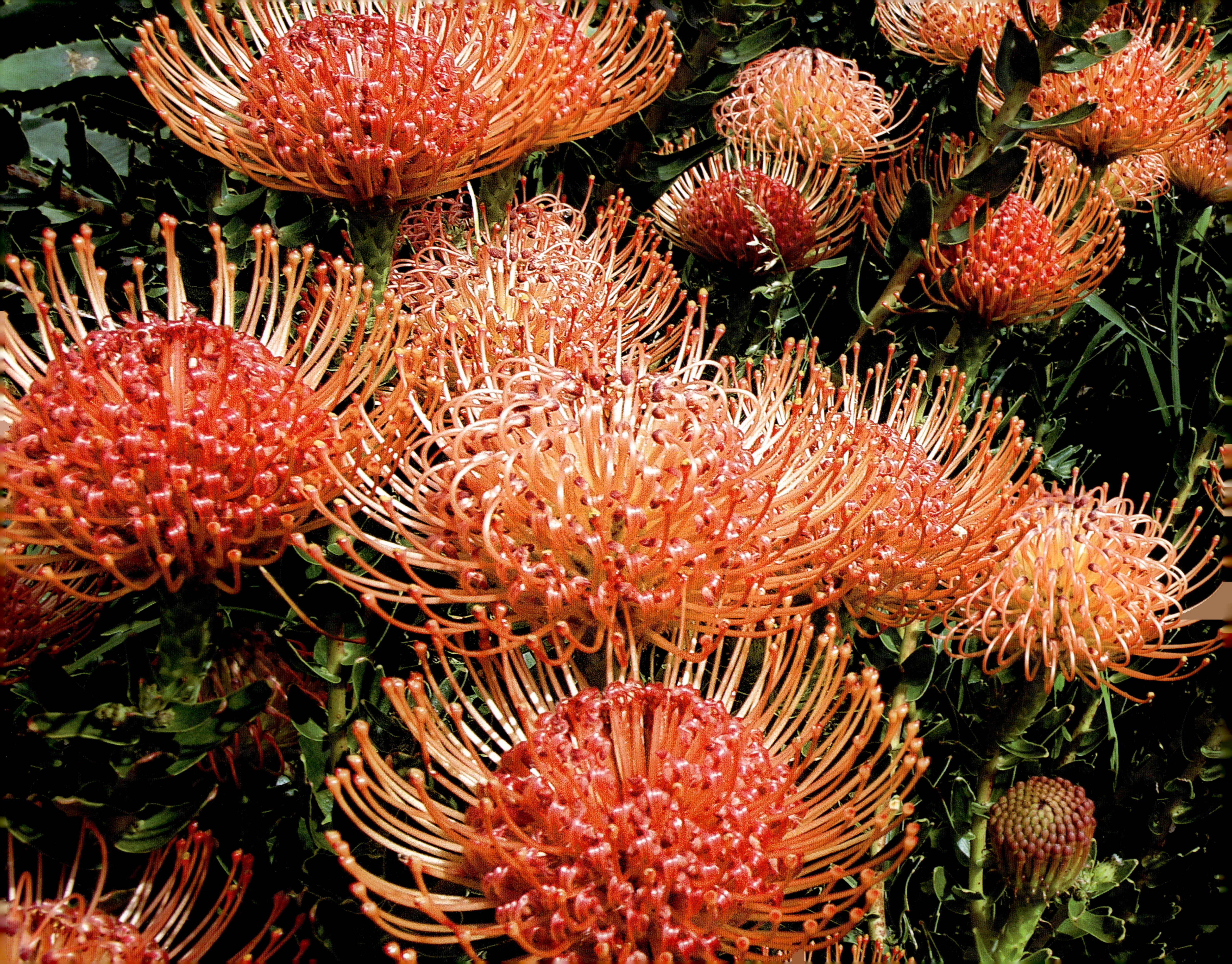

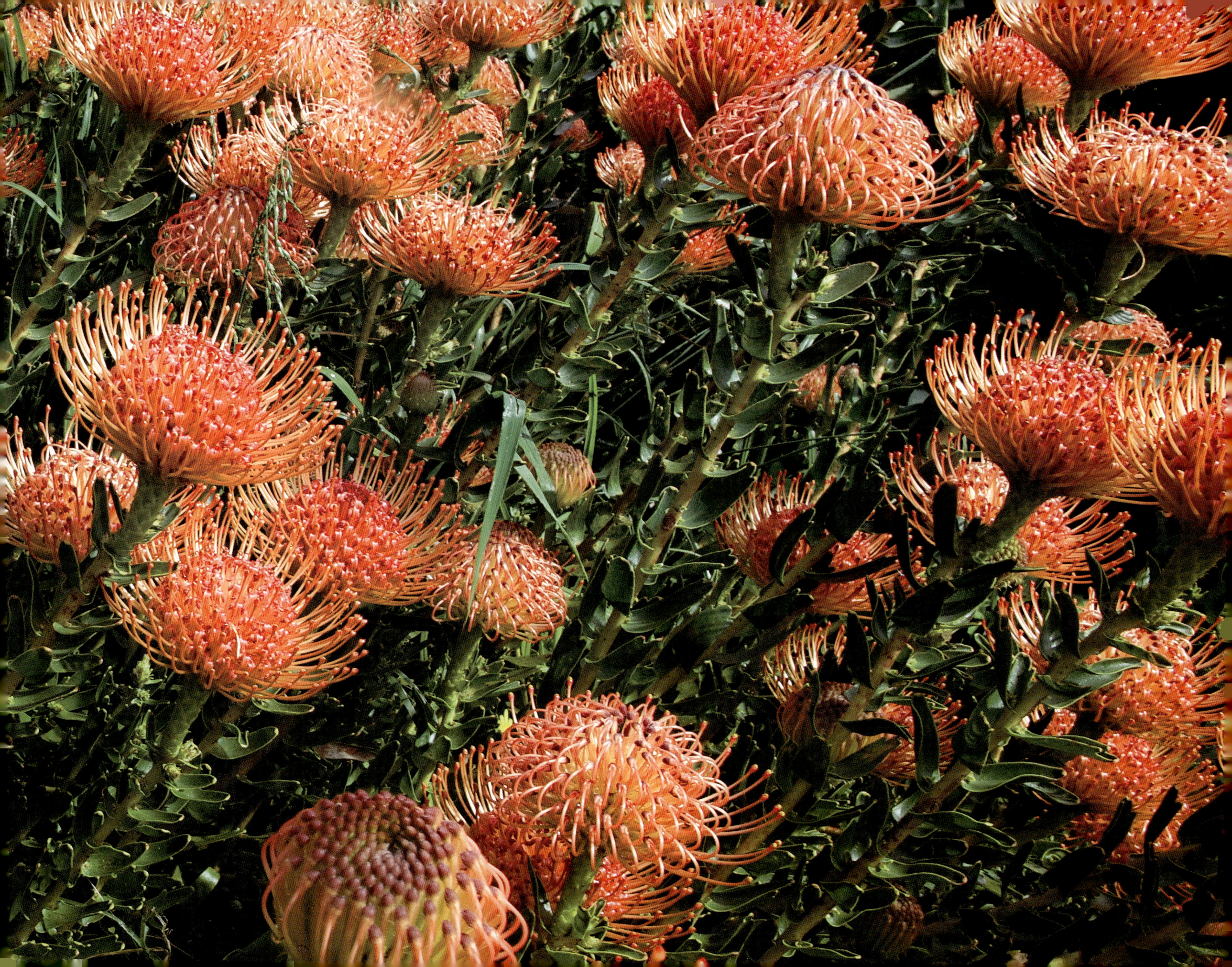

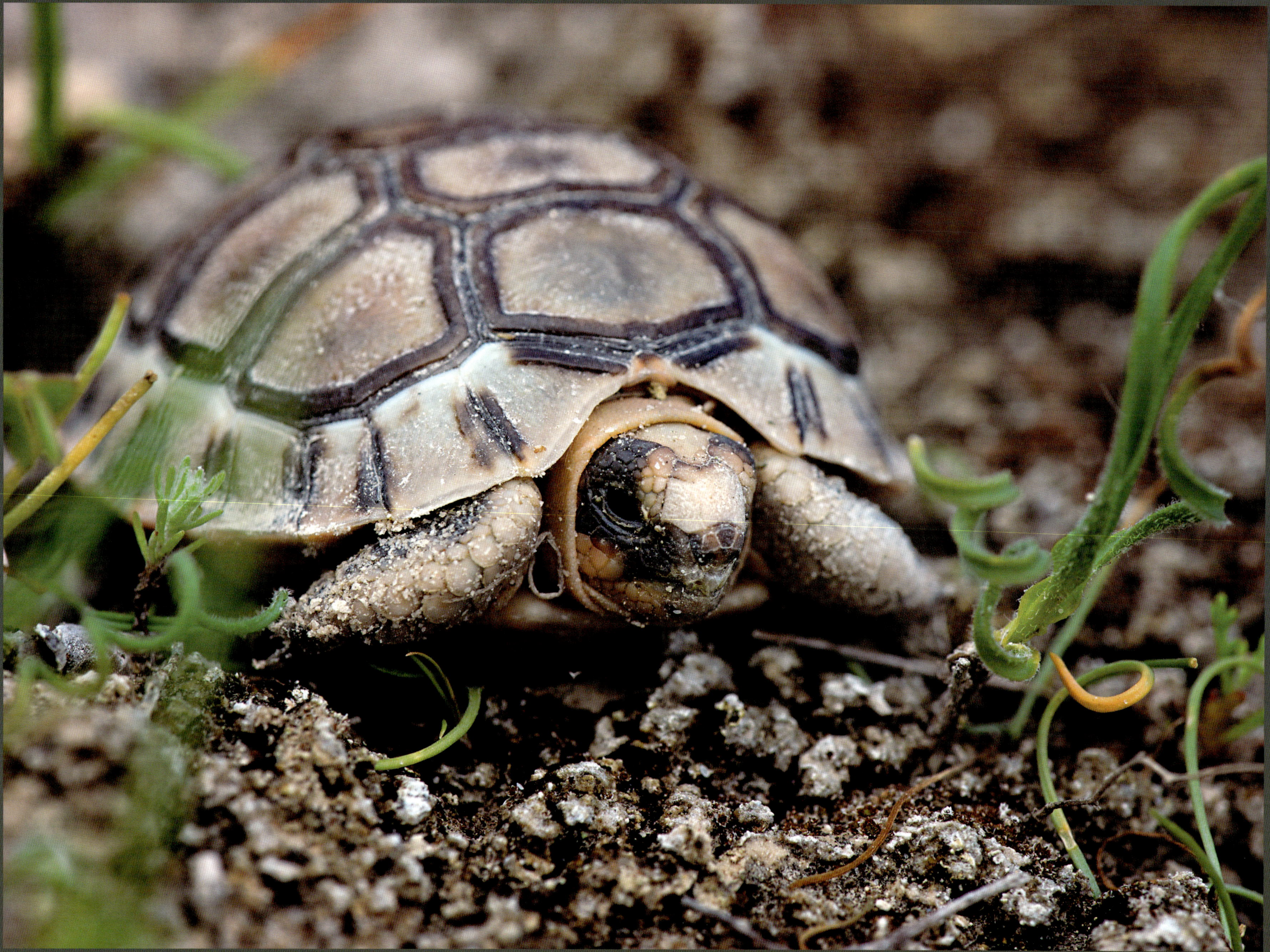

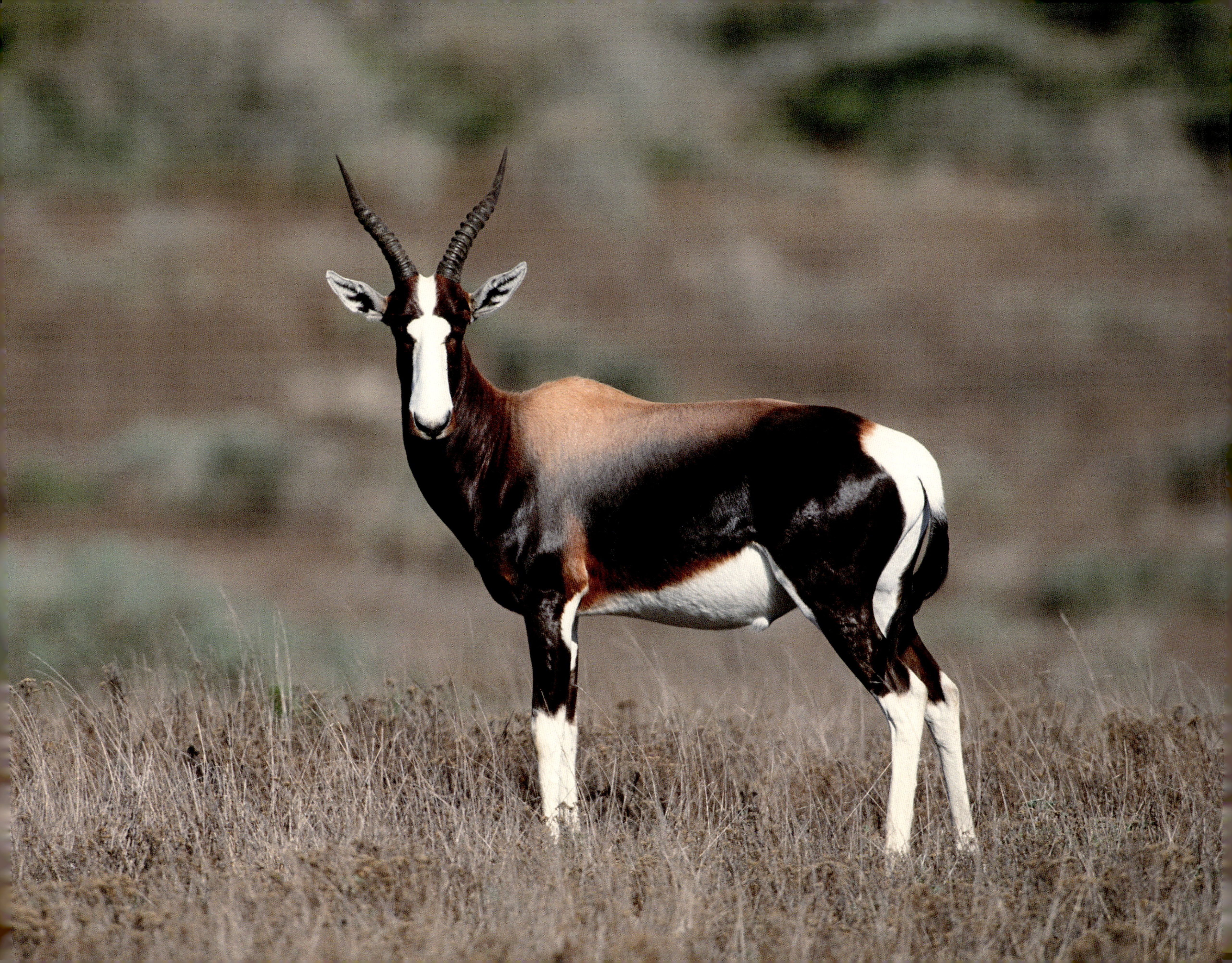

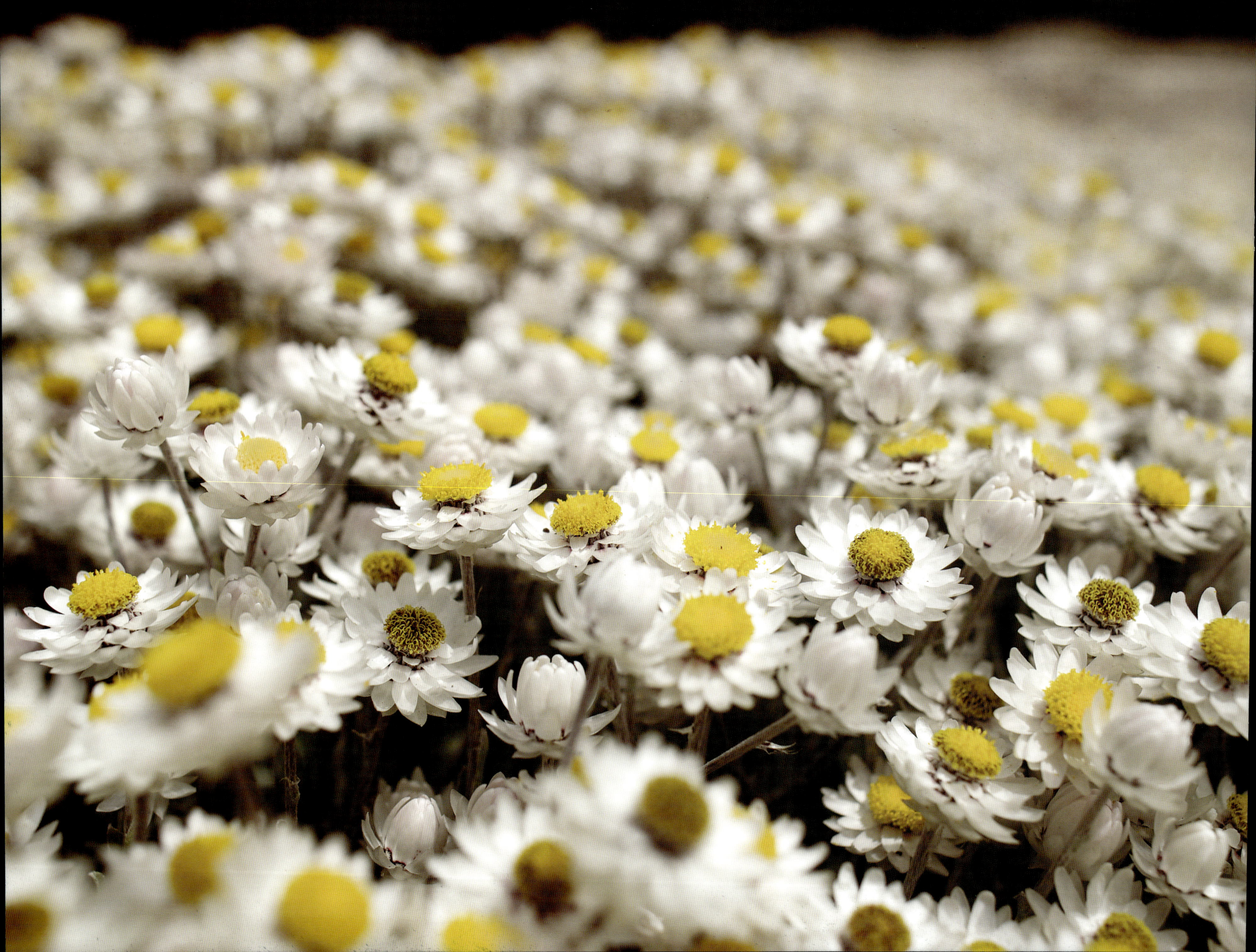

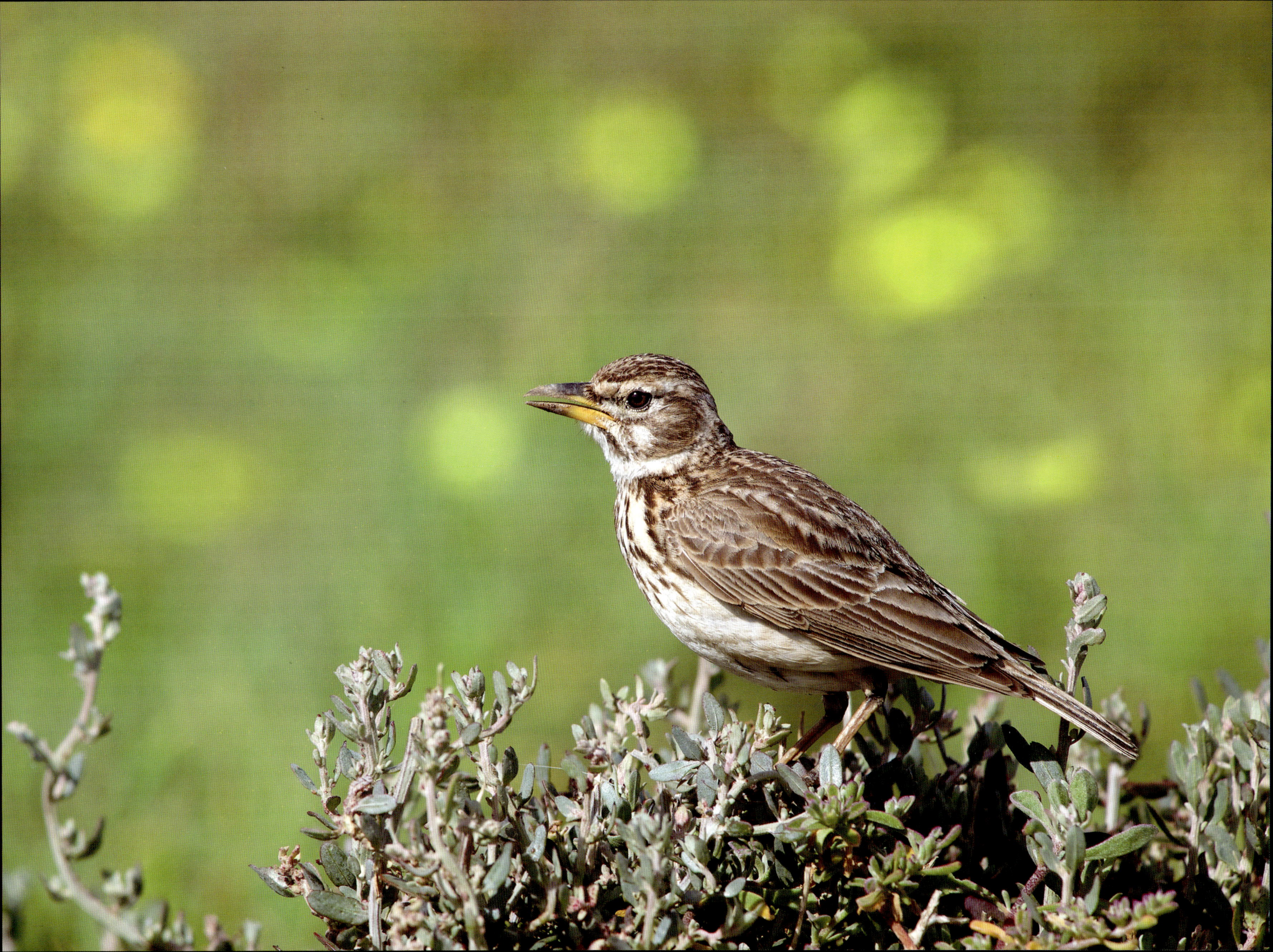

More about the Fynbos Biome

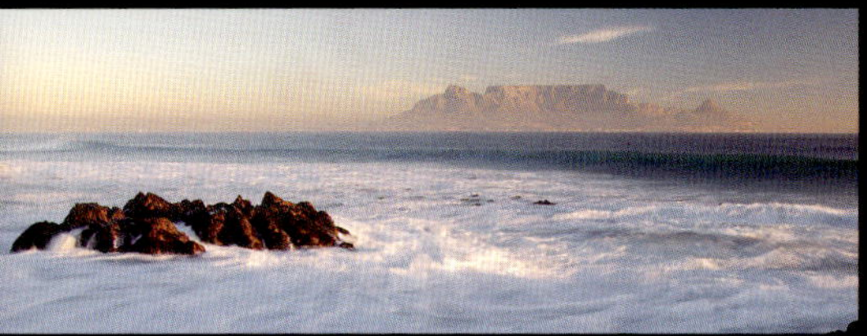

Table Mountain

The iconic status of Table Mountain is enhanced by the unique fynbos vegetation, which is found nowhere else in the world. Fynbos is classified as one of only six recognised floral kingdoms of the world, which in itself is worthy of a special accolade.

Even though fynbos is the smallest of the six floral kingdoms, it is considered to be the richest, with more plant species than those found in all of Europe. The Fynbos Biome has about 8 700 plant species, of which 68% are found nowhere else. Apart from the well-known protea family, fynbos is essentially a shrubland with 600 types of heath-like shrubs or ericas (most are endemic to this region), restios (a reed-like species), and many kinds of daisy and bulbous types. Unfortunately, more than 1 700 of these species are threatened or even on the verge of extinction due to several environmental threats.

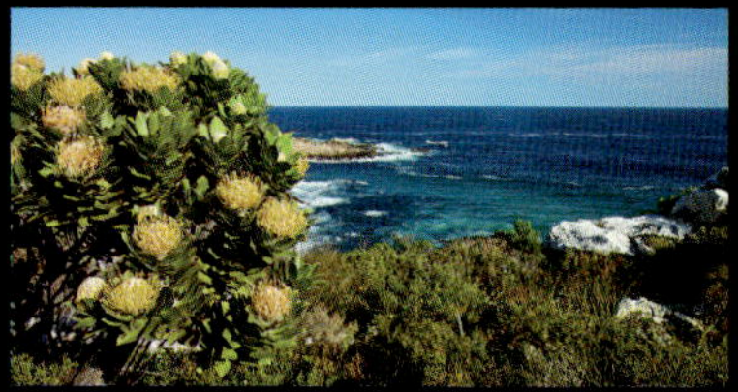

Leucadendron conocarpodendron

Descriptions from early travellers to the Cape of Good Hope tell us that the pincushion *Leucadendron conocarpodendron* was once abundant in the area now covered by Cape Town. It is still an abundant and also an endemic species to the Cape Peninsula. This specimen with its green foliage and bright, golden-yellow flowers grows into a small tree on the coastline at Betty's Bay. The windy cool conditions, wet winters and hot dry summers are extremely suitable growing conditions.

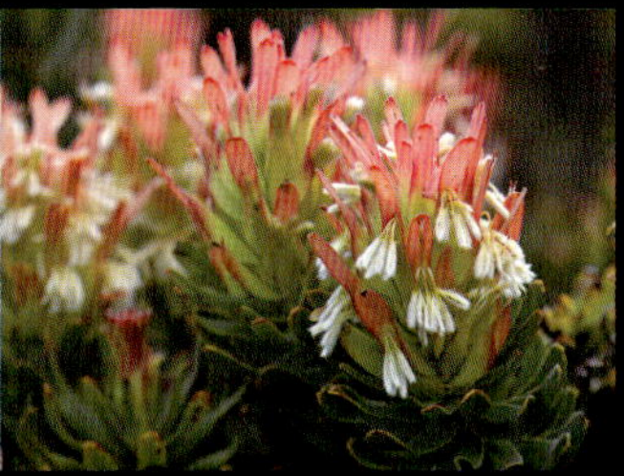

Mimetes cucullatus

Mimetes cucullatus is another abundant member of the protea family. The eye-catching, unusual flower heads are colourful all year round and are pollinated by sunbirds and the Cape Sugarbird. The seeds have a grey-white, oily protuberance, which is very attractive to ants. They collect the seeds and carry them off to their underground nests where the protuberances are eaten off. This does not damage the seed, which is left whole and safe in the ants' nest where it remains viable for many years. Seeds left lying at the base of the parent plant are often eaten by rodents.

About 430 plant species in fynbos are pollinated by birds. Of these, about 100 belong to the erica group and 79 to the protea group. Birds visit the flowers for the energy-rich nectar that is produced.

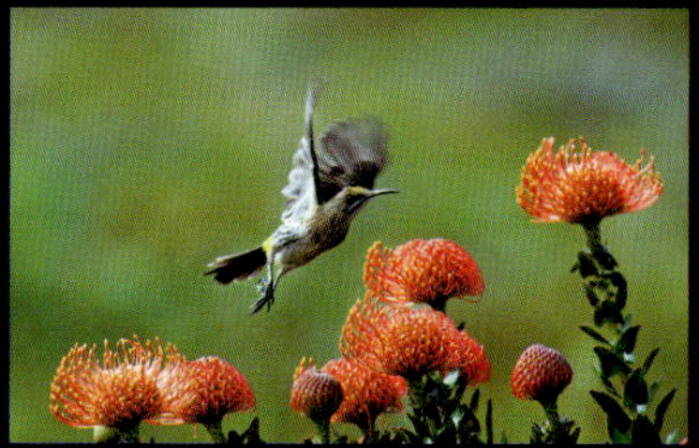

Cape Sugarbird

The Cape Sugarbird is one of the six bird species endemic to the Fynbos Biome and is a specialist nectar feeder when it comes to feeding off protea heads. The male has a very long tail (twice the length of the body) while the female's tail is about half that length. In both, the undertail is bright yellow and the top of the head dull brown. It reaches the nectar with its long, sharp beak and long, brush-tipped tongue. It may visit up to 300 protea flower heads per day. Although the staple diet of this sugarbird is nectar, it will also feed on spiders and insects. The characteristic strong winds in the Cape may make feeding off protea heads difficult, but the Cape Sugarbird has adapted to this with the development of sharp claws.

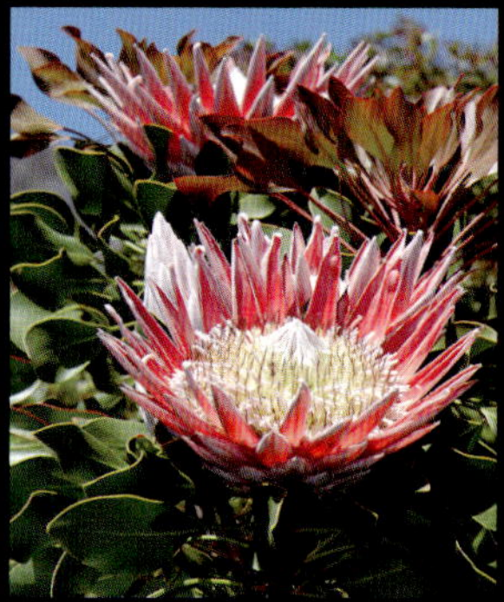

Protea cynaroides

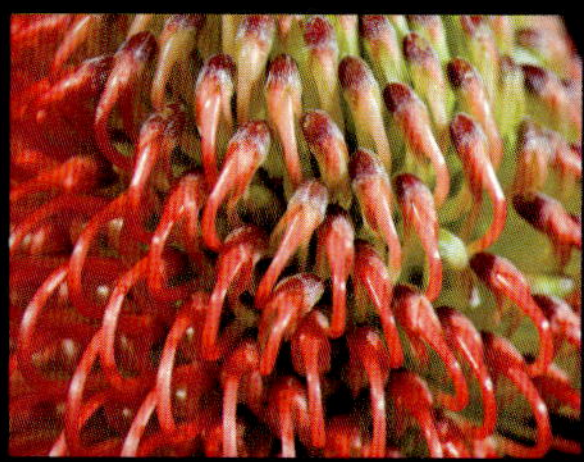

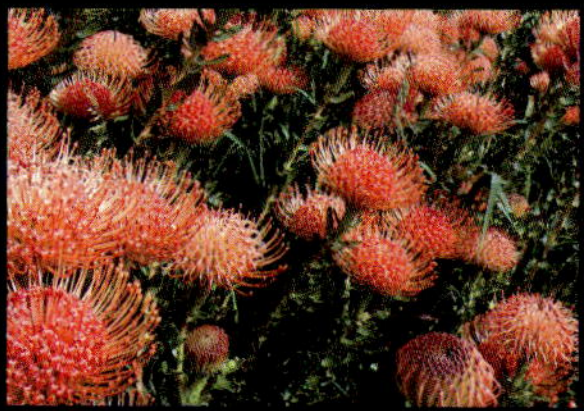

Pincushions

But it's not the only pollinator or the only visitor to these flowers. Beetles, flies, bees, moths, butterflies, thrips and earwigs may be found – some there for the nectar, some to prey on other visitors to the flowers.

Fire is a natural and normal occurrence in fynbos, and the life of fynbos plants is actually shaped by fire. Fire is a key factor, preventing fynbos from reverting to thicket by providing new opportunities for plants to regenerate. Many species, such as the yellow brunia, *Brunia alopecuroides*, survive fires by resprouting from a large, woody, underground rootstock. Other species, like the *Mimetes sp*, are re-seeders, i.e. they are killed by fire but seedlings germinate when conditions are suitable after the fire. Although damaged, even those with thick, corky bark do sometimes survive fires.

All members of the protea family have small flowers grouped in conspicuous flower heads. The unfolding florets of the pincushion *Leucospermum cordifolium* show the fascinating spiral arrangement of the florets within the flower head. The best known and most spectacular of them all is the king protea or *Protea cynaroides*. No wonder it is honoured as South Africa's national flower.

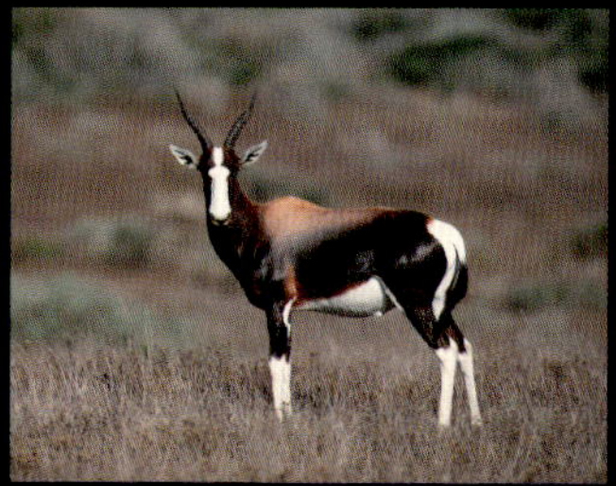

Bontebok

The bold splendour of the many kinds and sizes of proteas and pincushions – in shades ranging from yellow to pink – are prized worldwide as magnificent cut flowers. Although much harvesting is done from natural areas in the mountains, commercial protea and fynbos farming is undertaken on a huge scale.

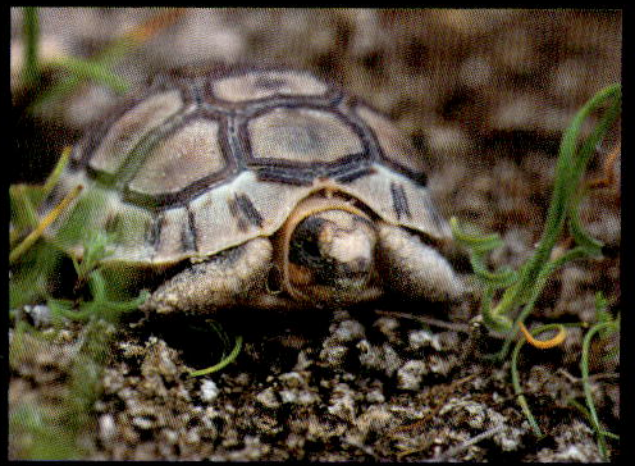

Angulate tortoise

Large game animals are not abundant and at most places completely absent in this biome. There is historical evidence that mountain zebra, quagga, bluebuck (endemic and extinct), roan antelope, red hartebeest, eland, bontebok, elephant, black rhino and Cape buffalo occurred in the area, as did predators such as lion, hyena, wild dog and leopard. It is probable that the herds were confined to the more fertile Renosterveld regions of the Fynbos Biome. The beautiful bontebok is the only antelope endemic to fynbos and so the Bontebok National Park was proclaimed to save this species from extinction. Other larger mammals commonly found in this park and

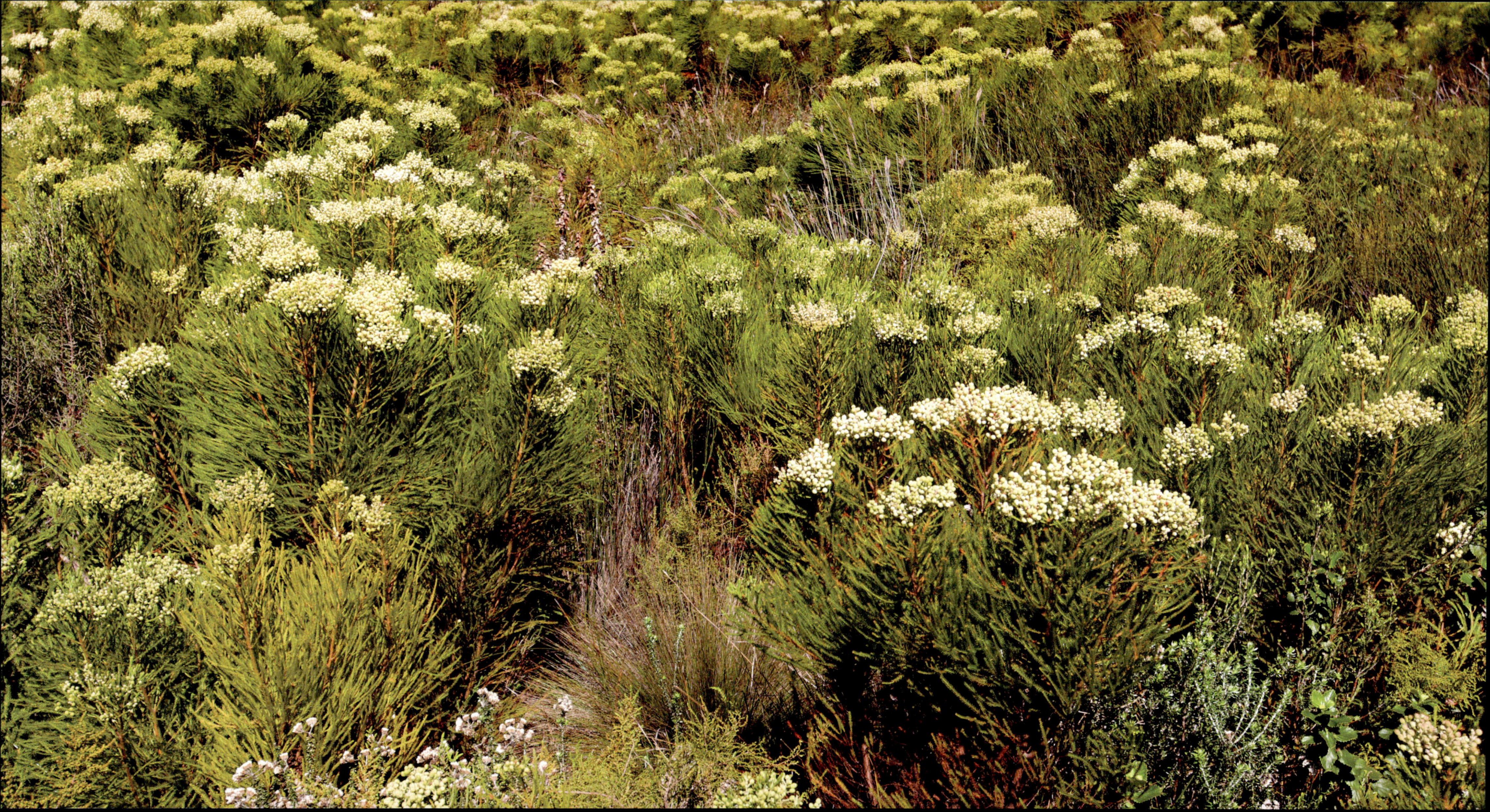

Brunia alopecuroides

in fynbos in general, include steenbok, grysbok, baboon and leopard.

Of course, the Fynbos Biome hosts many smaller species, including insects, frogs, reptiles and birds. The rare geometric tortoise is the only one endemic to this biome, but another species, the angulate tortoise, is fairly common and widespread.

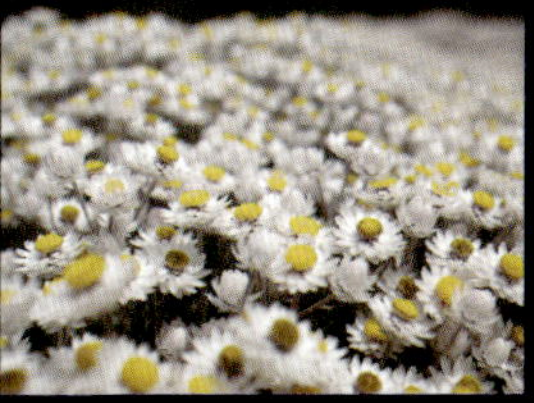

Helichrysum sp

The endangered Renosterveld is characterised by the daisy family. Of all the daisies, the everlastings or Helichrysums are particularly beautiful. Grasses are also abundant but overgrazing leads to an increase in shrub cover. Most of the shrubs in the Renosterveld have typically small, tough, grey leaves.

The Large-billed Lark is one of the typical so-called 'lbj's' (little brown jobs) using the Renosterbos shrubs as a lookout. It feeds on seeds and insects, and it nests on the ground.

The Fynbos Biome is indeed a wonder of the world and places the southern tip of Africa in a league of its own. When he sailed round the world in 1580, Sir Francis Drake, viewing the mountains of the Cape Peninsula from his ship, called it 'the fairest Cape in all the circumference of the Earth'. Not only the dramatic scenery, but also the Fynbos Biome – which is indeed one of the botanical wonders of the world – truly justifies the name Fairest Cape.

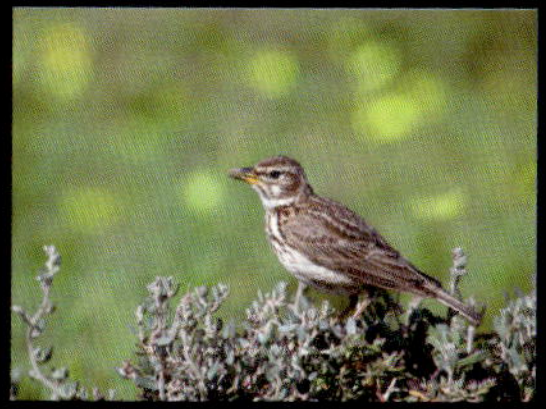

Large-billed Lark

Succulent Karoo

The changing face of a semi-desert

As the richest semi-desert in the world, the area known as the Succulent Karoo is extraordinary. Although it covers a relatively small area (only 6.5% of the subcontinent), it surprisingly boasts a plant diversity of about 5 000 different recorded species. Of these, at least 4 000 are succulents, of which half are endemic – occurring only there and nowhere else. Succulents are mostly low perennial shrubs with fleshy moisture-storing leaves, branches and stems. No other vegetation type in the world includes more species of succulent plants than this biome in southern Africa.

The Succulent Karoo occupies mostly low-lying country to the west of the western rim of the central escarpment in Namaqualand and southern Namibia, and extends into the rain shadow of the Cape Fold Mountains and into the Little Karoo of the Western Cape. In stark contrast to the low-lying areas are the rugged mountains of the Richtersveld in far northern Namaqualand.

The soils are rich in lime but weakly developed and shallow. This makes the region particularly vulnerable to wind and water erosion. Although the vegetation is sparse, it does to some extent protect the soil. Overgrazing, however, is a real threat.

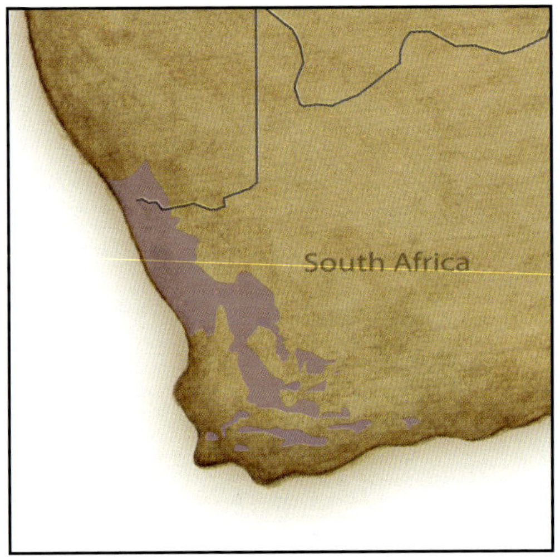

Rain is scarce at between 20mm and 290mm per annum. In some areas there is year-round rainfall, but most of the biome is restricted to winter rainfall. The summers are hot and extremely dry due to berg winds and temperatures soaring above 45°C. Consequently, the Succulent Karoo is a drab and almost colourless environment for most of the year.

For a few weeks in spring, however, the region has a completely changed face. Masses of a great diversity of succulent, ephemeral annuals and bulbous plants burst into flower and charm both visitors and locals alike. Fields of orange, purple, pink, white and yellow flowers dazzle the eye. But the beauty is elusive since the spring flowers only appear after sufficient winter rains and only reveal their true splendour in full sunshine.

Vertebrates associated only with the succulent landscapes are relatively rare. From the 88 desert vertebrates, 25 are endemic with reptiles in the majority.

Places to experience the Succulent Karoo

Spring, after good winter rains, is the best time to plan a visit to any of the nature reserves in the Succulent Karoo. Van Rhynsdorp and especially Nieuwoudtville are the best places to go for bulbous plants, while the Namaqua National Park and Goegab Nature Reserve close to Springbok in Namaqualand are excellent for annuals and mesems (i.e. the *Mesembryanthemum* group). The |Ai-|Ais/Richtersveld Transfrontier Park that straddles the far Northern Cape and the southern part of Namibia across the Orange River is a favourite destination for botanists and adventurers, but it is not suitable for ordinary sedan vehicles. In the Little Karoo, small nature reserves can be reached from Worcester and Robertson, and along the R62 road through Bonnievale, Barrydale and Ladismith to Oudtshoorn.

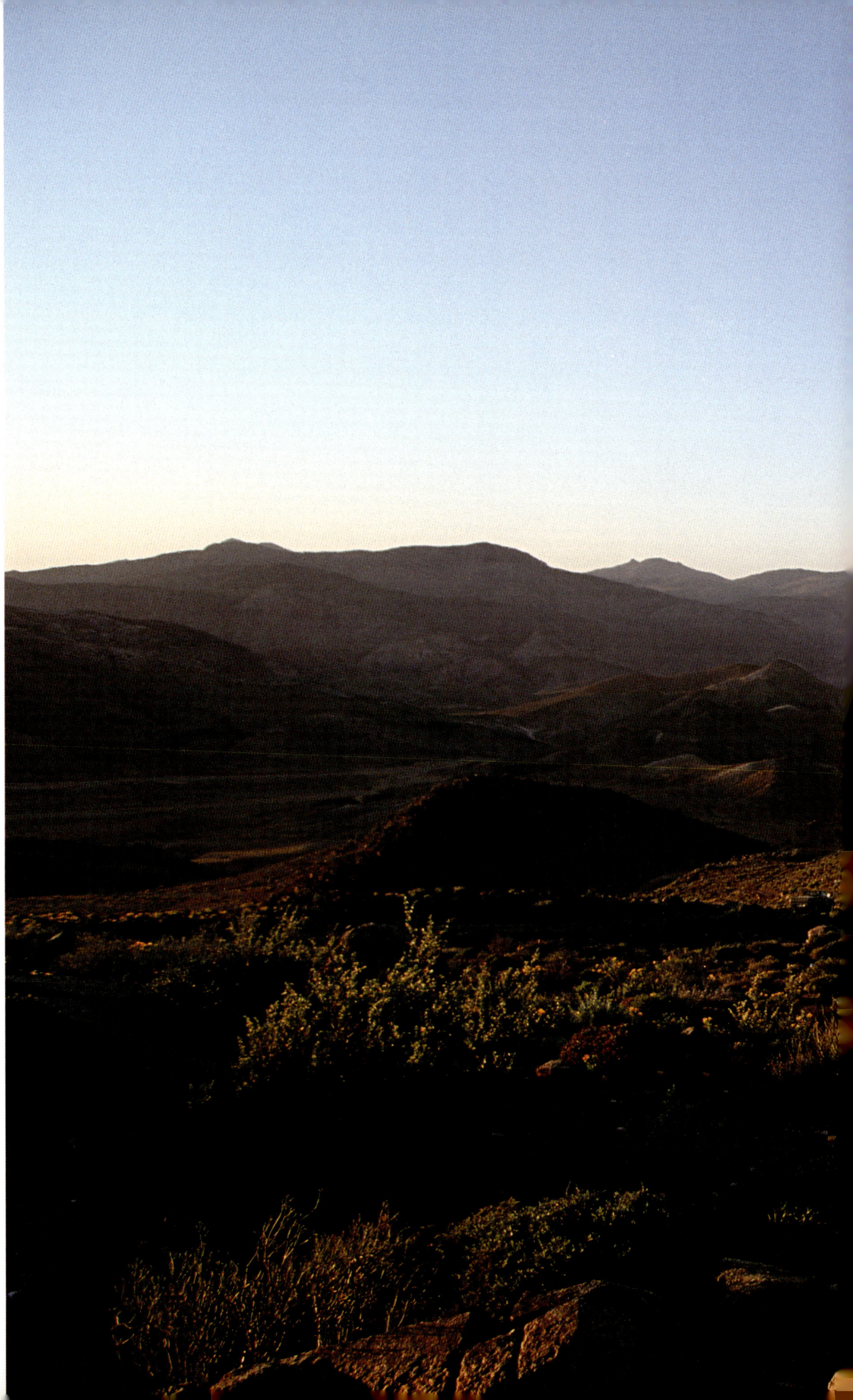

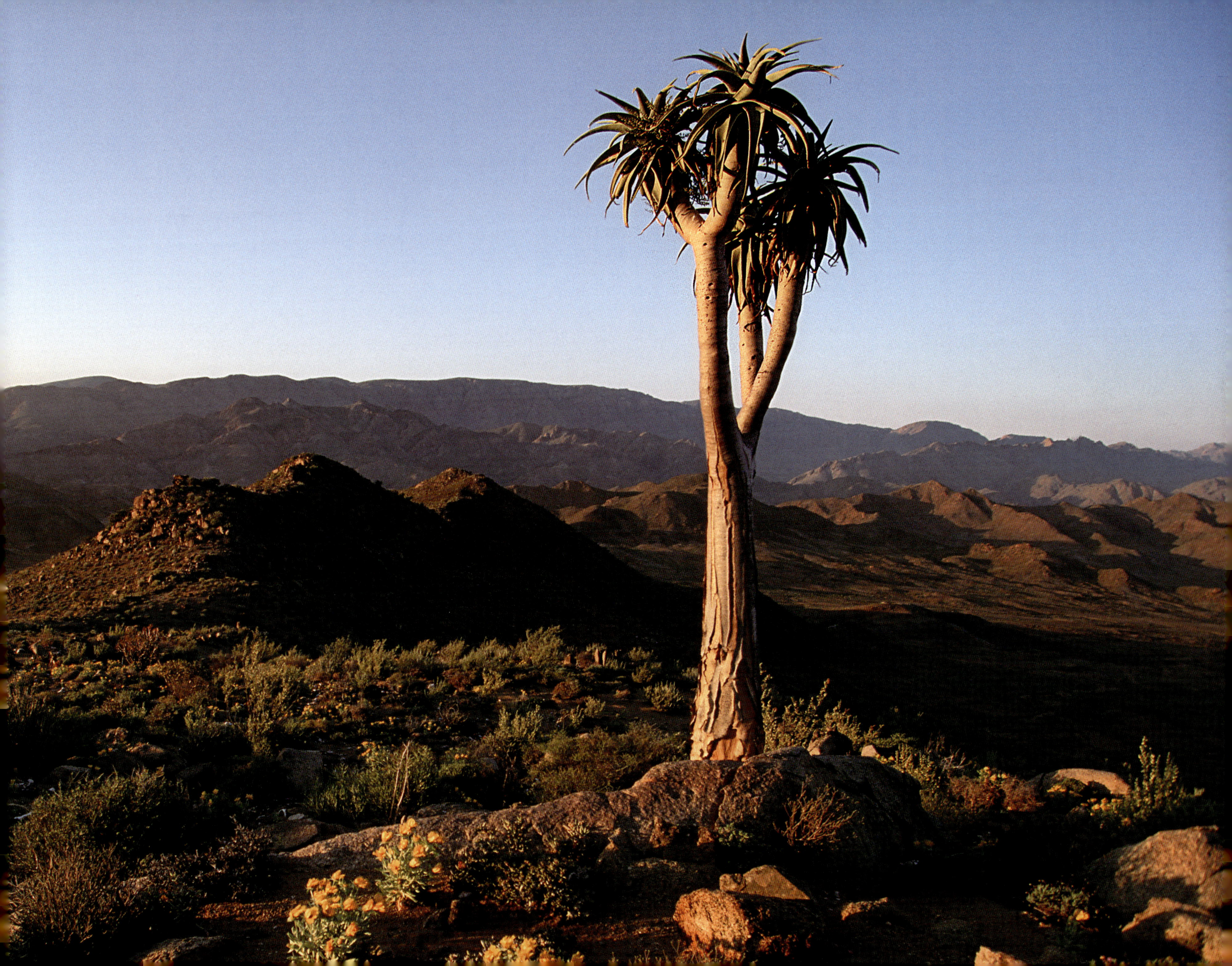

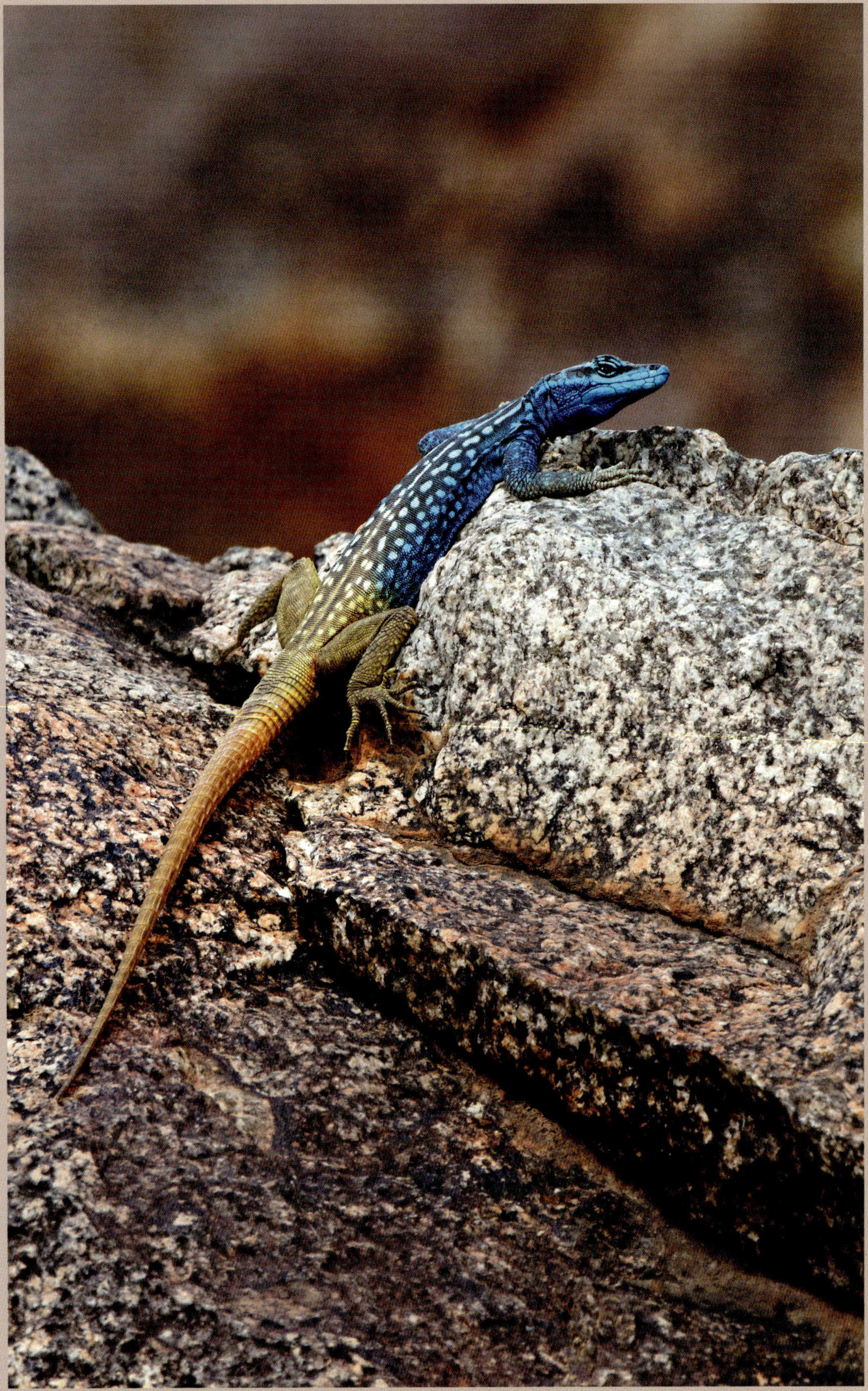

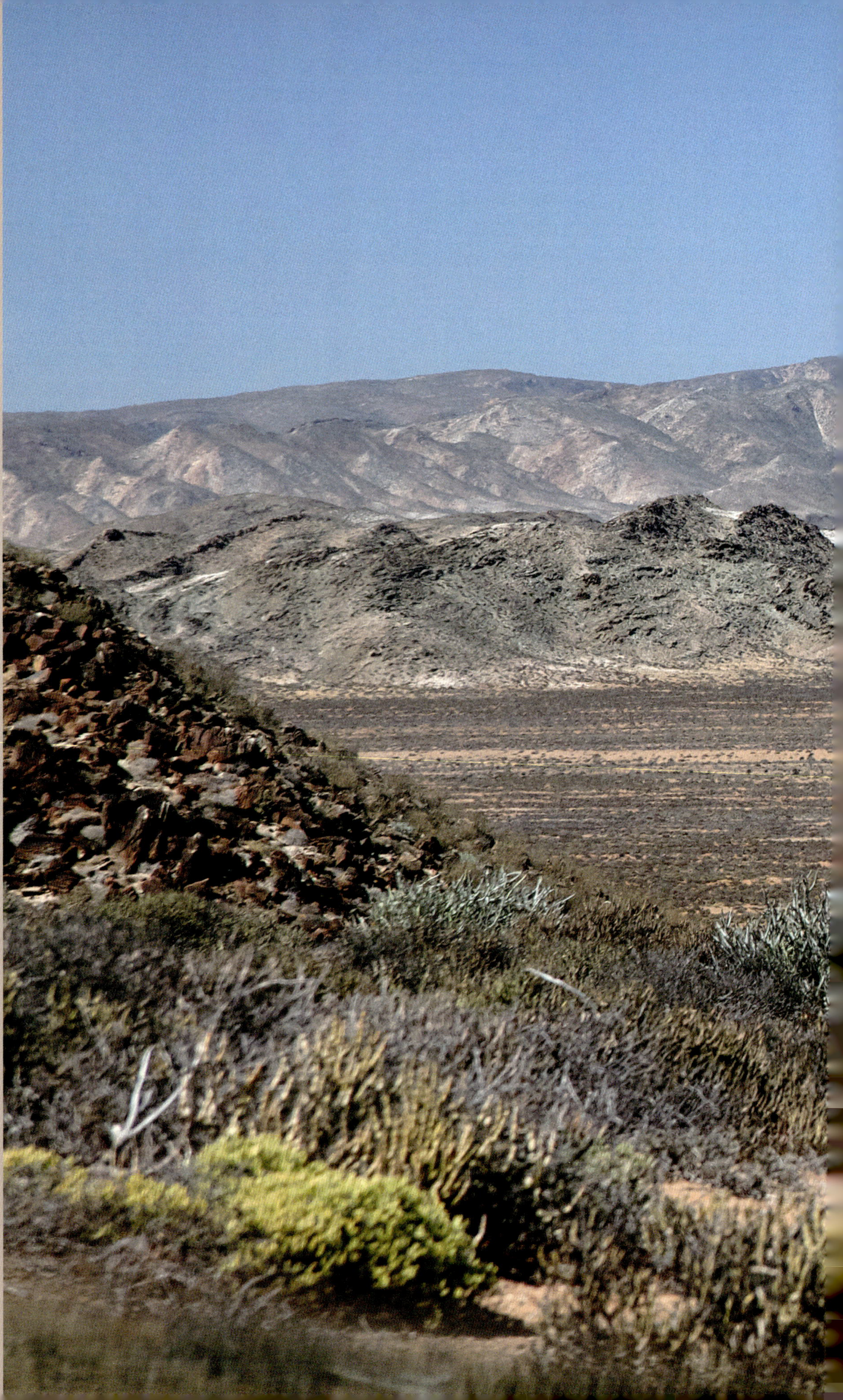

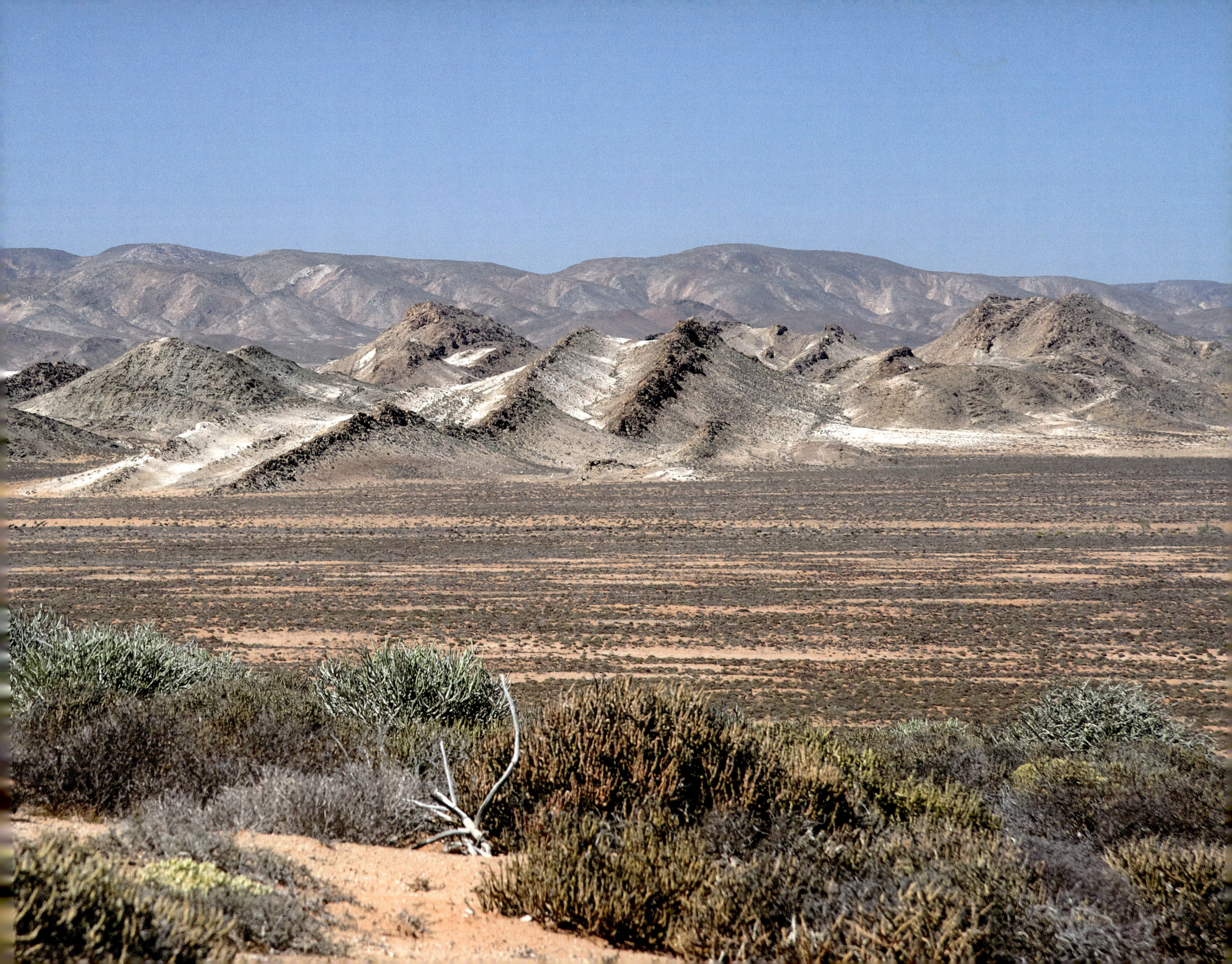

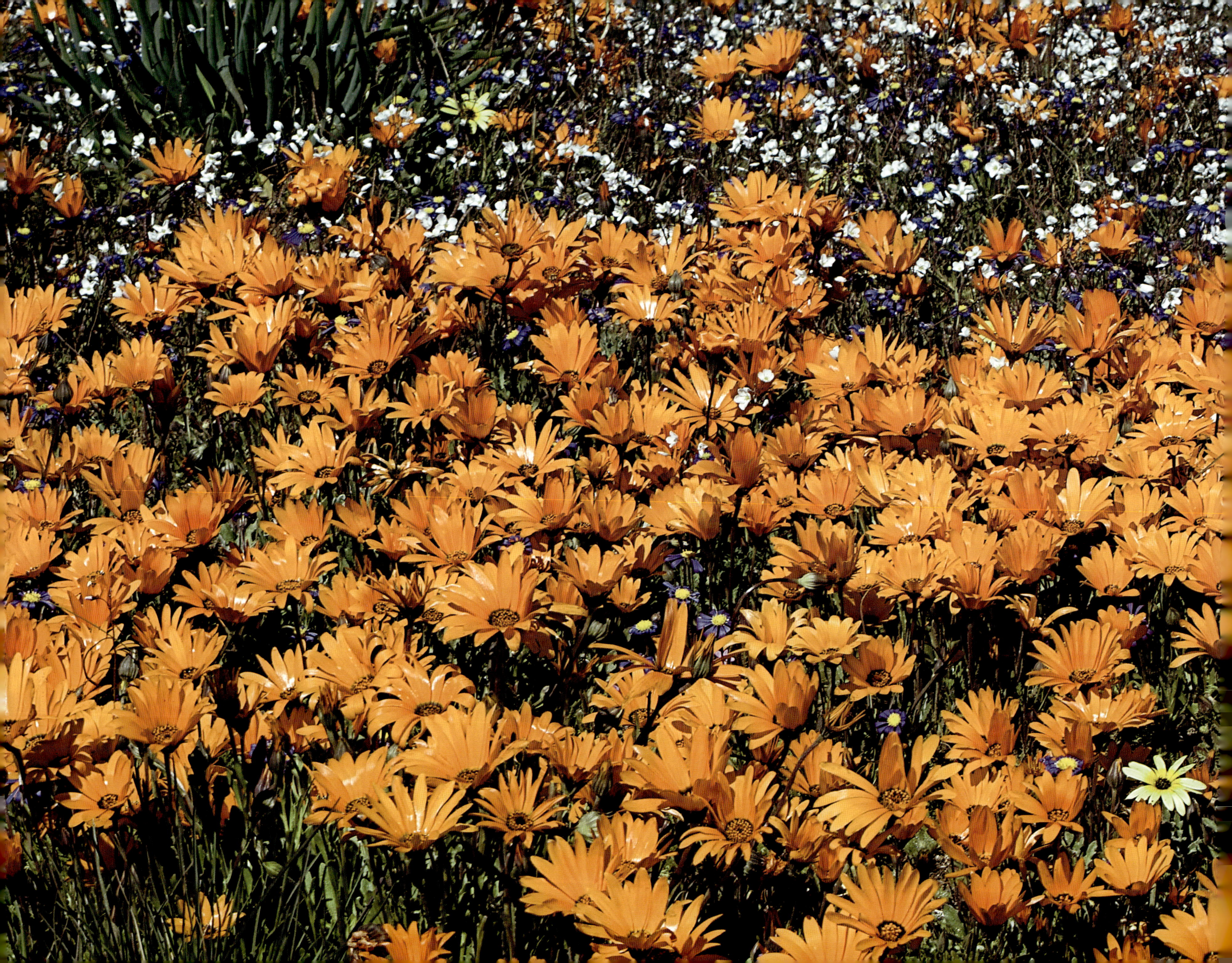

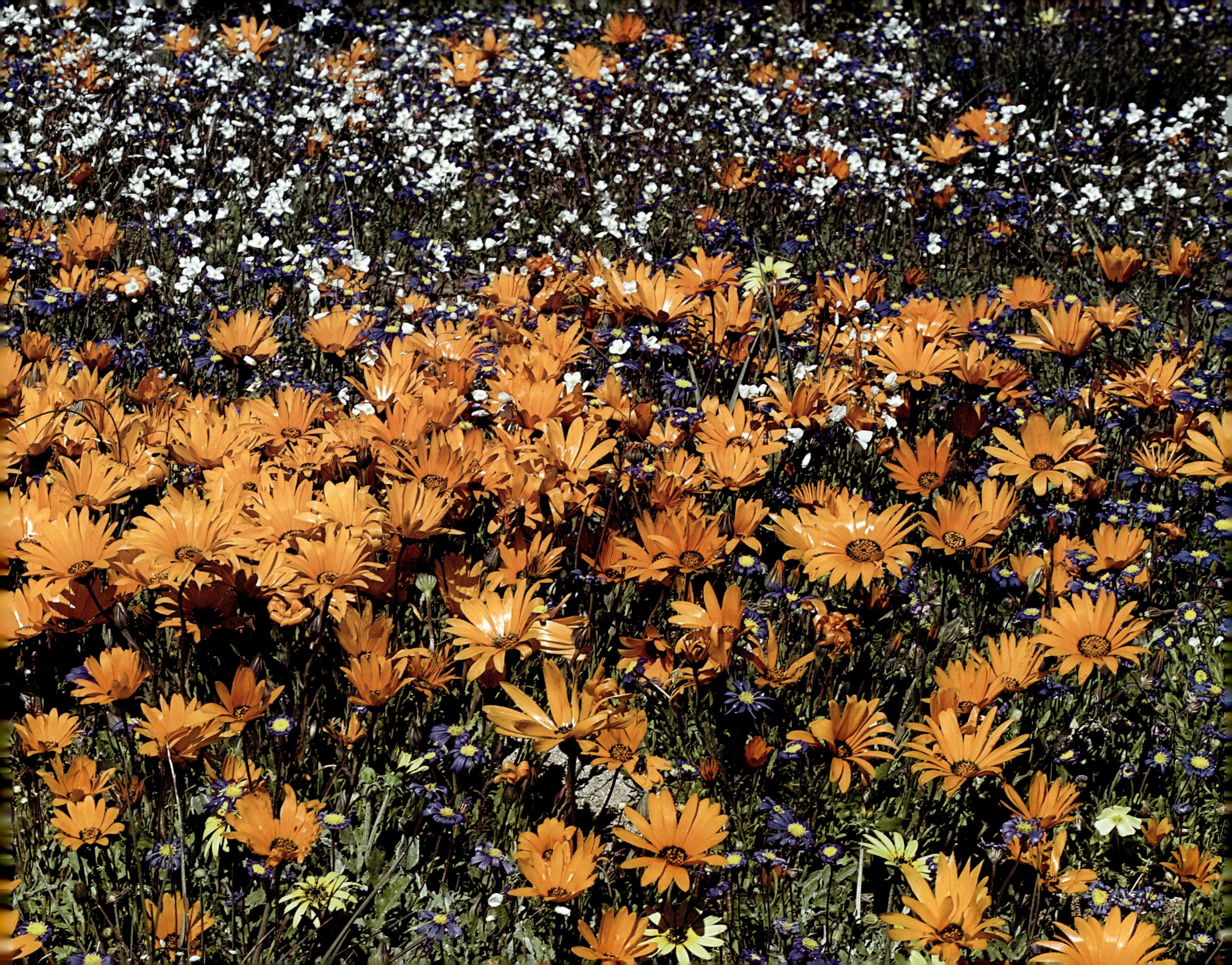

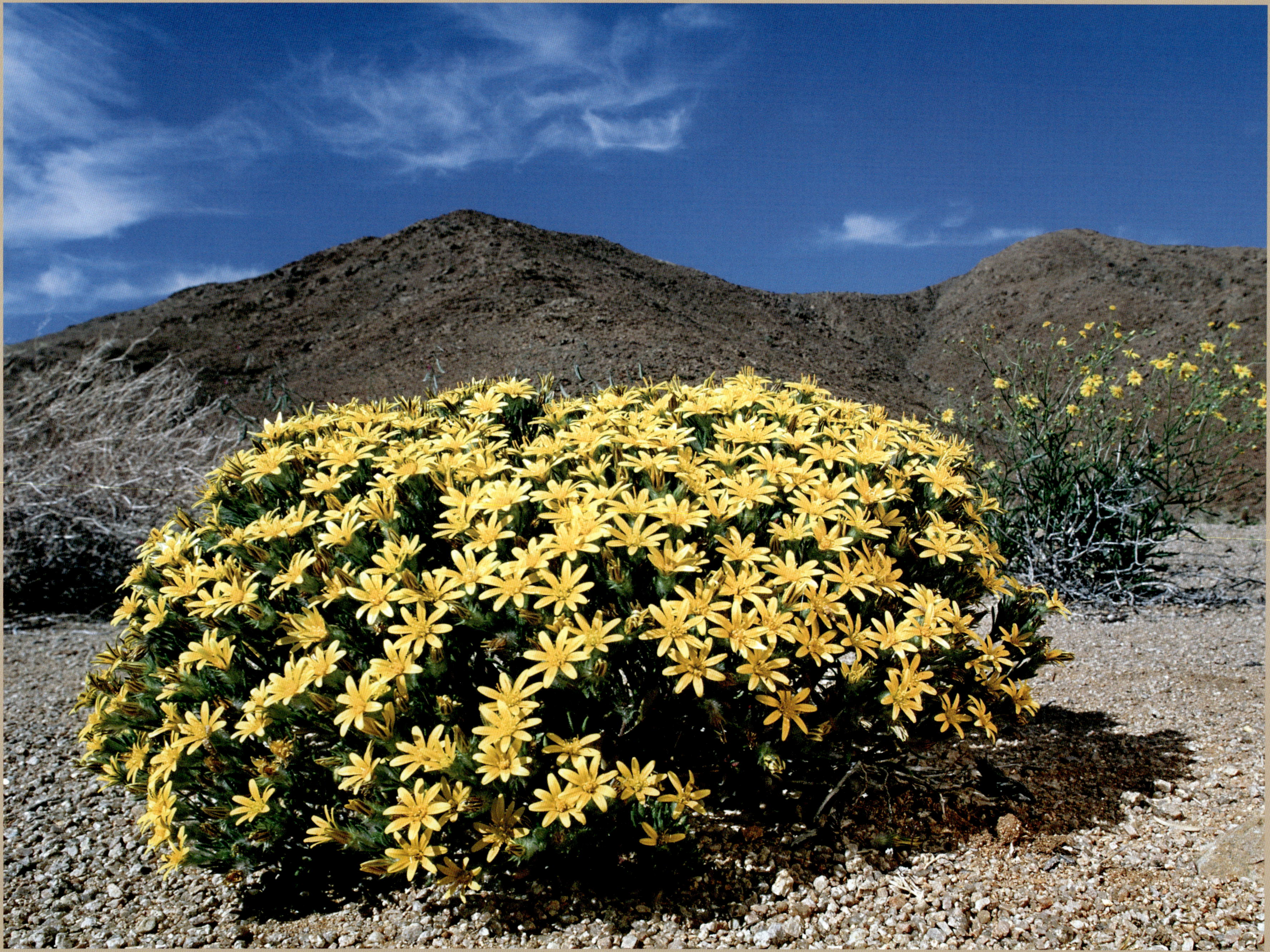

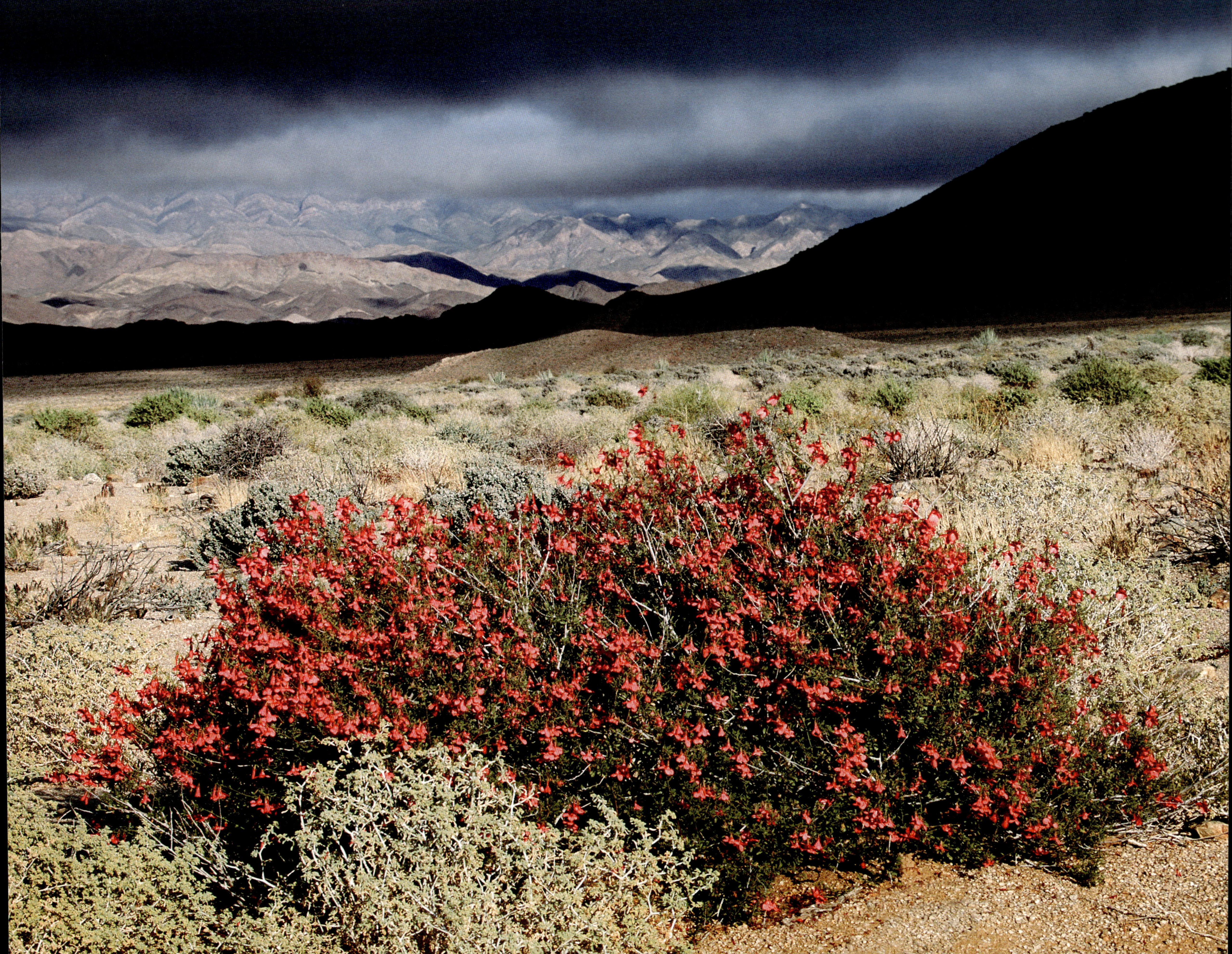

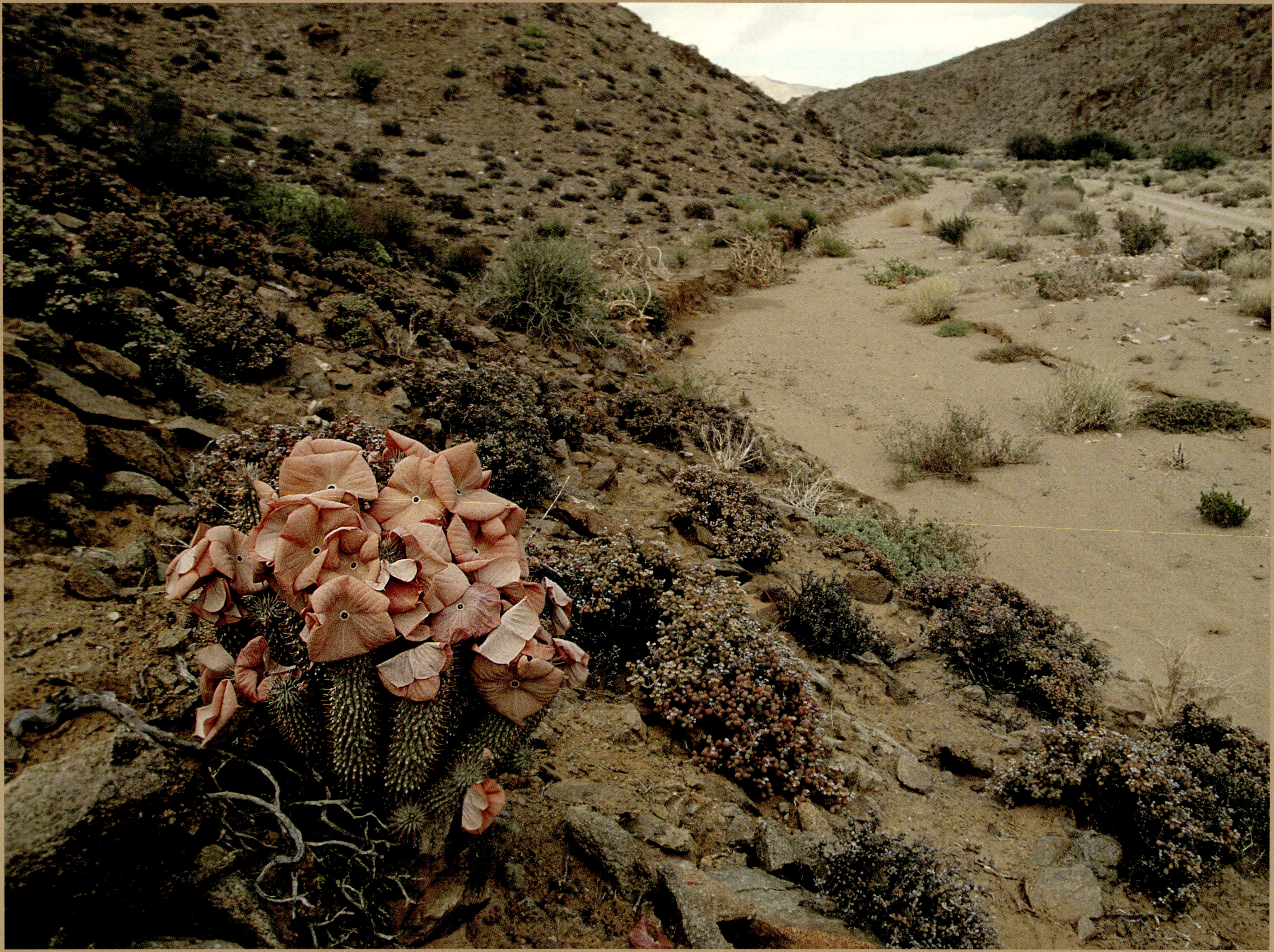

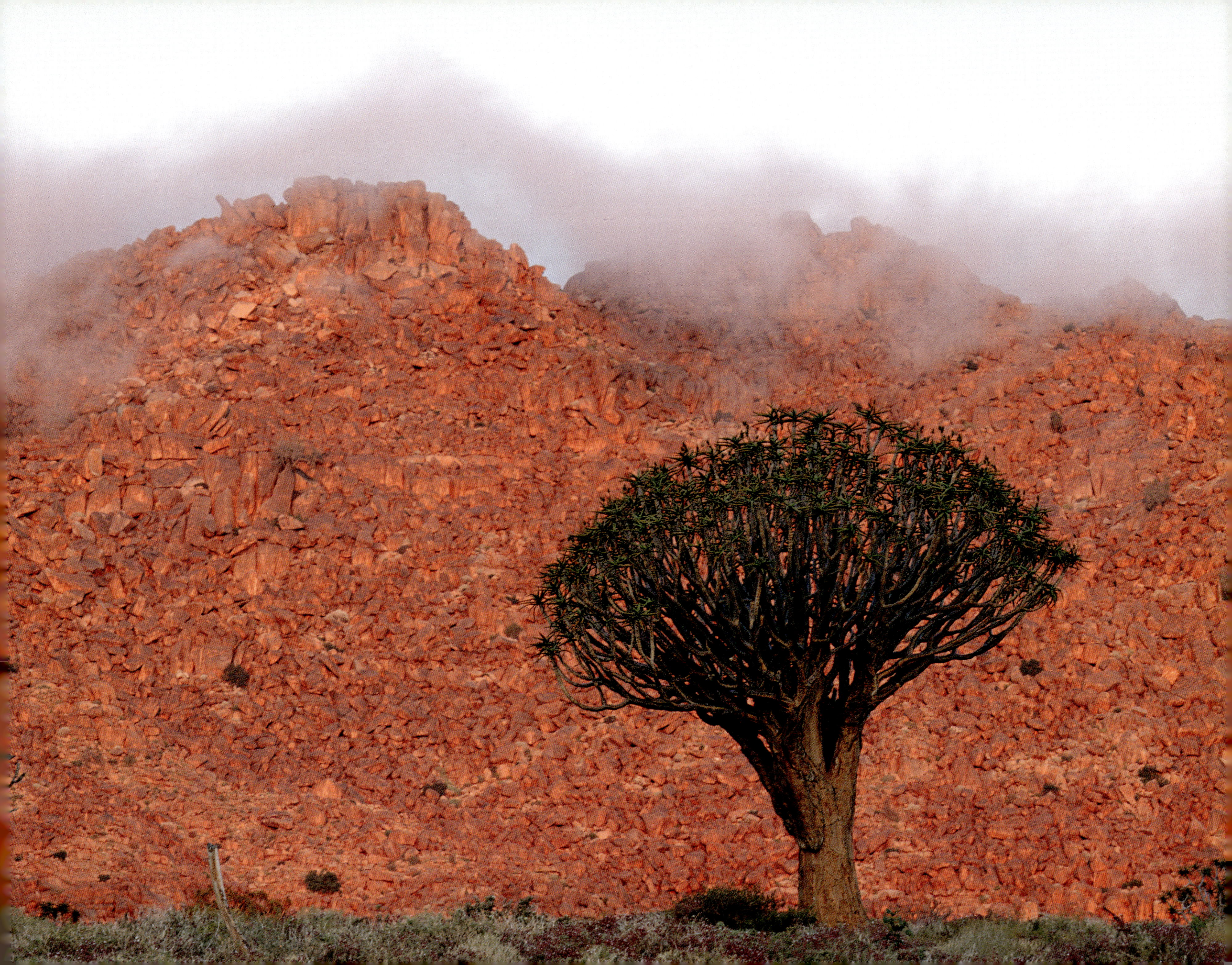

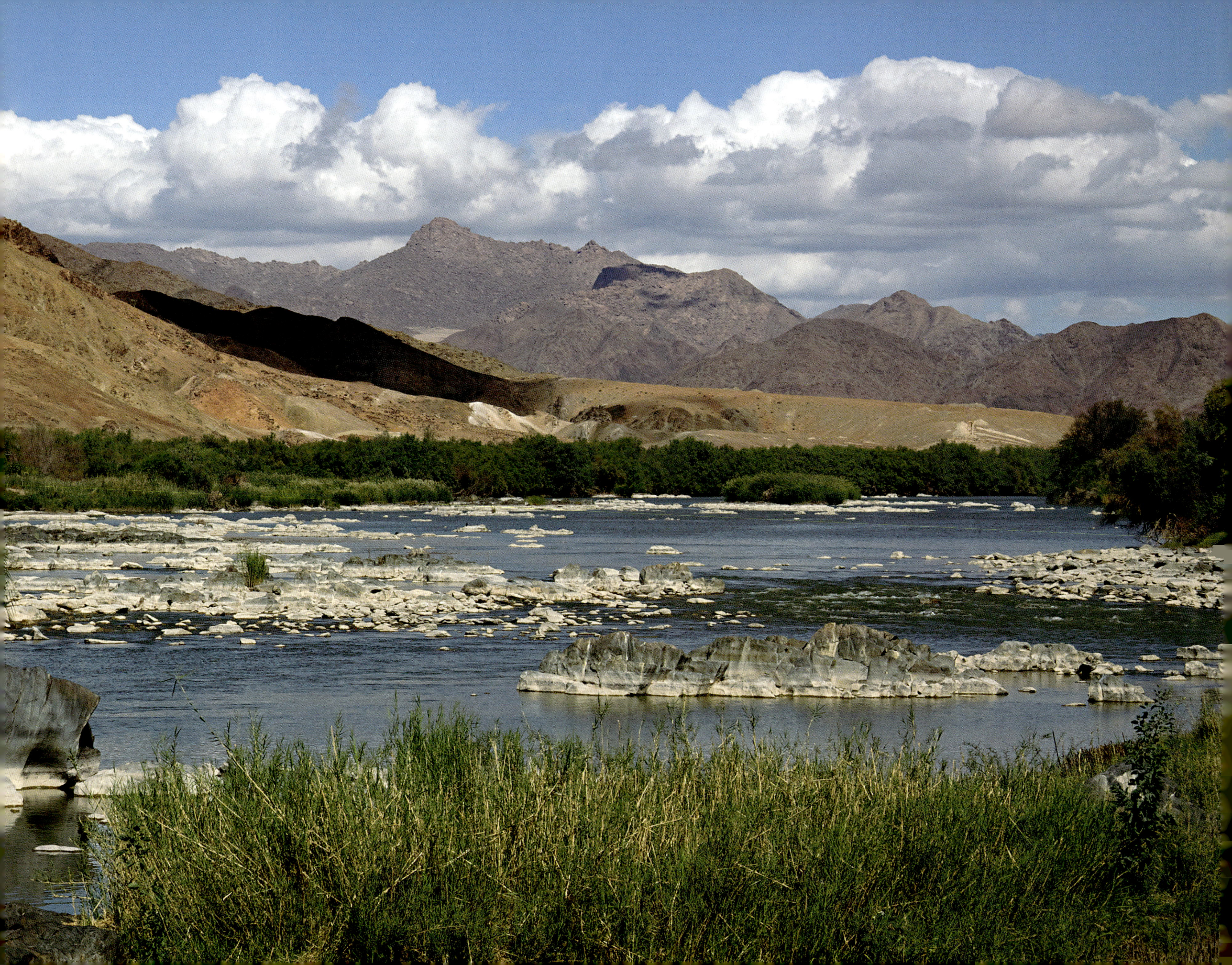

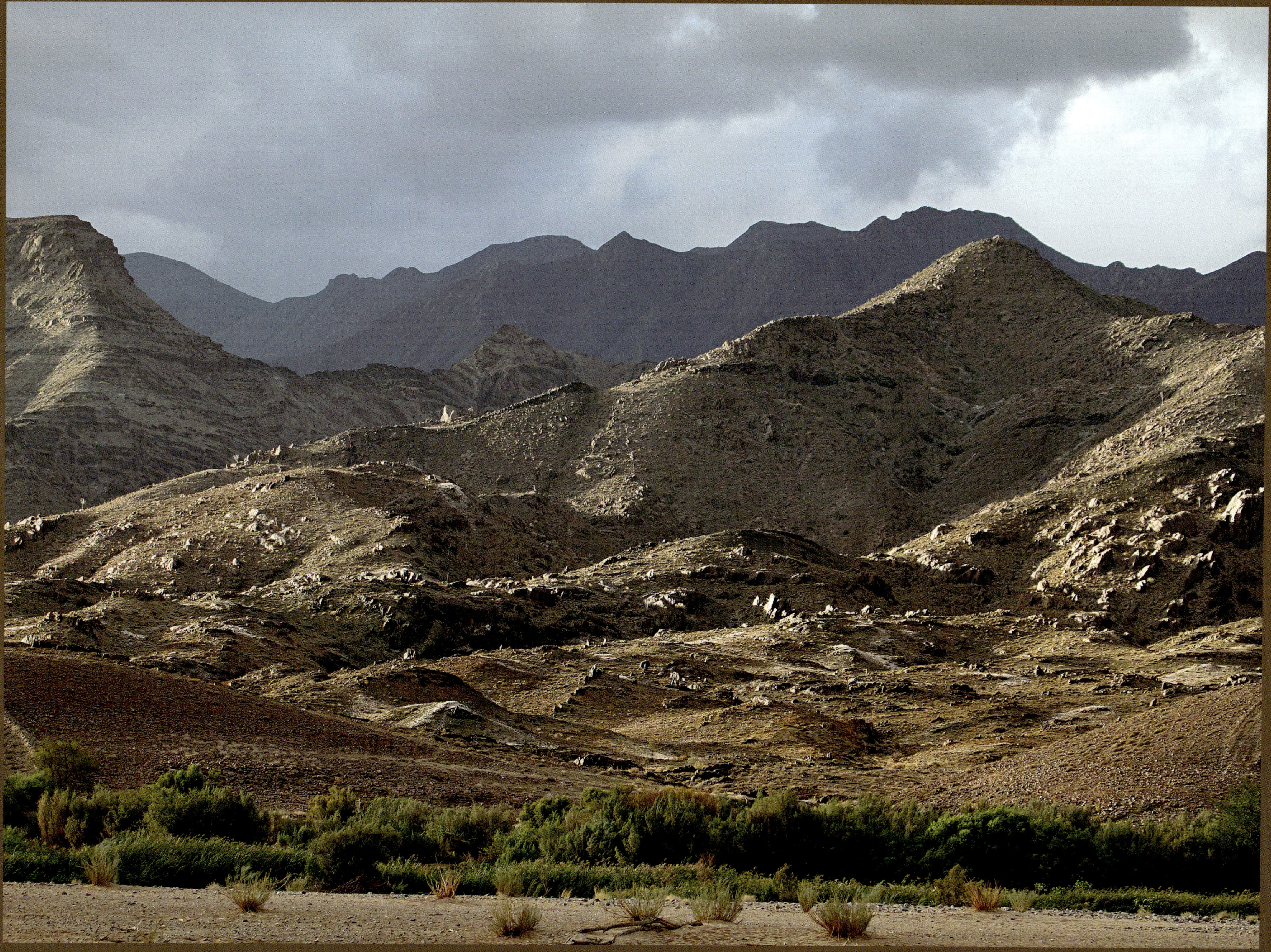

More about the Succulent Karoo Biome

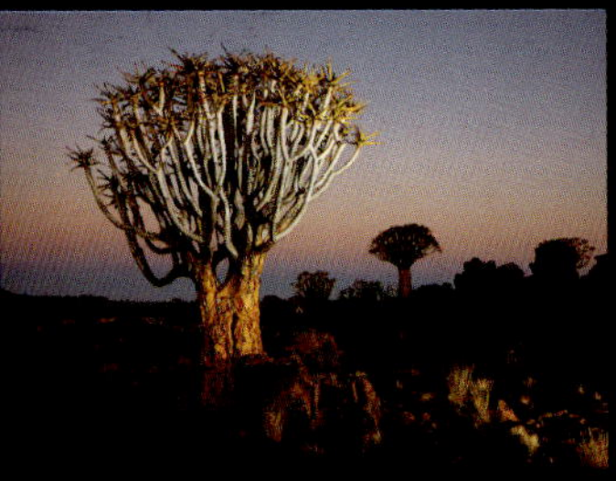
Aloe dichotoma

There are hardly any trees in the Succulent Karoo, but those that do occur are extremely hardy, conspicuous and typical of arid areas. Most spectacular are the quiver trees of the northern section of this biome. With its dense, rounded crown and large golden-brown bark scales that reflect the sun, the common quiver tree, *Aloe dichotoma*, is striking and occurs from Nieuwoudtville all the way north into Namibia. They grow on rocky outcrops, sometimes in dense stands where they are popular tourist attractions. The best places to see such dense stands or 'forests' are between Nieuwoudtville and Loeriesfontein in South Africa, and north of Keetmanshoop in Namibia. Slow-growing, the quiver tree soaks up water to store in its fibrous trunk. The San people used the hollow stems of the branches for their quivers, hence the name. Like most aloes, the leaves are spirally arranged in rosettes. The bright yellow flowers are borne high up on erect flower heads, which is a distinguishing feature of this species.

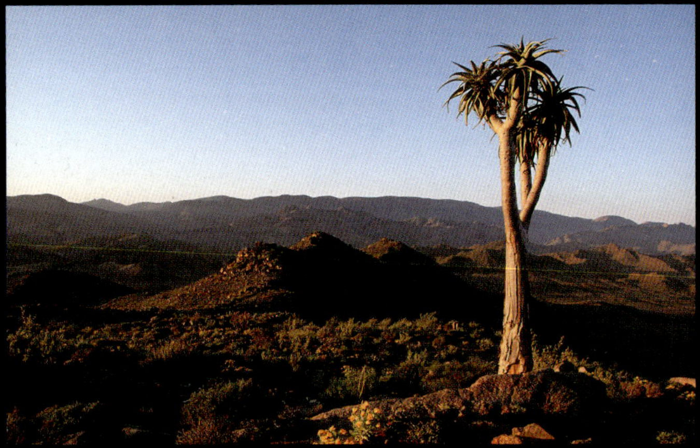
Aloe pillansii

Extremely endangered and endemic to the Richtersveld, is the bastard quiver tree, *A.pillansii*. This scarce giant can grow up to 10m high. Characterised by fewer but larger leaf rosettes, the flower heads (inflorescences) arise almost horizontally from the lower leaves and the bark lacks the scaly appearance of *A.dichotoma*.

The craggy land called the Richtersveld is a place like no other. A sense of true wilderness prevails in this inhospitable, lunar landscape. Yet it is hauntingly beautiful. The unfolding vistas of rough mountain ranges reaching into the far distance offer profound experiences of wonder and amazement. Mountains of decomposing grey-brown granites, crumbling weathered rocks of black dolorite, quartzites in shades of white to pink, conglomerates of various geological origins and the most prevalent colours of grey to black, all give the impression of space and desolation.

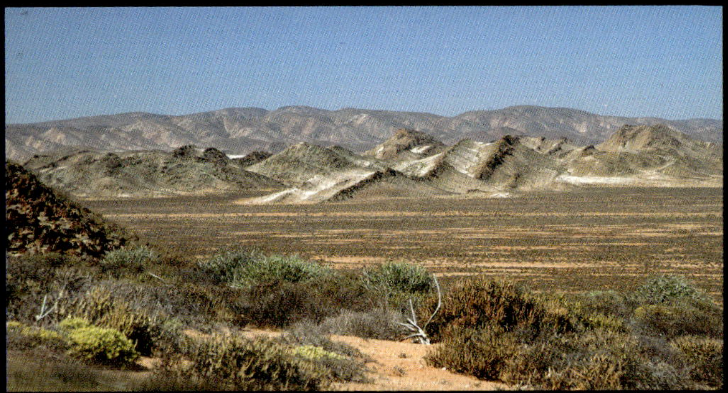
Typical Richtersveld landscape

When early morning light moves over the mountain tops into the deep valleys, the colour of the different rock faces change from mauve to pink and yellow. But once the sun sits high in the sky, the magic colours have all vanished.

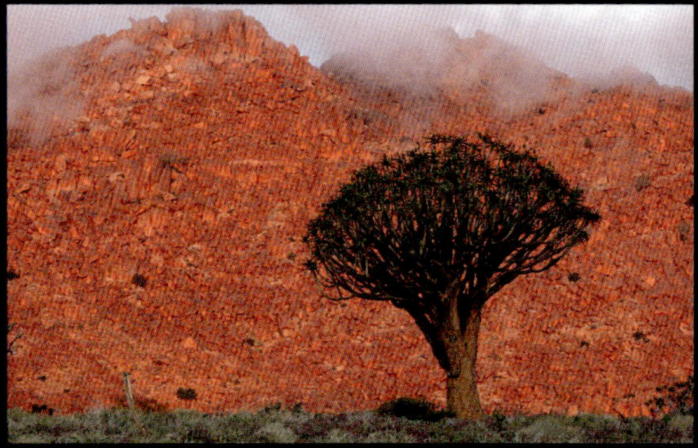
Morning glow on rocky slope in the Richtersveld

The Richtersveld is a place of hidden jewels. In the midst of all the rocky rubble, a succulent hoodia, *Hoodii gordonii*, spills its leathery flowers for a few days and so sends its putrid smell to attract the flies that pollinate them. The thorns on the succulent stems protect them from being grazed and the spiny angles help to reduce evaporation by scattering and reflecting the sun's rays.

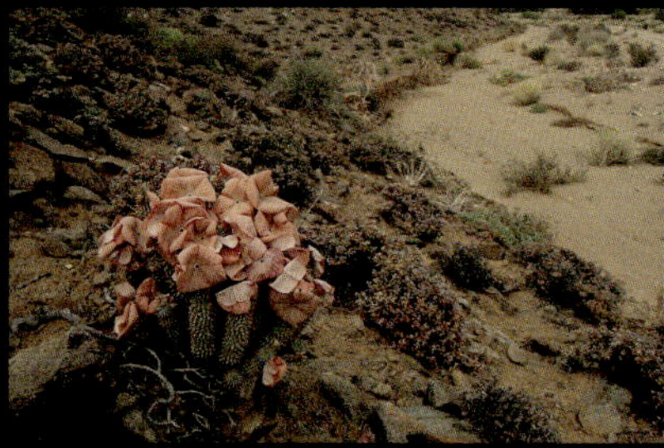
Hoodii gordonii

The shy Cape flat lizard favours rocky areas in the Richtersveld. It usually lives in small groups and is often seen sunning itself on top of a rock. When it feels threatened, it runs for cover. While it is a predator that hunts by ambush, it will also eat flowers and berries.

Namaqualand is famous for its spring flowers. Each year thousands of tourists visit the region between the end of July and early September to witness this extraordinary floral splendour. In the long, dry summers air temperatures are often over 40°C and the land has a rather bleak aspect, but after the first winter rains the veld undergoes a swift and extraordinary change as carpets of flowers cover the earth, often as far as the eye can see. Fields of orange, purple, pink, white and yellow dazzle the eye. These ephemeral plants survive the dry season in the form of seeds and most belong to the daisy family, though many succulent perennial mesembs also contribute to the colourful display.

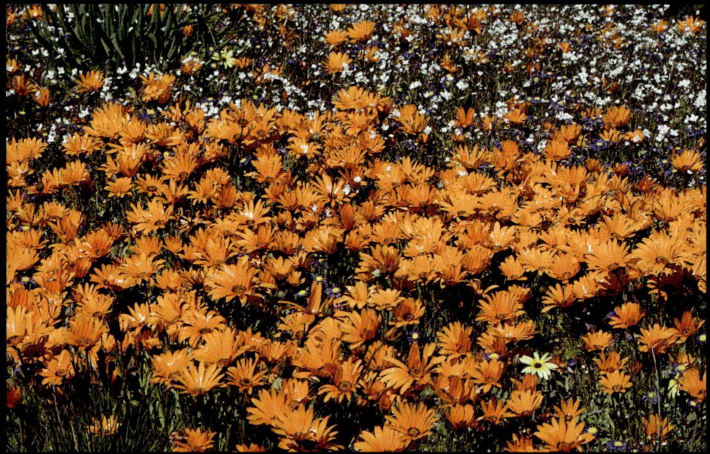
Spring scene at Namaqua National Park

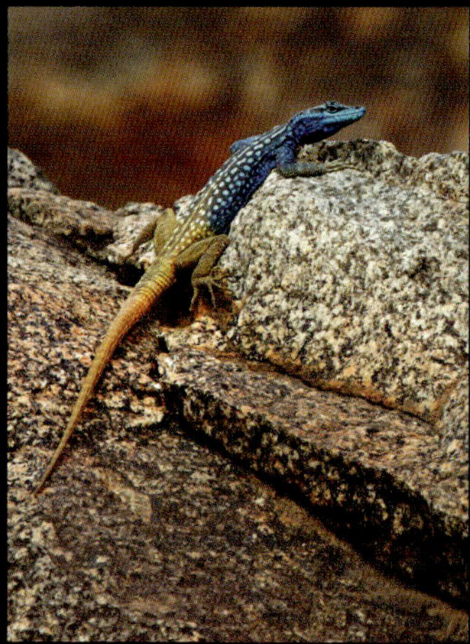
Cape flat lizard

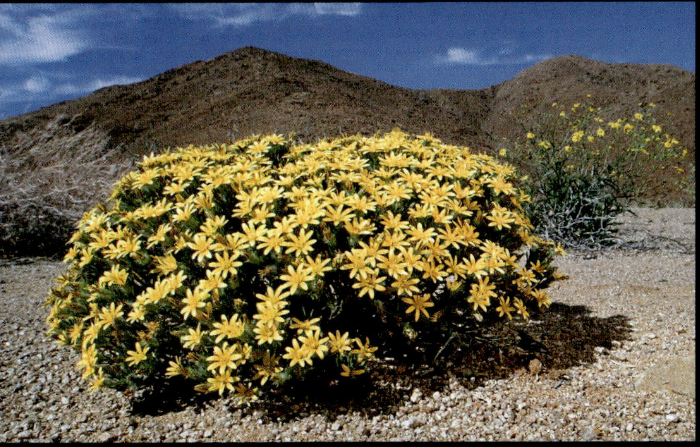

Perennial yellow daisy bush

After rain the seeds germinate quickly, flower in a few days, set seed and die back again, taking full advantage of the short-lived moisture in the soil.

Most of the succulent plants are leaf succulents, which means they store moisture in their leaves to withstand the extremely dry times of the year. Largest of these families are the mesembs comprising of about 2 500 species. Since they all have a noonday flowering habit, they only open their blooms when the light is sufficiently strong. They range in size and growth form from compact perennials such as the famous stone plants with one or two leaf pairs to many others like the tufted or cushion-forming perennials with mostly light yellow flowers, the sprawling annuals called ice plants due to small bladder cells in their leaves, the creeping annuals with flowers that range from white to red and mauve (bokbaai vygies), and the dwarf perennial shrubs (vygies). Most mesembs have brightly coloured flowers to attract pollinators, yet none of them are blue.

Not all the plants store water in their leaves or stems above ground: many do so in underground organs, such as bulbs, corms or swollen roots that provide a valuable food source to many animals, especially digging and burrowing mammals. Termed geophytes, these plants are not strictly regarded as succulents. The striking desert rose, *Hermannia stricta*, with its nodding pink-red tubular flowers, provides a splash of colour. It flowers for up to two months even during a prolonged drought. The woody stem is covered with small toothed leaves, limiting water loss and reflecting light.

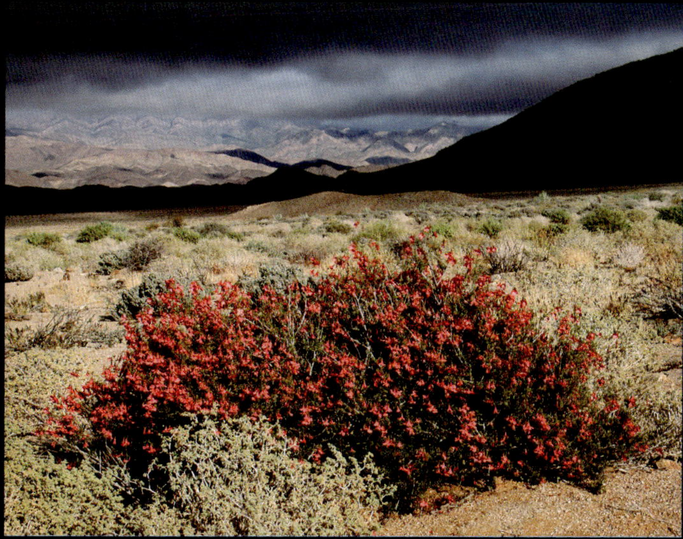

Desert rose or Hermannia stricta

At the place called the Giant's Playground, soft decomposing granites form weird weathered sculptures that invite imaginative naming. The ones called 'chicken drumsticks' illustrate this.

The area is dramatically bisected by the only perennial river in the region. The Orange River winds its way down from far inland to empty its waters into the Atlantic Ocean. In summer its presence gives the visitor some faint sense of relief from the searing heat; in winter it is a reassuring sign of life. But despite the temperature extremes and the constant dry and strong winds that sculpture the landscape, the Richtersveld in the Succulent Karoo is a magic place.

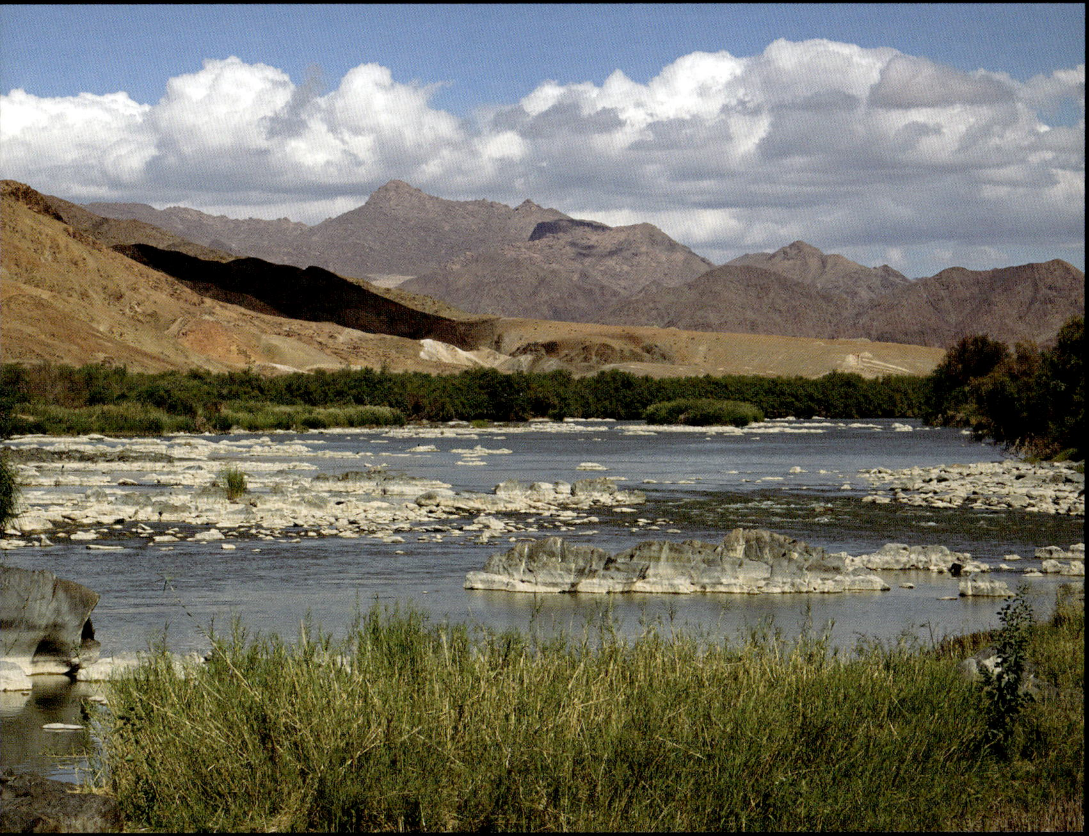

The Orange River in the Richtersveld

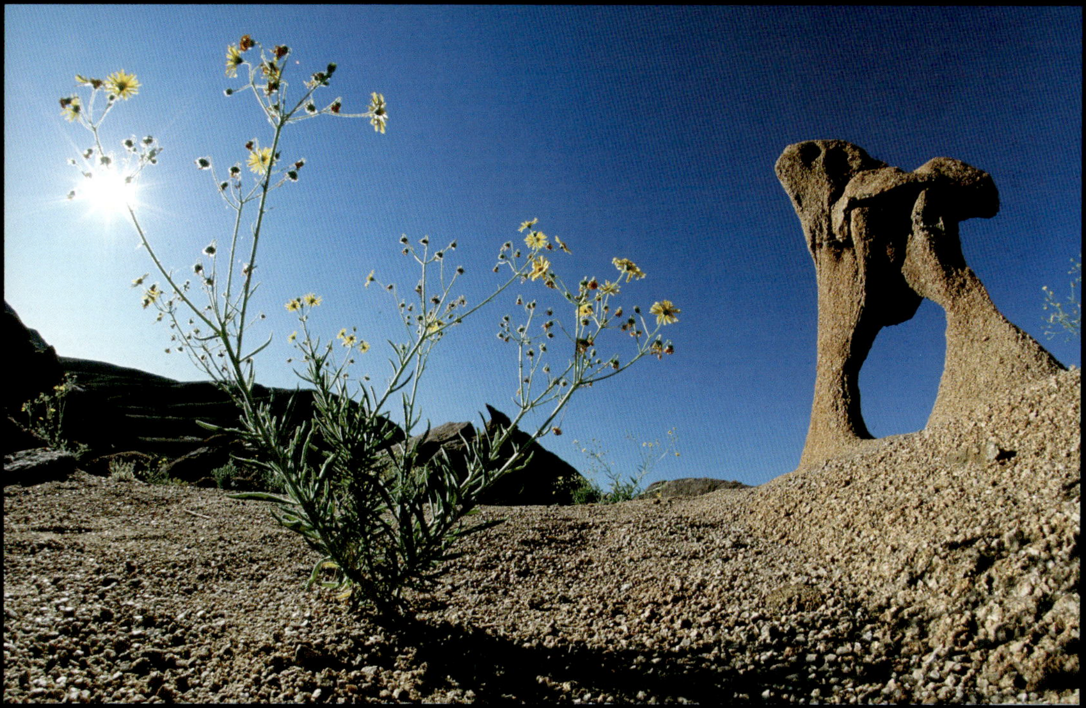

Giant's Playground

Nama Karoo

The central southern African plateau

The vast Nama Karoo is the second largest desert biome in southern Africa and covers most of the dry western central plateau. It stretches from the rain shadow of the Cape Fold Mountains, over the central low-lying basin of the Great Karoo and northwards into Namibia – an area of 605 000km^2.

Millions of years ago, the Karoo was a vast inland basin with seas, lakes and swamps where prehistoric reptiles and amphibians prospered in wet lush forests. Dry and wet stages alternated, forming a crust of sand and mudstone layers. Gradually an ice age set in. Then, about 180 million years ago, severe and large-scale volcanic eruptions took place. Dolorite intrusions pushed through cracks of the earth's crust. Molten lava covered the entire basin and beyond. By wind and rain, heat and cold, erosion gradually chiselled the land into shape. But most of the volcanic rocks were hard and resisted the action of the elements. As the softer sand- and mudstone were eroded away, the volcanic rock remained and protected the softer rocks below, hence the characteristic flat-topped hills or 'koppies' so typical of the Karoo. Other dolorite outcrops are exposed simply as low ridges of tumbled boulders. A few fairly low mountain ranges occur at the rim of the escarpment.

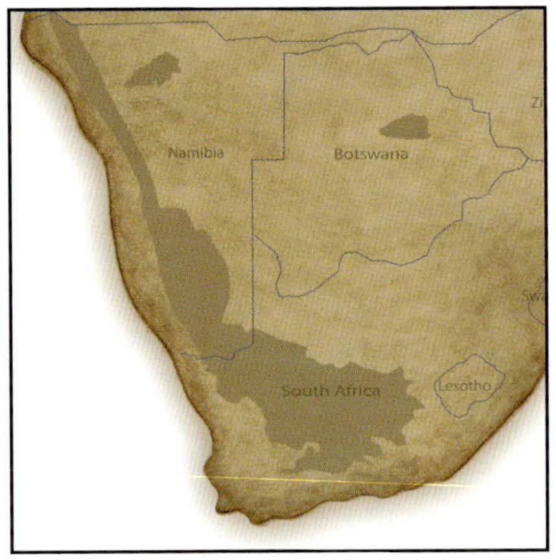

Rain falls in summer and autumn and varies from 200mm per annum in the west to 400mm in the east. Not only is the rainfall unreliable, but also patchy, with sporadic thundershowers. The vegetation is so sparse that wildfires are rare and play no significant role.

The prehistoric animals have all vanished over the millennia, but their remains have made the Karoo a famous destination for palaeontologists. Today the most conspicuous mammal in the Karoo is the well-known springbok. It used to occur in large migratory herds following the rain, but today one finds only smaller herds on farms and game reserves.

Places to experience the Nama Karoo

Within South Africa, the Tankwa Karoo National Park near Sutherland is small and difficult to reach but very special to birders and botanists; the Karoo National Park near Beaufort West is a good place to see fossils, a variety of Karoo game and experience Karoo silence; Camdeboo National Park near Graaff-Reinet offers fantastic vistas over the plains of the Camdeboo and good sightings of typical Karoo fauna; the Mountain Zebra National Park near Cradock is famous for the conservation of that rare animal; at the Augrabies Falls National Park the landscapes along the Orange River and the magnificent Augrabies Waterfall are very special and the Kgalagadi Transfrontier Park also partly falls within the Nama Karoo Biome. Within Namibia, most of the south-eastern part falls within the Nama Karoo and there are several smaller conservation areas with typical elements of this biome, e.g. the eastern Namib-Naukluft Park, the Hardap Dam Reserve and in the Keetmanshoop area, as well as parts of the Ais-Ais and Fish River Canyon Park.

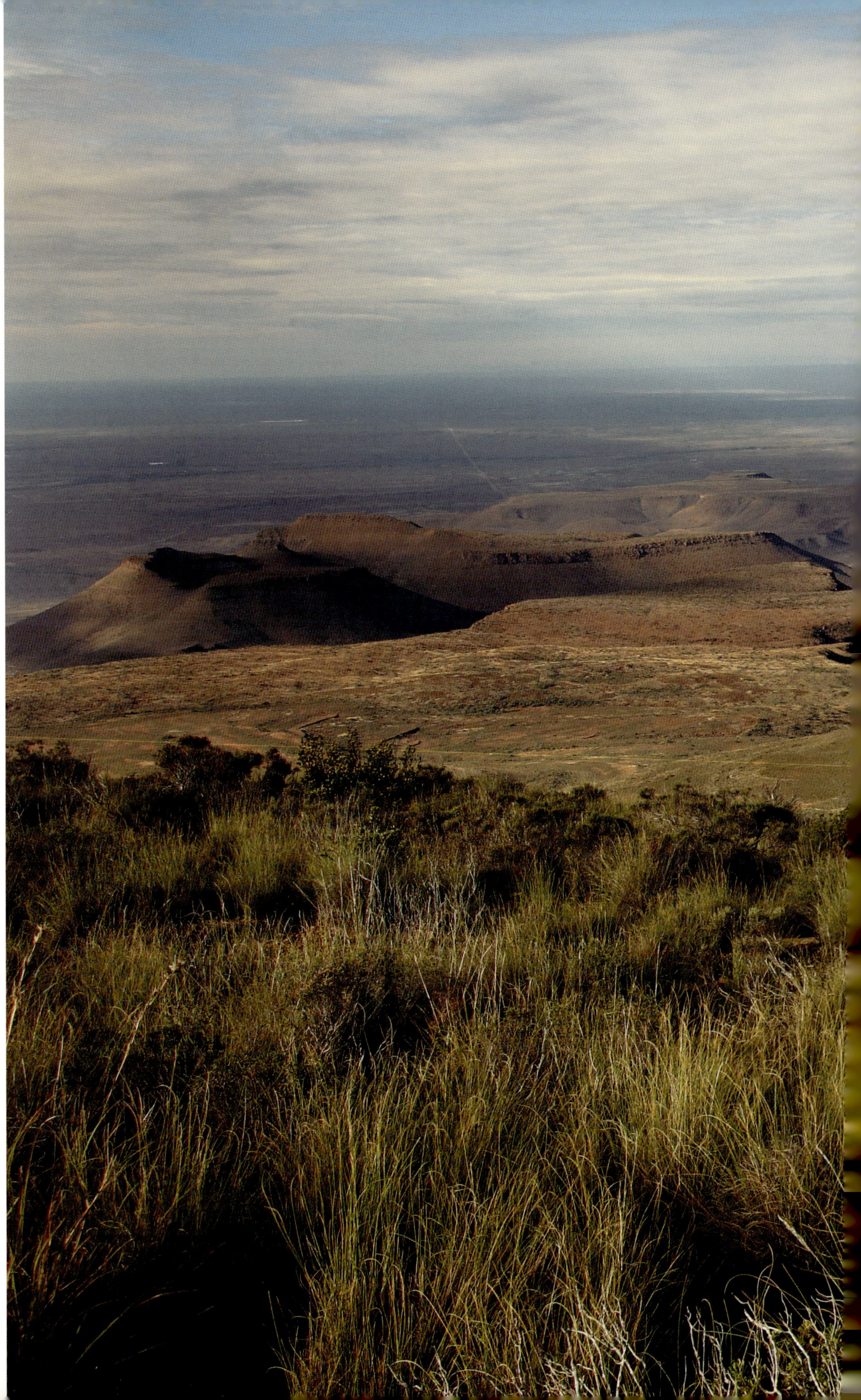

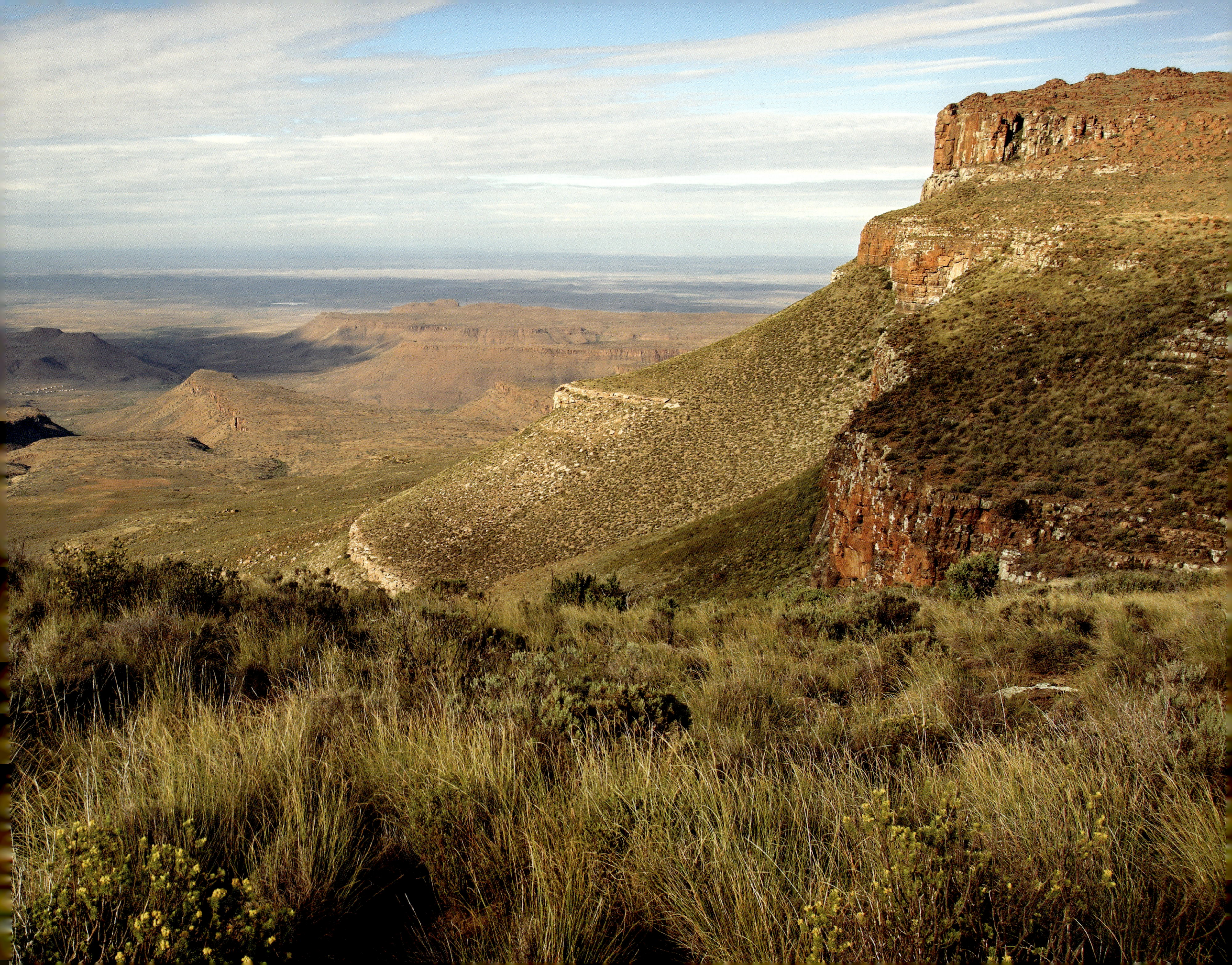

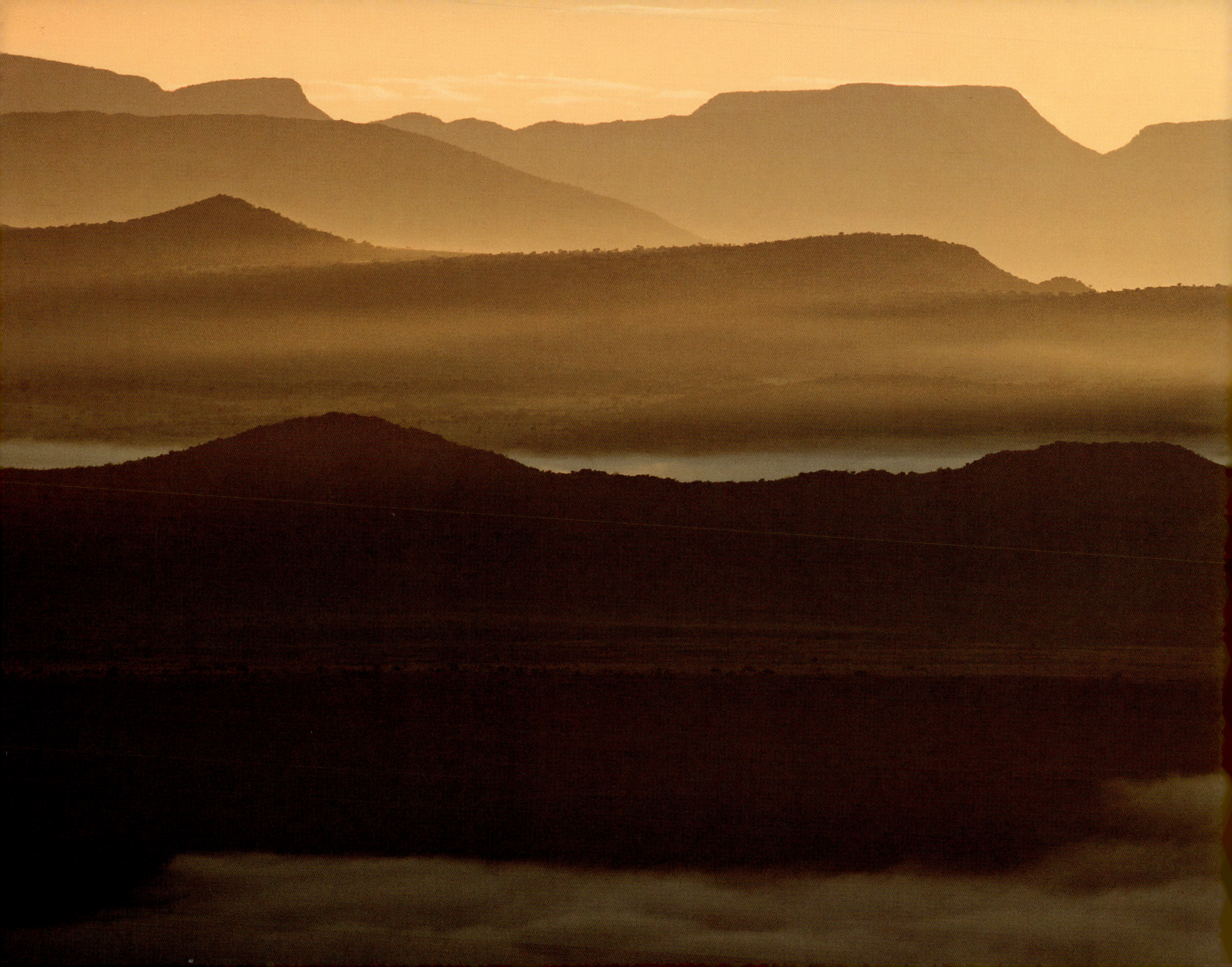

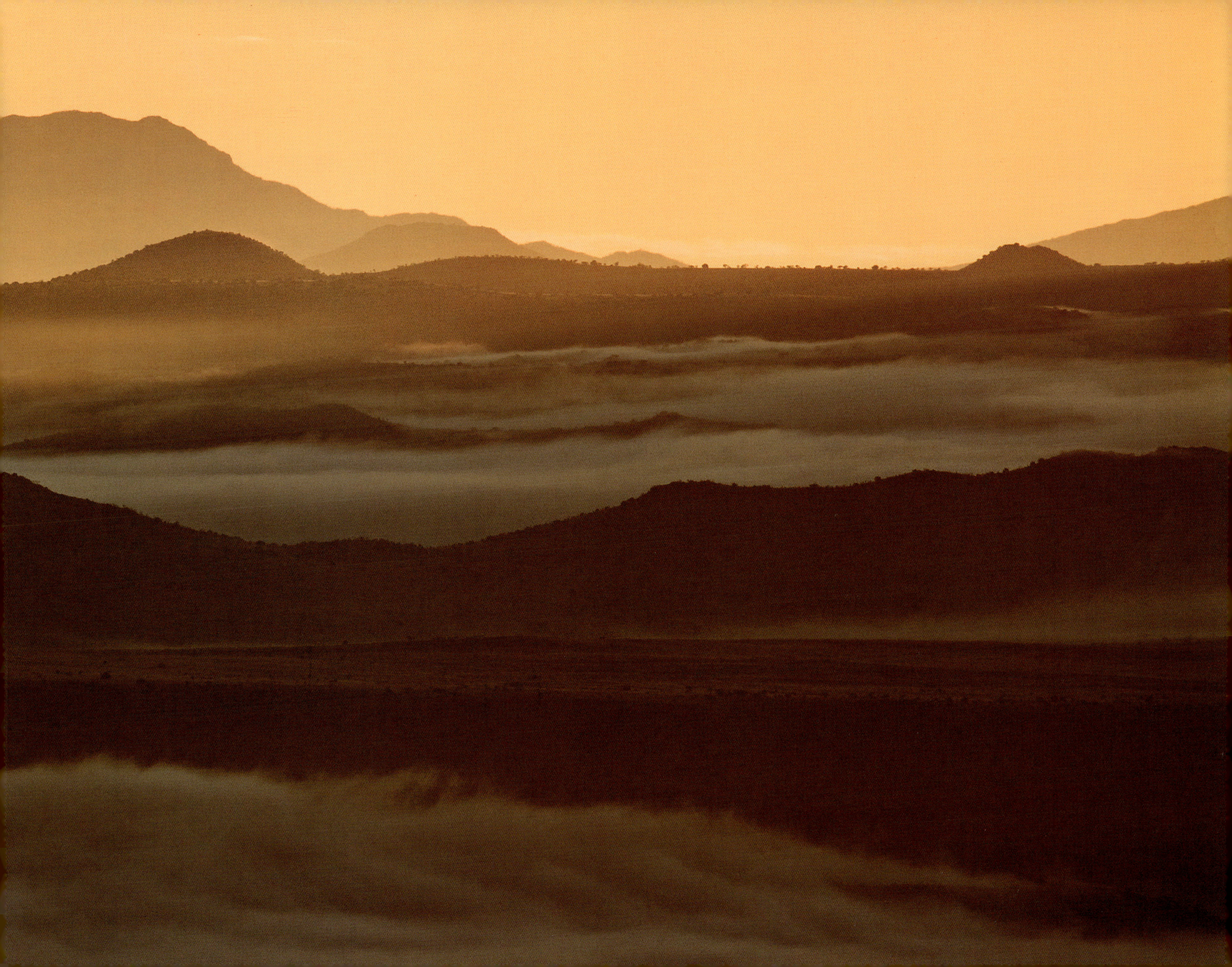

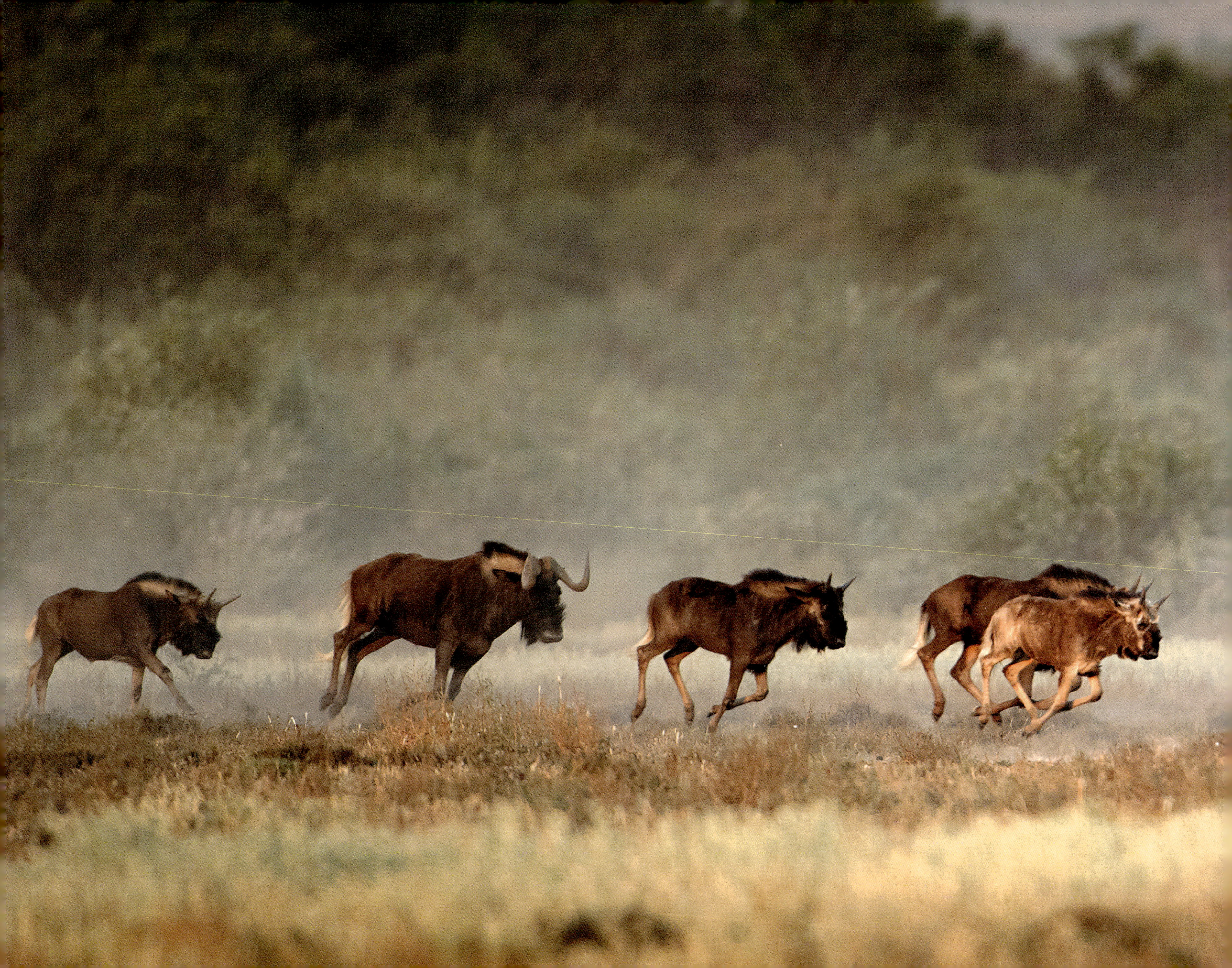

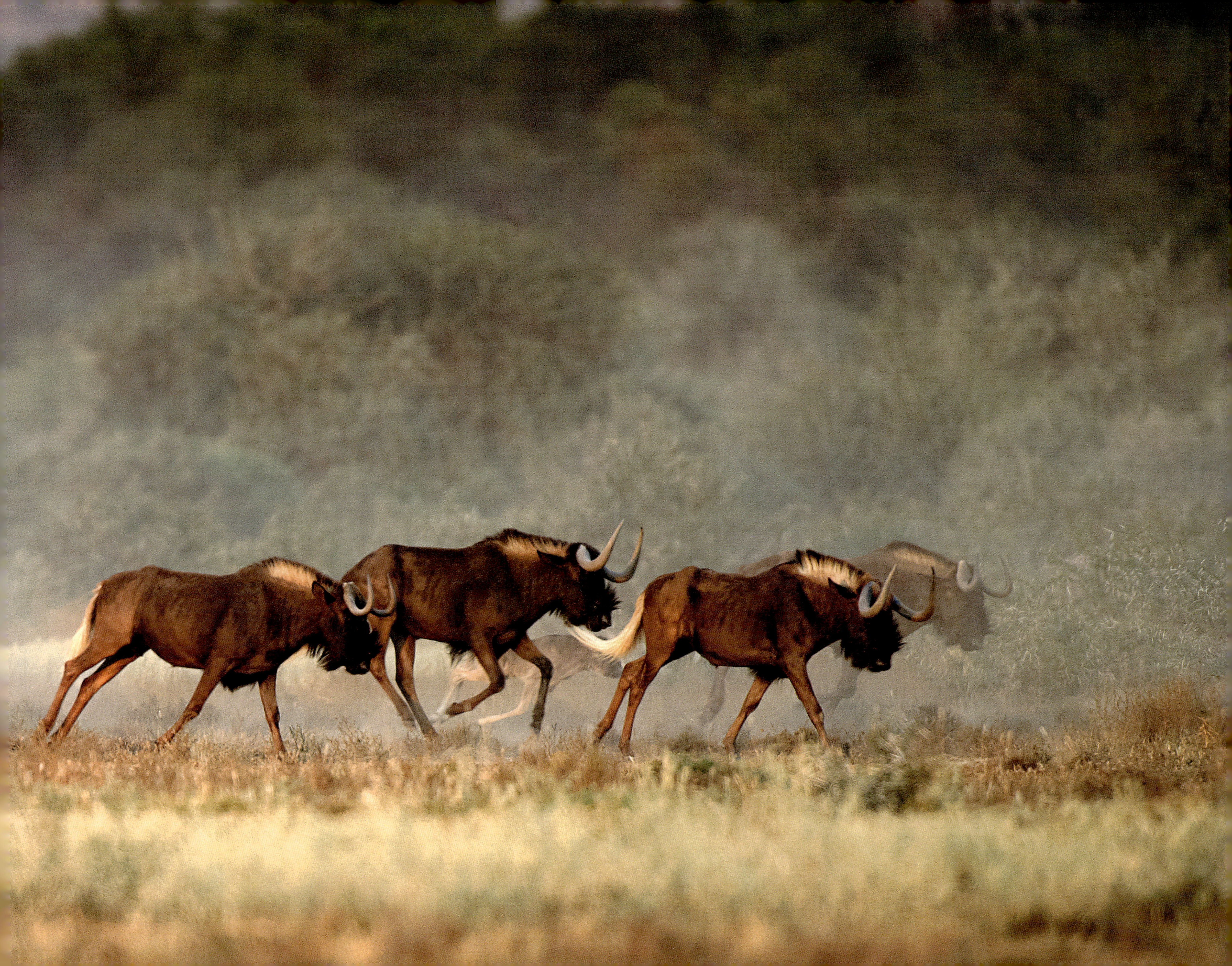

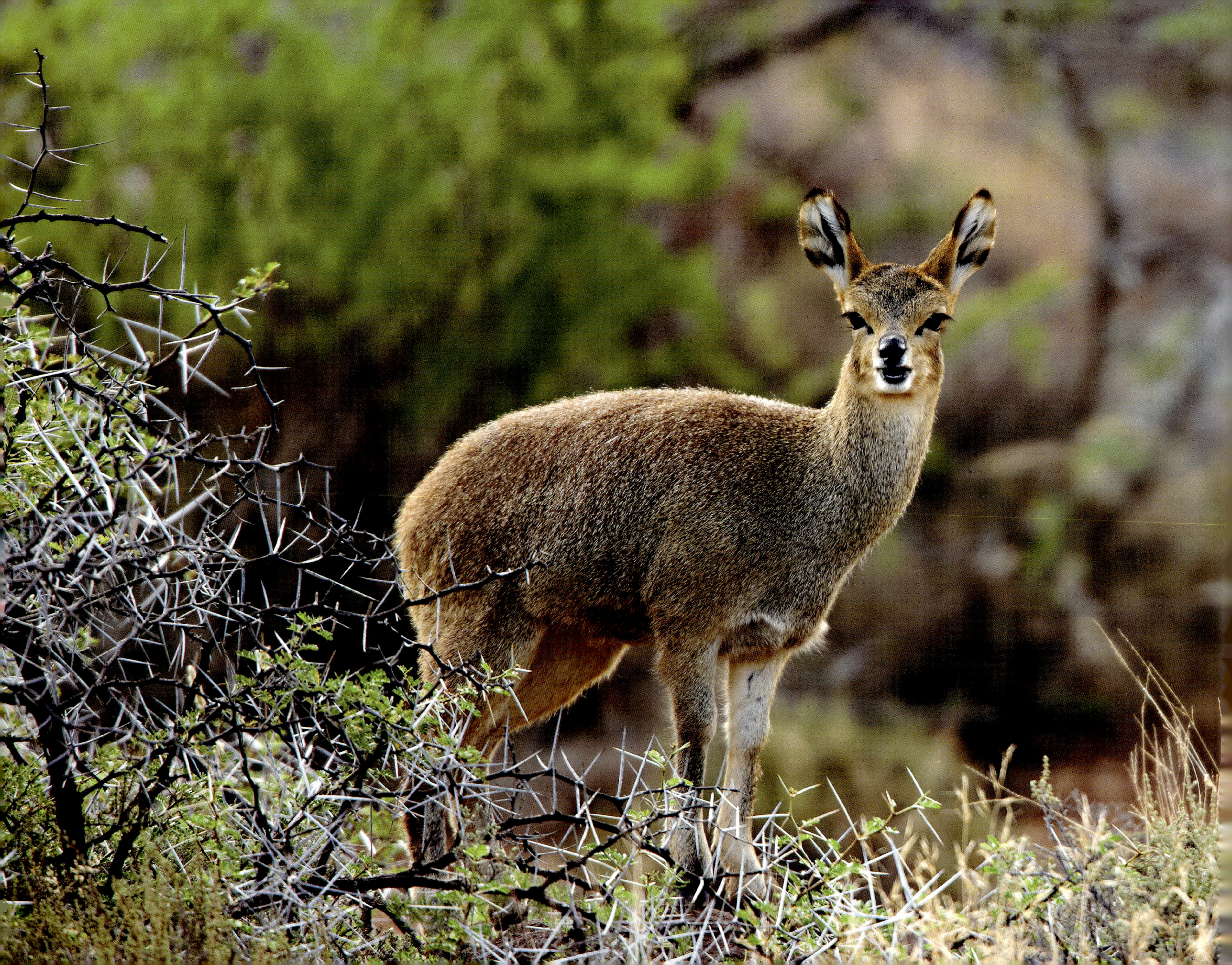

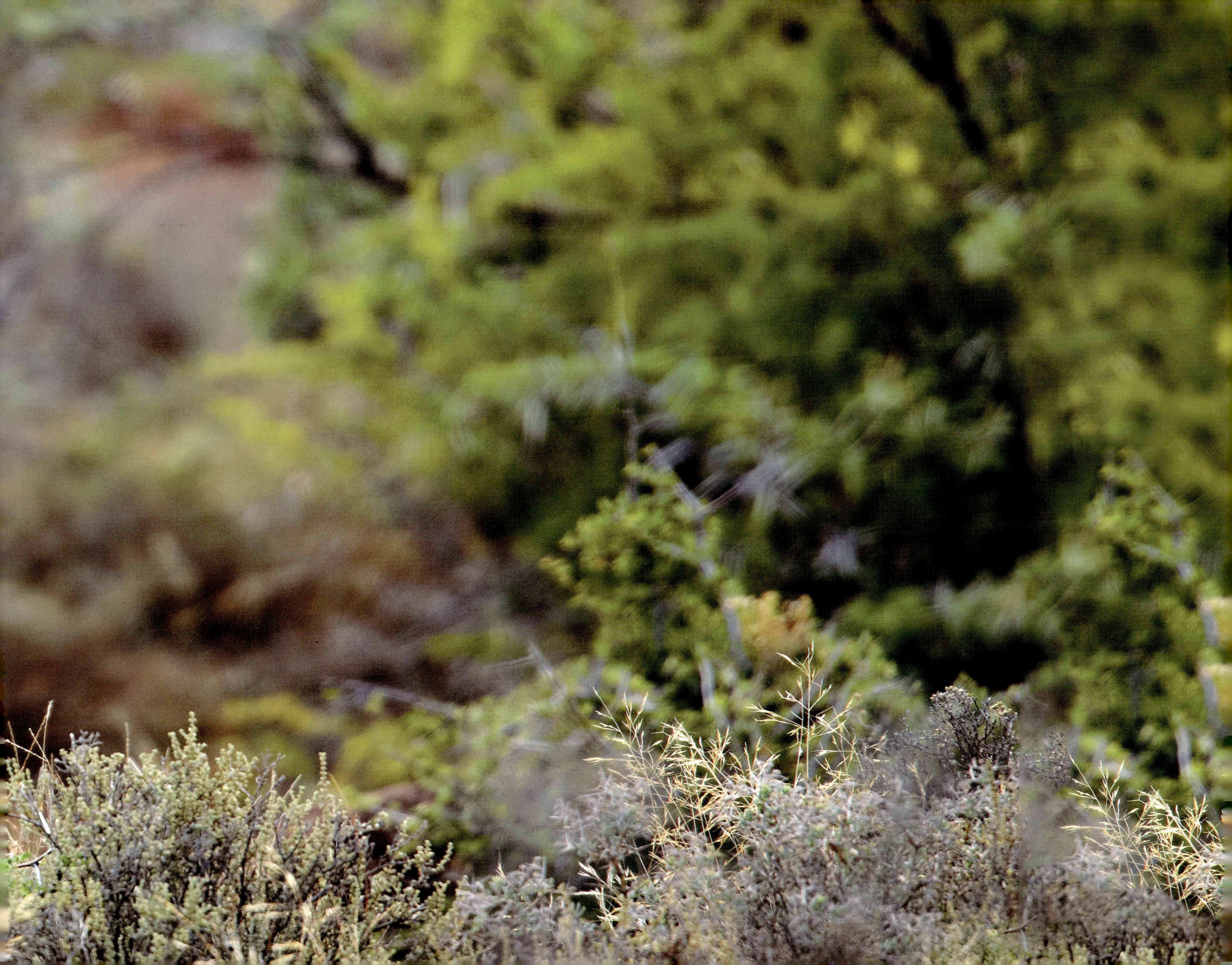

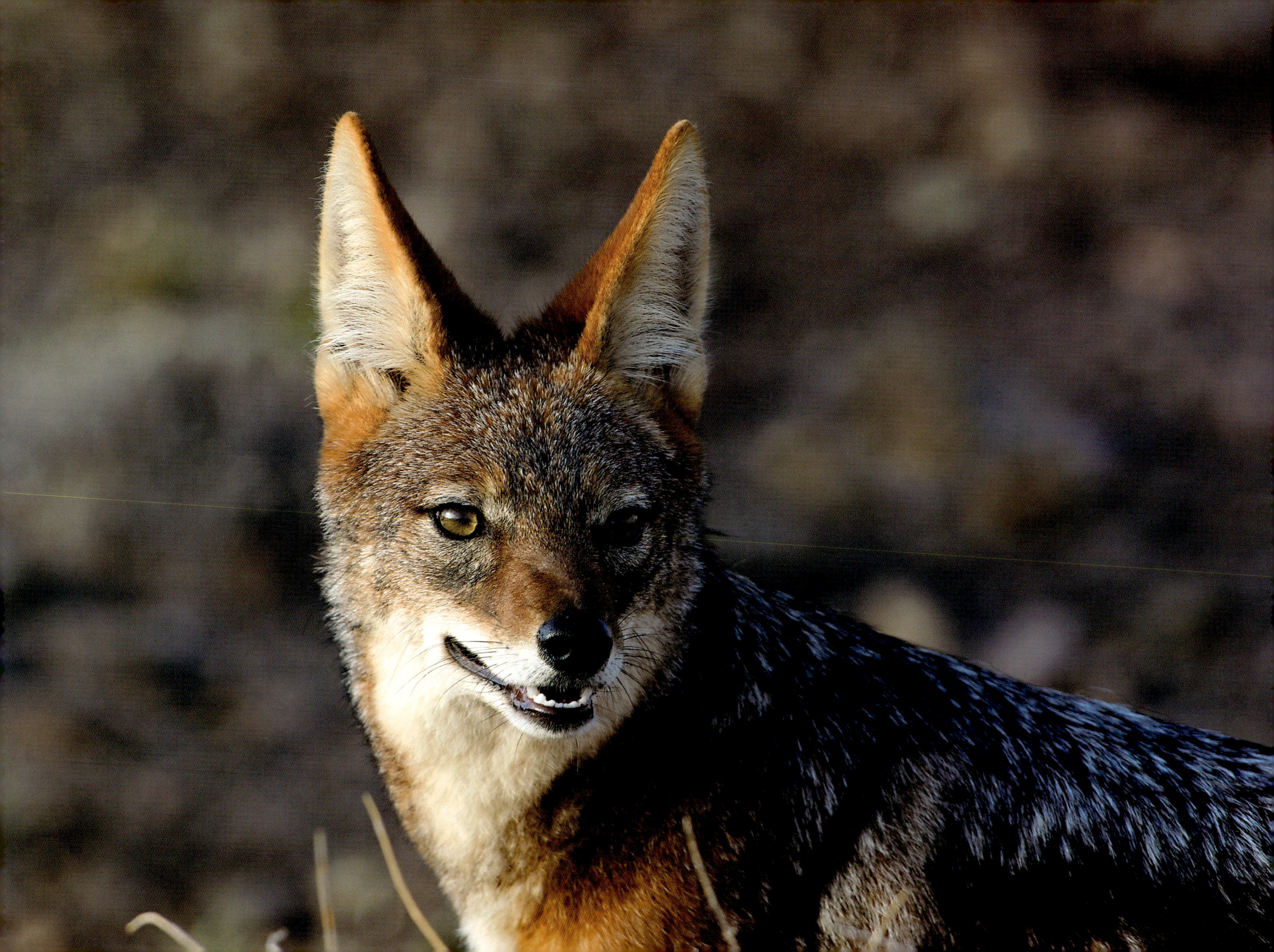

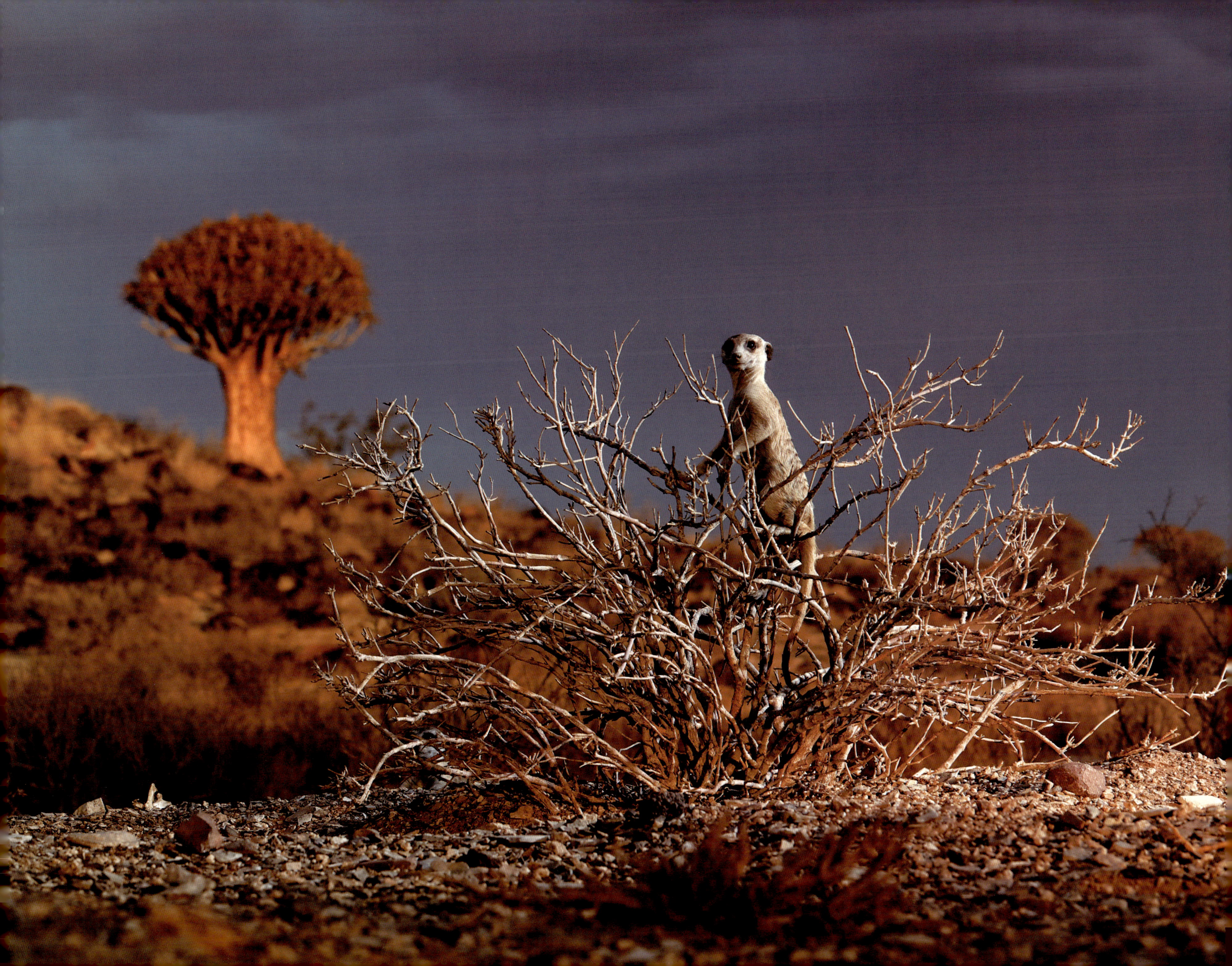

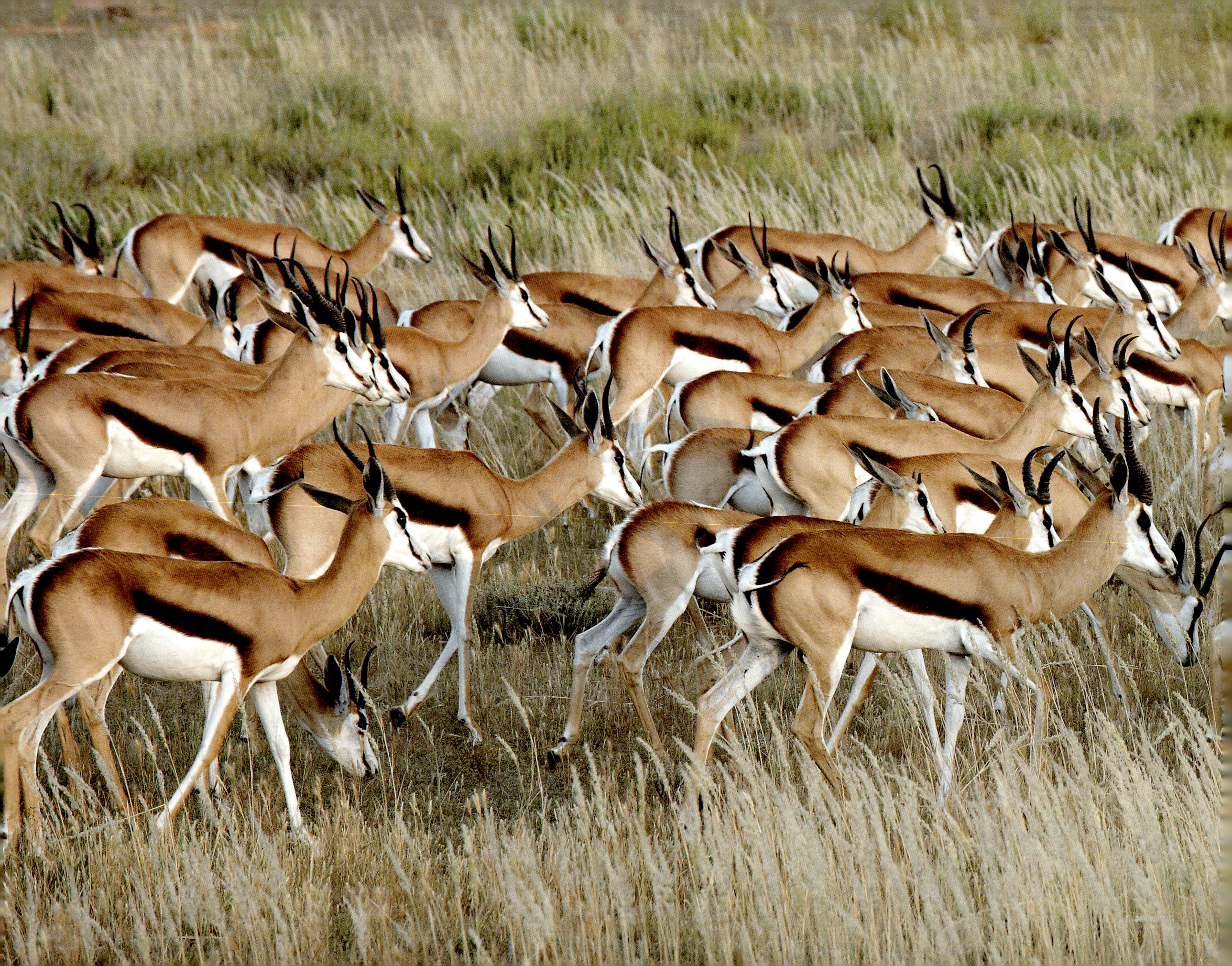

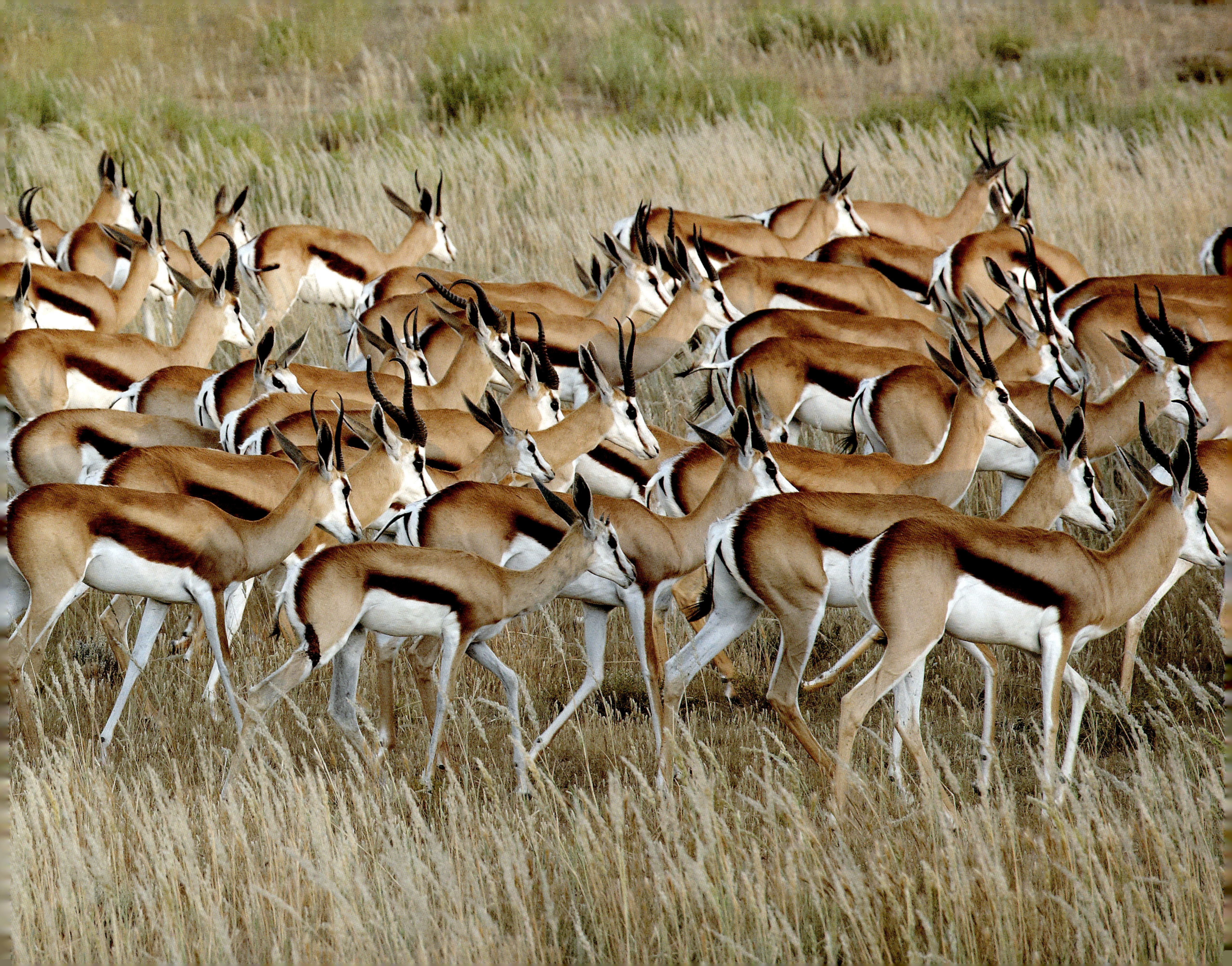

More about the Nama Karoo Biome

The Karoo has a special charm. To the casual observer the plains may seem monotonous, dry and never-ending. Yet nowhere else in southern Africa do the skies seem so immense and the vistas so vast and alluring. The view from the top of the Nuweveld Mountains near Beaufort West seems to transport you to infinity.

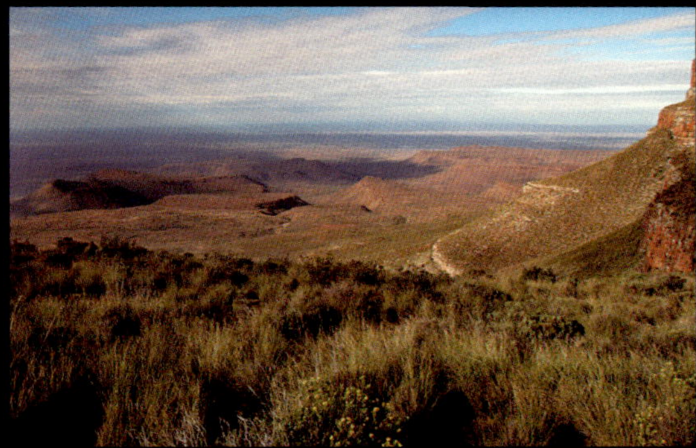

View from the Nuweveld Mountains – Karoo National Park

From the mountain plateau at the summit of the Bankberg in the Mountain Zebra National Park, the distant Karoo hills are bathed in rich morning light as sunshine is about to spill over the flat-topped peaks into the misty valleys.

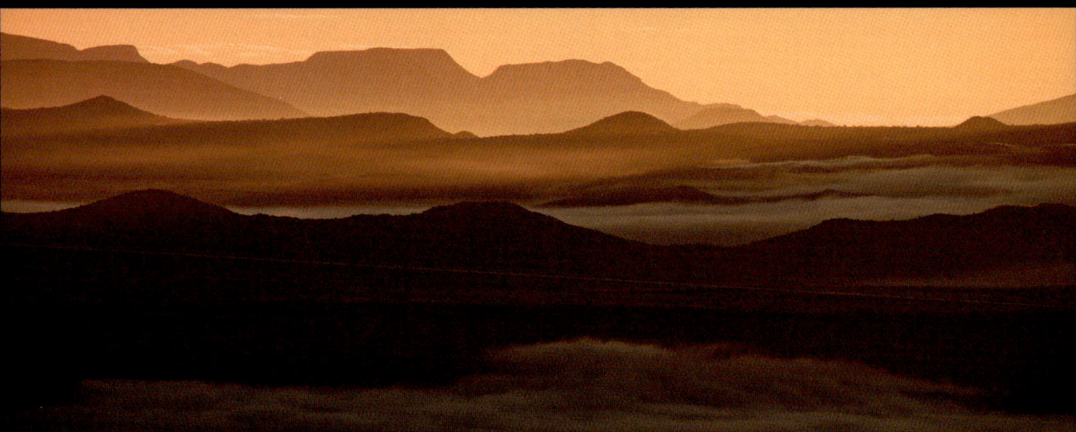

View from the summit of the Bankberg in the Mountain Zebra National Park

Trees, mostly thorny, are scarce and often confined to dry watercourses. The plains at the foothills of the Sneeuberge (Snowy Mountains) in the Camdeboo National Park are a grassy shrub land that supports several types of large grazers such as the black wildebeest. Historically, large migrating herds of these and other antelope species traversed the Karoo plains, continually moving; now they are confined within the borders of the 19 405-hectare park, which practically surrounds the town of Graaff-Reinet in the Eastern Cape.

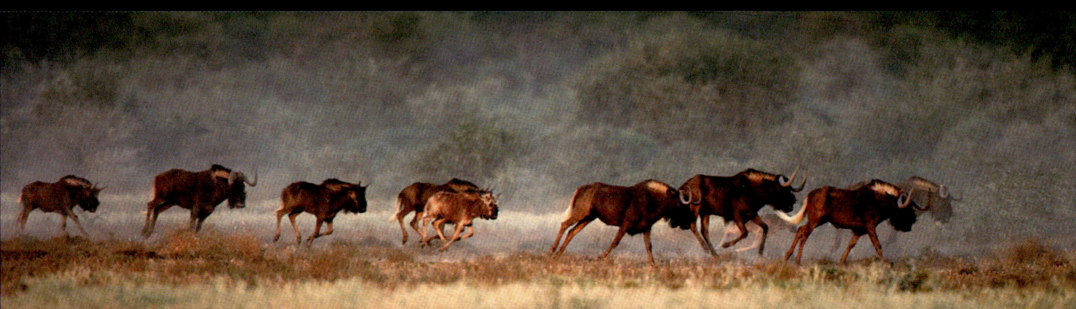

Black wildebeest in the Camdeboo National Park

Klipspringer antelope are well adapted to a life on the rocky dolorite outcrops of the numerous flat-topped Karoo hills, also called 'koppies'. They are predominantly browsers and they nibble the fresh green shoots of the Karoo scrub. Confined to rocky habitat, they daintily hop from rock to rock, negotiating deep crevices and precipices. They are plentiful and regularly seen in the Karoo National Park.

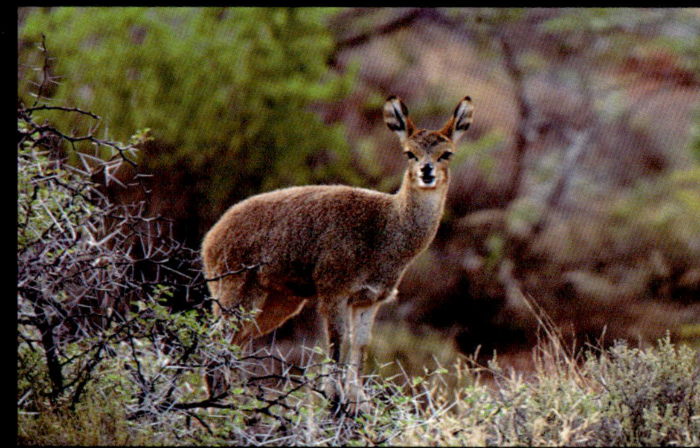

Klipspringer in typical Karoo vegetation

Where herbivores are plentiful, predators follow. Although lion have been reintroduced into the Karoo National Park, the more common and widespread predators on the farms and in the parks are smaller ones, such as the black-backed jackal and the caracal. Farmers are not very fond of them since they cause huge stock losses. Despite their bad reputation as sly and destructive predators, when observed from close by in a vehicle, they are actually beautiful creatures. Leopard used to be widespread in the secluded kloofs and gullies of the mountain ranges, but sadly they are also ruthlessly hunted by farmers determined to reduce stock losses.

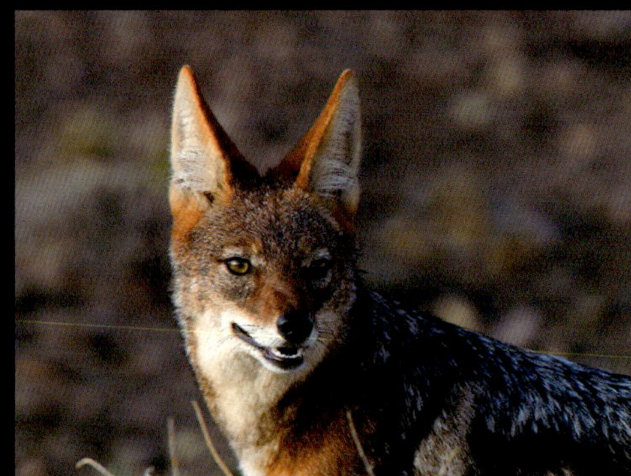

Black-backed jackal

The lucky traveller may come across a group of suricates. Spotting a sentry keeping watch on top of a shrub may be the first sign that they are around. These small desert mammals live in social groups. The advantage of their lifestyle is that members of the group can take turns to stand sentry while the rest forage by digging for insects and scorpions. The sentry watches the skies and scans the horizon for predators while uttering a continuous repertoire of calls to keep all members of the group informed about what is happening around them. When a potential predator is spotted, a shrill, sharp bark is enough to send everyone diving into the nearest burrow for protection. This group was spotted close to a rocky outcrop covered with quiver trees near Keetmanshoop in Namibia.

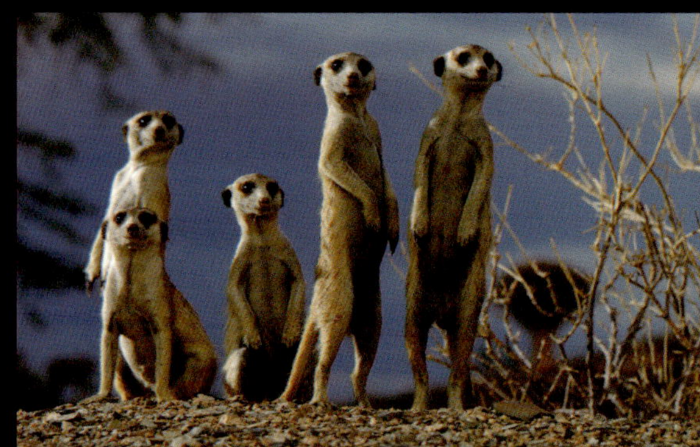

A family of suricates

The dainty springbok are the only gazelle species in southern Africa. Before the arrival of farmers with their guns and fences, travellers reported huge mass migrations of these antelope within the Karoo. Today herds of springbok seldom number more than a few hundred, compared with the millions reputed to have formed the 'trekbokken' herds a few centuries ago. These antelope are partial to the short grass on the fringes of pans and ephemeral river courses, even if water is only seasonally available. Amazingly, they are able to survive on the moisture they derive from succulent karroid vegetation and even by digging out succulent roots. Being gregarious by nature, and living in areas of sporadic and localised rainfall, small herds will move to areas where rain has fallen in order to feed on the fresh green vegetation. In this way much larger aggregations form, which can quickly run into thousands.

There are two sub-species of the mountain zebra, namely the Cape mountain zebra in the southern regions of the Nama Karoo and the slightly larger Hartmann's mountain zebra found in Namibia. As both sub-species are grazers, they can also be placed under the Grassland Biome. They do, however, prefer mountainous areas with kloofs and ridges where they can find shelter, but they also need access to water every day. The best place to see Cape mountain zebra is in the Mountain Zebra National Park. These zebra differ from the plains zebra in that the black stripes on their heads and bodies are narrower but more numerous, while on the rump the black stripes (often deep chocolate-brown) are broader and have no 'shadow' stripes between them. The white belly and distinct dewlap is another distinguishing characteristic.

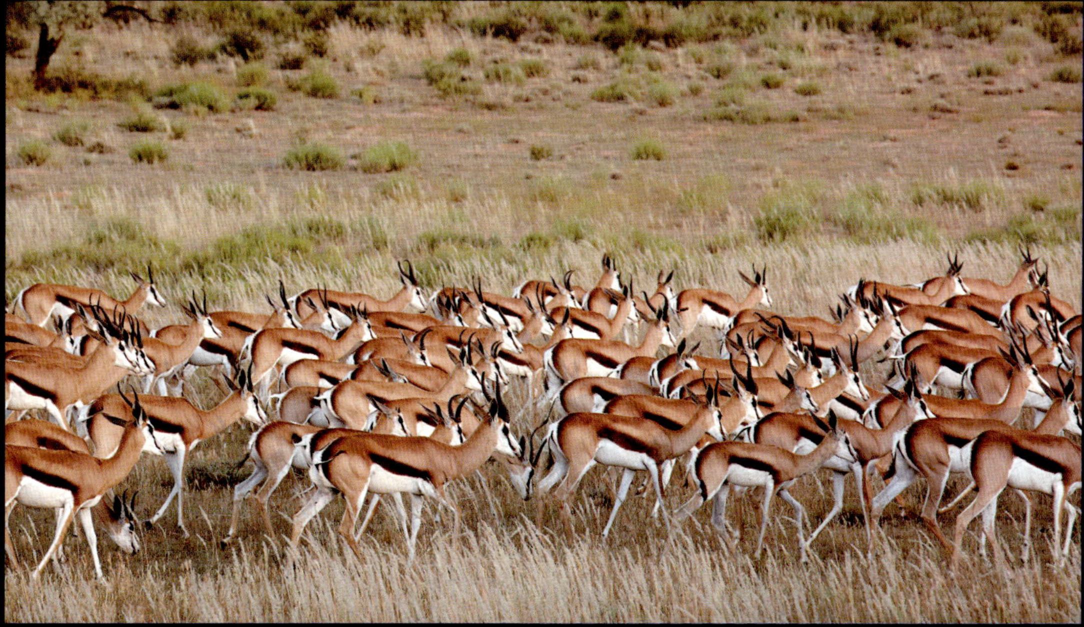

Springbok

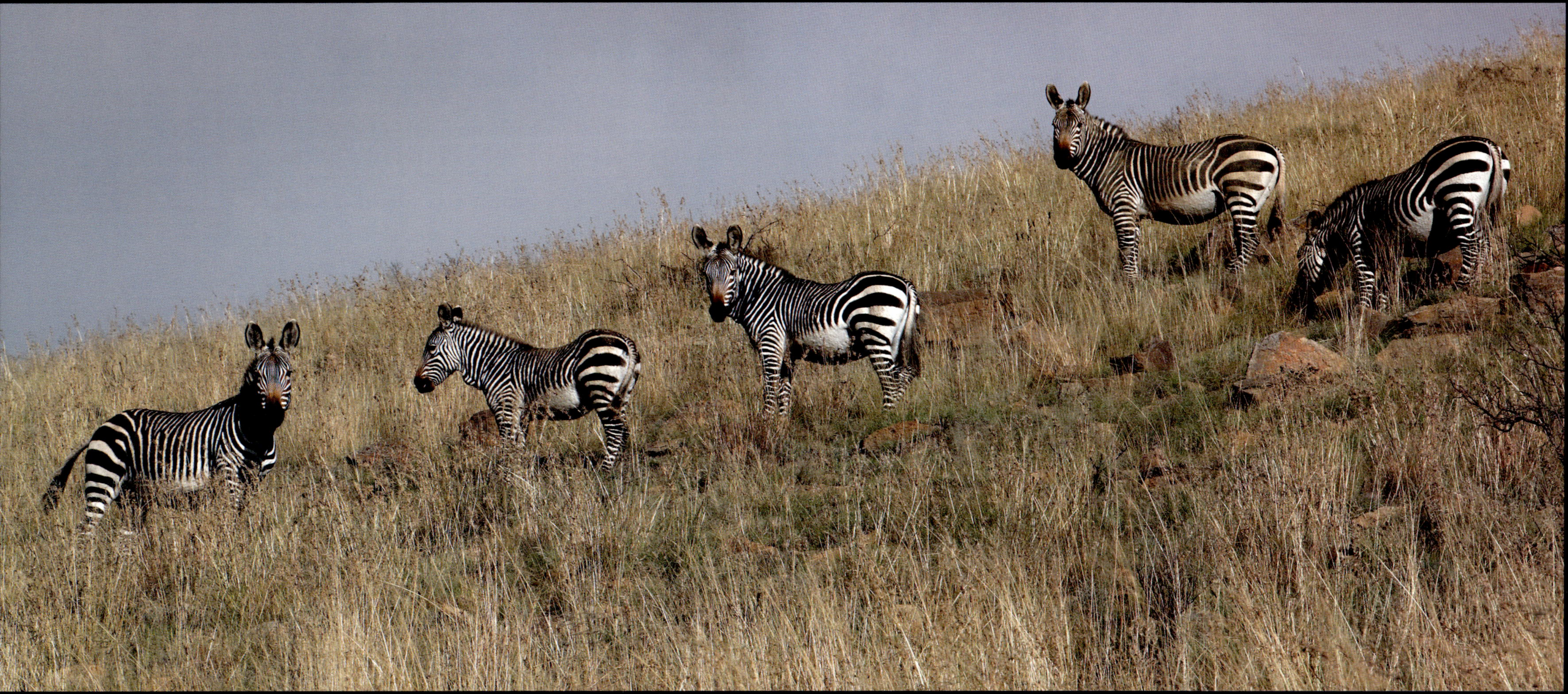

Mountain zebra in the Mountain Zebra National Park

Desert

Fascinating landscapes

Considering all deserts of the world, there is none like the Namib. Not only is it ancient, but due to climatic and other interacting factors, it is spectacularly beautiful, alive with unusual animals and plants, diverse in origin and intrinsically fascinating.

The Namib Desert stretches north along the entire coast of Namibia and is termed a true desert since the region receives very little rain (between 5mm and 85mm per annum). In some years no rain may fall at all. Summer days are searingly hot and temperatures frequently rise above 45°C. On the other hand, winter nights are icy cold with severe frost and temperatures may drop below –15°C.

A few days each month, particularly in winter, midday temperatures rise quite high when the berg wind blows from the interior. This wind brings very dry air which is heated as it descends to the coast – this wind may blow unabated for three to eight days. Notwithstanding these severe conditions, the Namib is surprisingly viable. Closer to the coast, temperatures are kept low as moist air from the cold Benguela Current spills over the desert in the form of fog and thus provides life-supporting moisture to the sparse vegetation and animal life. Many insects, which include some tenebrionid beetles, can effectively utilise this fog water.

The desert landscape is diverse. Apart from the inland sea of towering sand dunes glowing in pink and gold with blinding white claypans in between, there are the endless pale greyish-mauve gravel plains interrupted by inselbergs of stark black or dark brown rock. Across these vast areas and etched into the desert floor are several dry watercourses lined with sparse vegetation and even trees such as camelthorn acacia and wild ebony. The well-known *Welwitschia mirabilis* occurs along drainage lines on the gravel plains. But most of the plants that occur are annual grasses which appear only after rare rains.

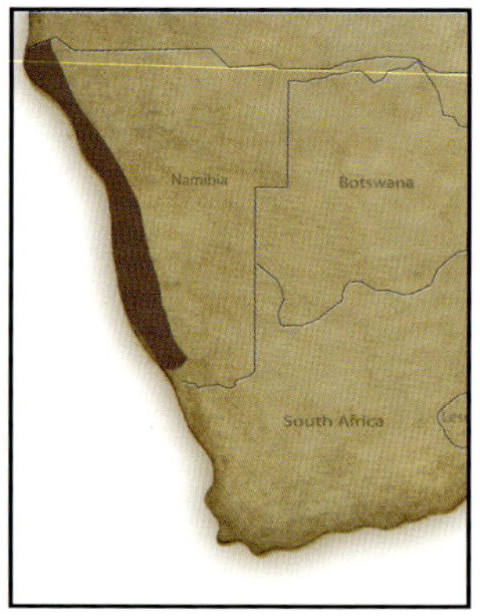

Animals such as the gemsbok, springbok, jackal and brown hyena are key species of the desert area. The Sperrgebiet (a German word meaning prohibited area) north and south of Luderitz has more biodiversity than anywhere else in the Namibian desert. This may be due to the lack of human intervention for at least a hundred years, and consequently it is now termed an arid biodiversity hotspot, the only one in the world. Only about five percent of an area of 26 000km^2 is being mined for diamonds. Despite the recent creation of a national park along the coastal parts, most of the area still serves as a barrier to humans and thus as protection for the fragile land.

Places to experience the desert

The remarkable Namib Naukluft Park and the Skeleton Coast in Namibia, as well as the rest of the Namibian coastal region, fall within the desert zone. Sossusvlei, close to Sesriem, with its spectacular sand dunes south of the ephemeral Kuiseb River, is favoured by tourists. The welwitschia and gypsum plains east of Swakopmund reveal unexpected life forms and the Skeleton Coast north of Henties Bay offers unforgettable desert experiences, including seal colonies. The Sperrgebiet south and north of Luderitz was declared a national park in June 2004 and is under the control of the Namibian Ministry for Environment and Tourism.

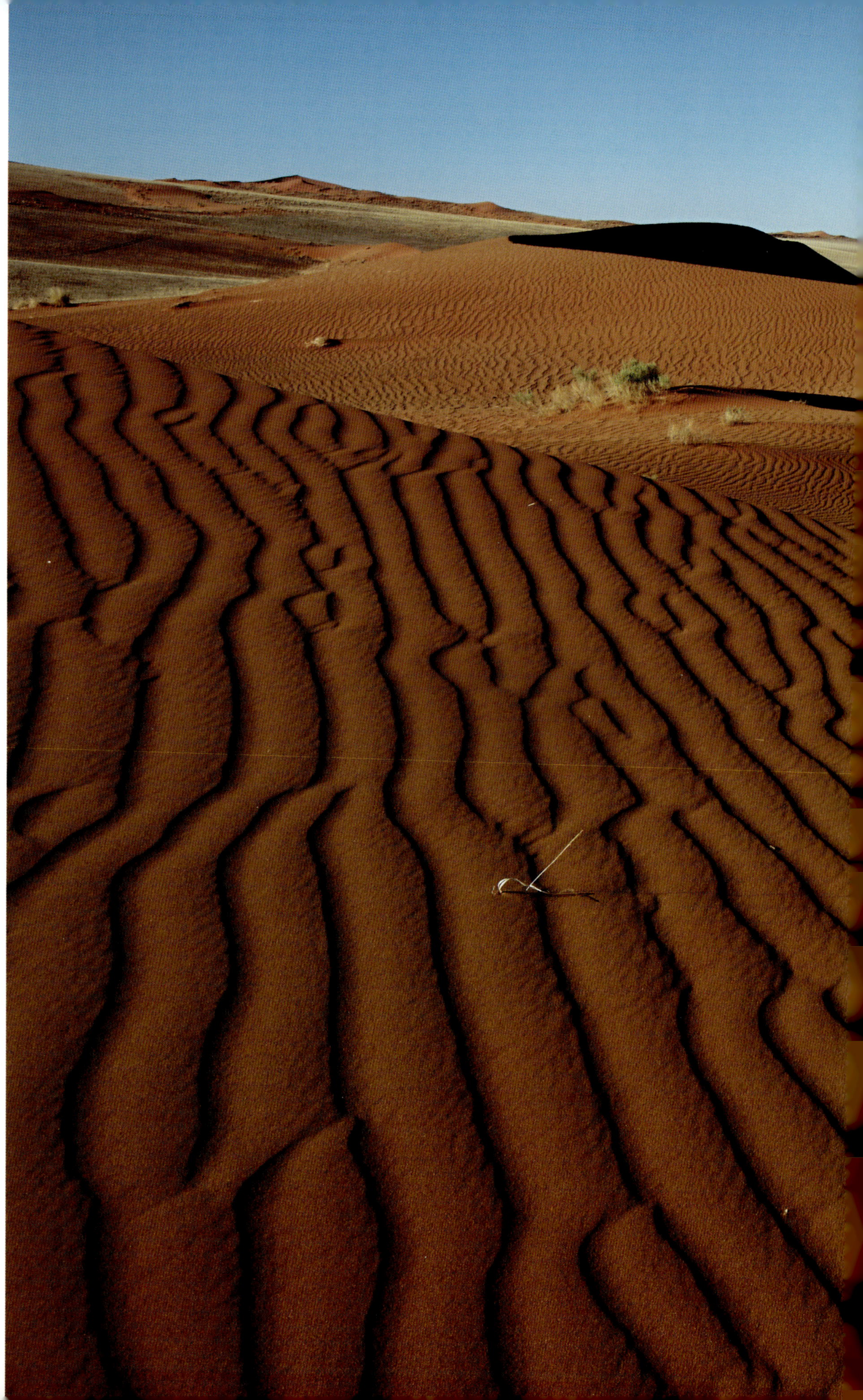

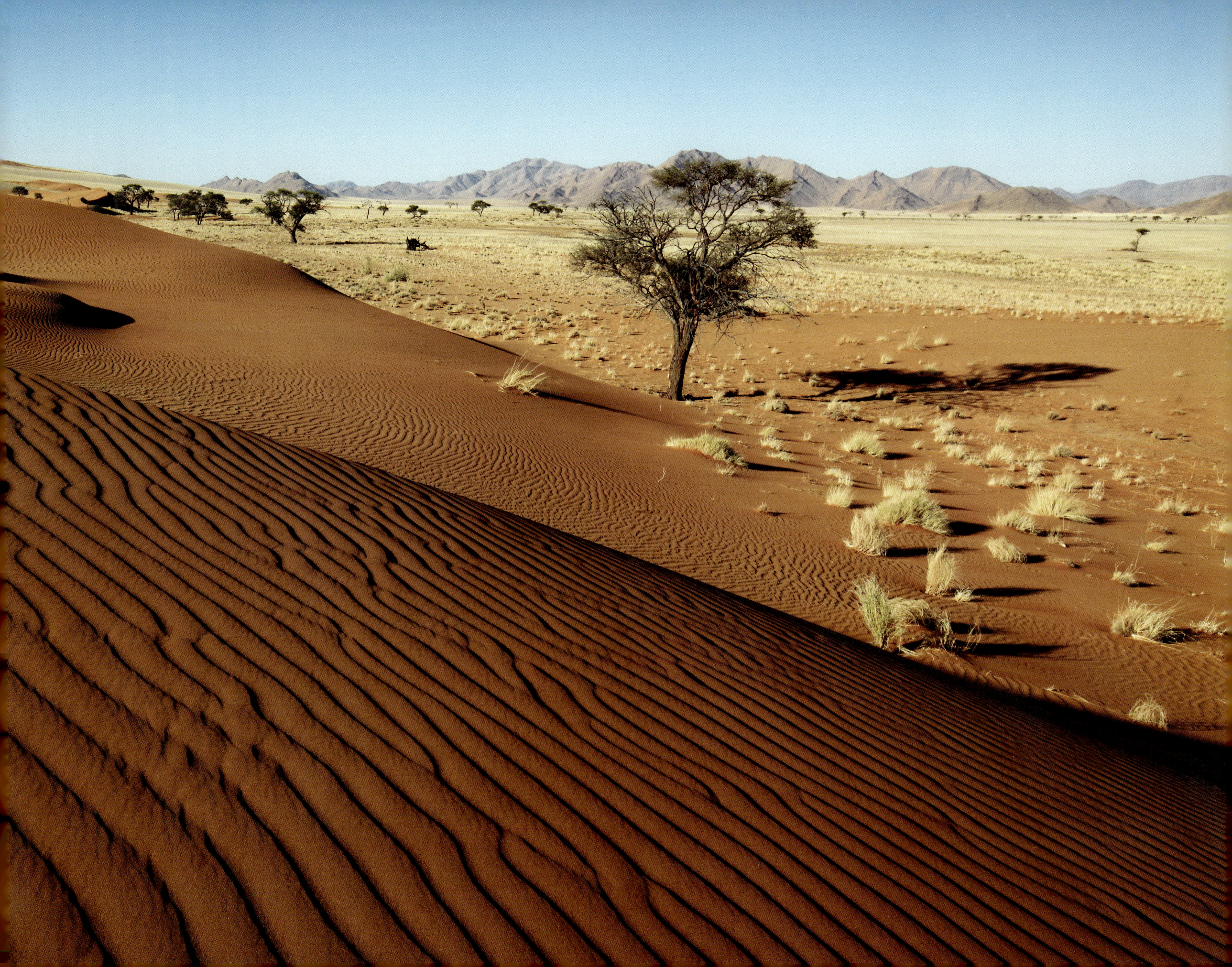

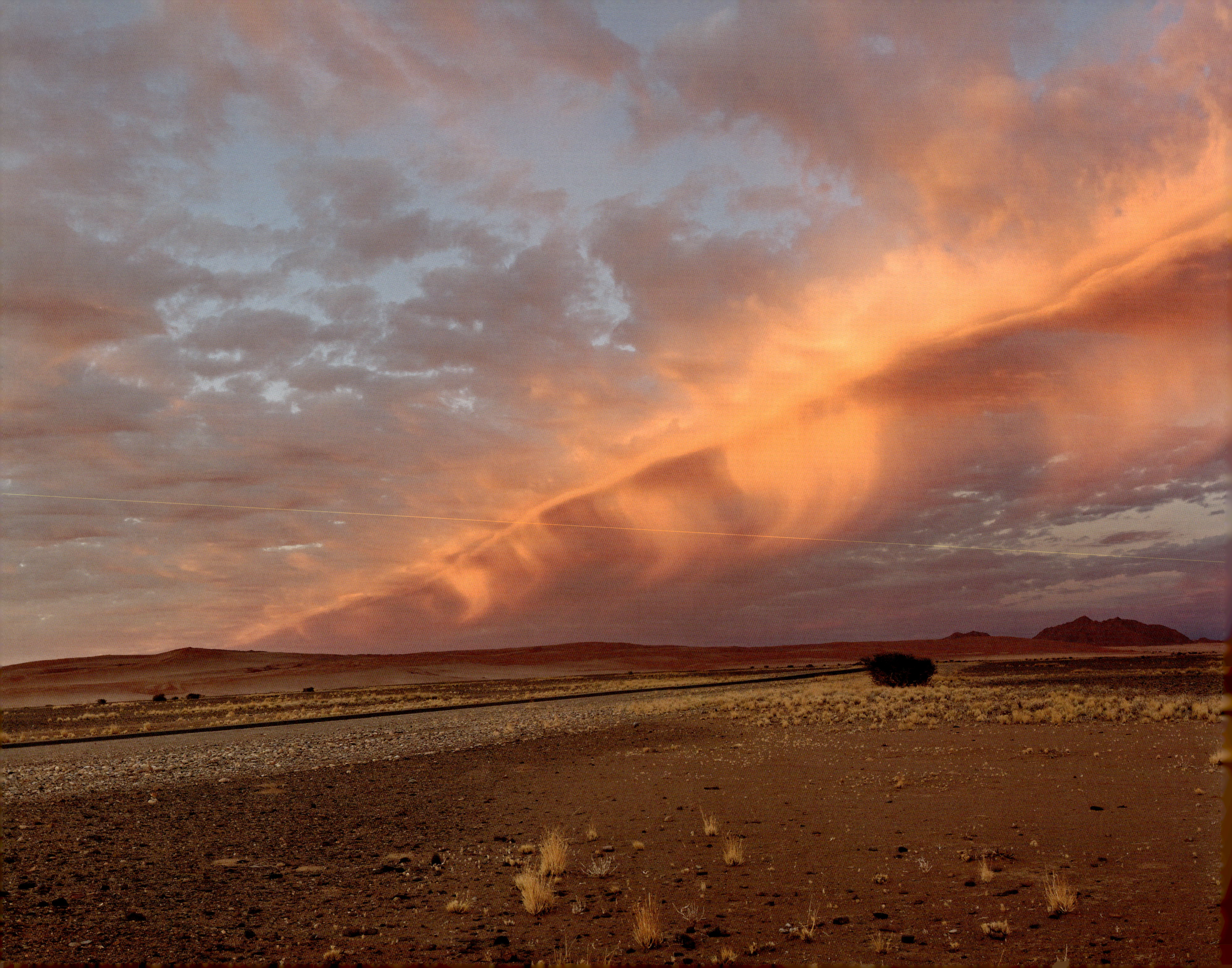

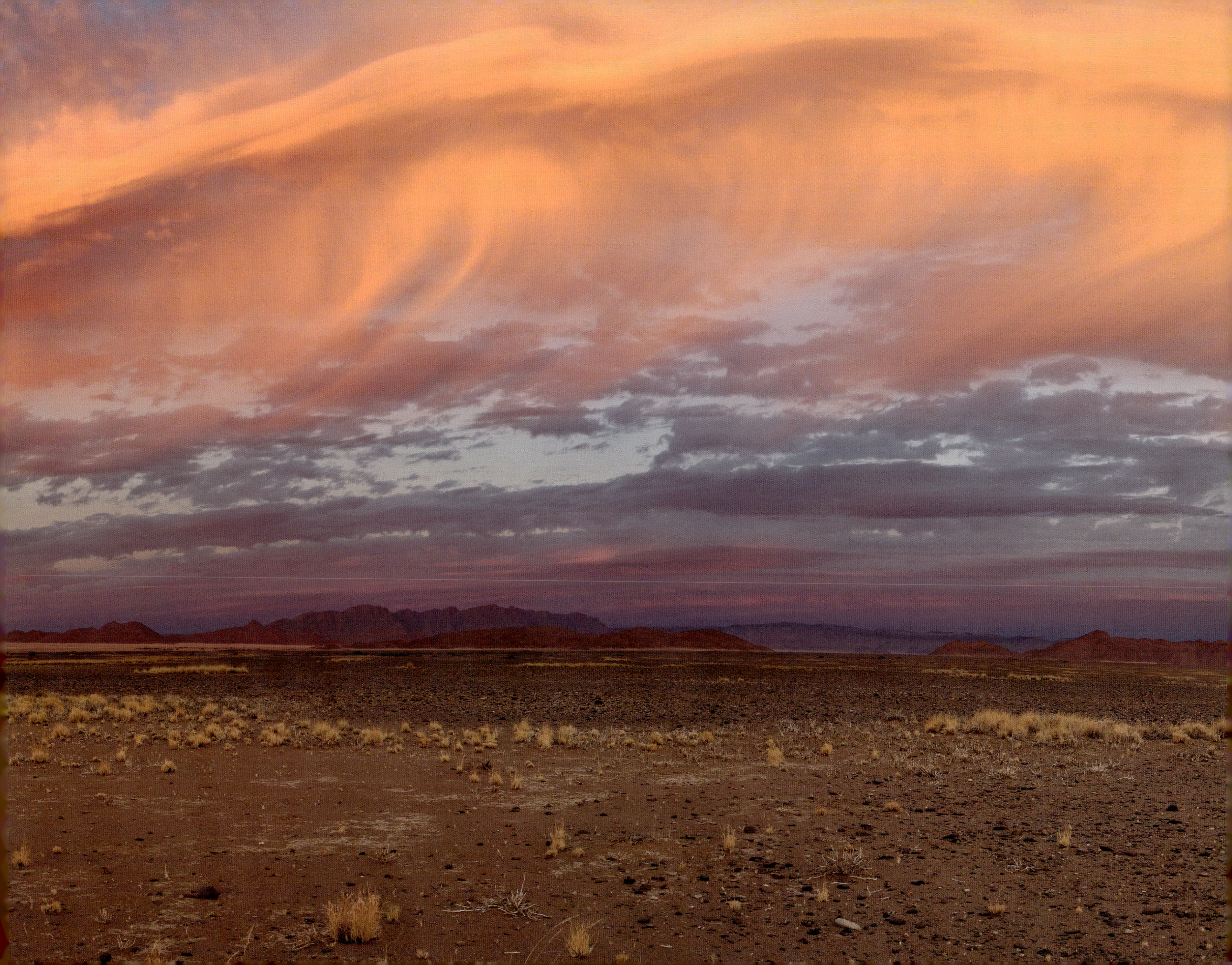

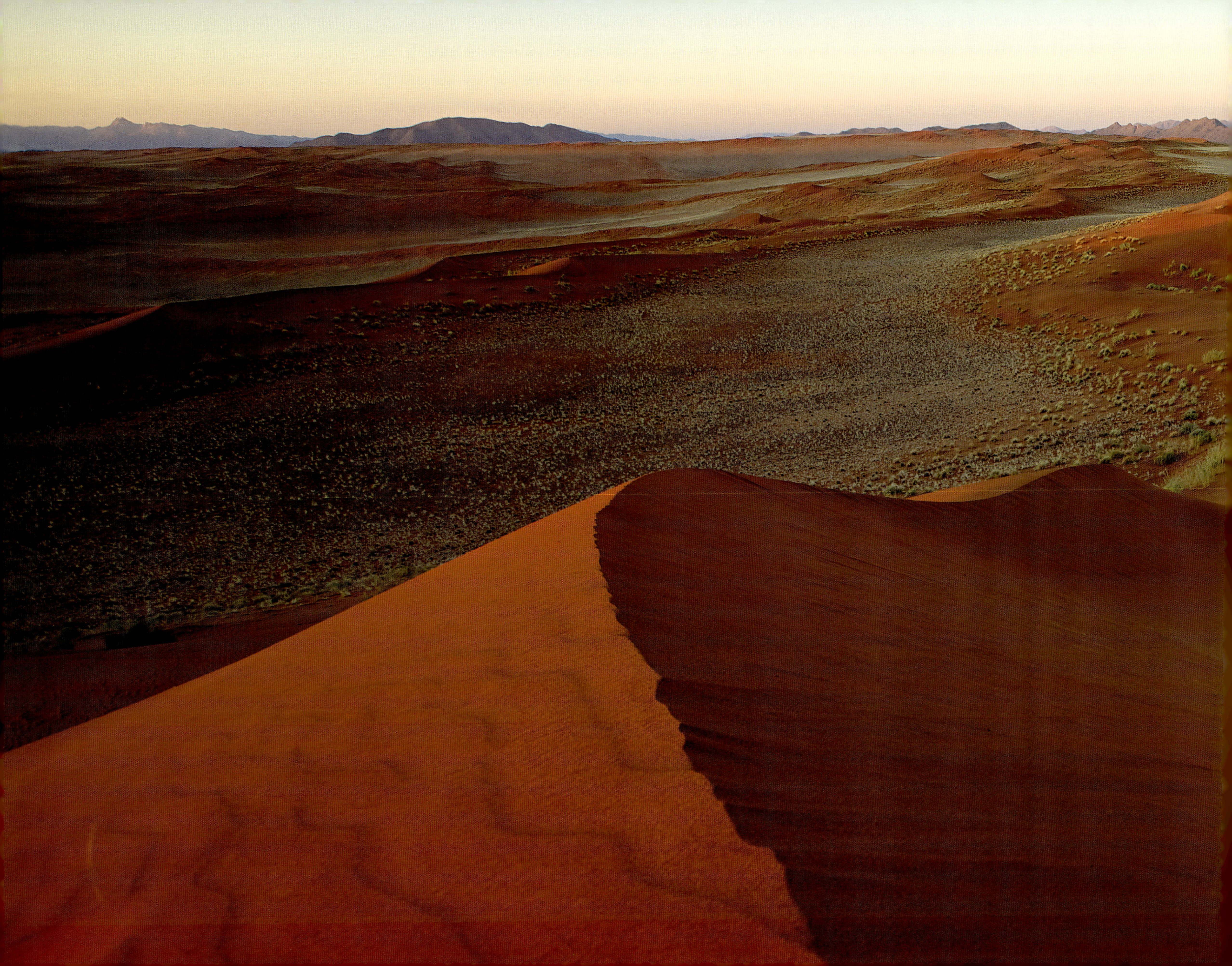

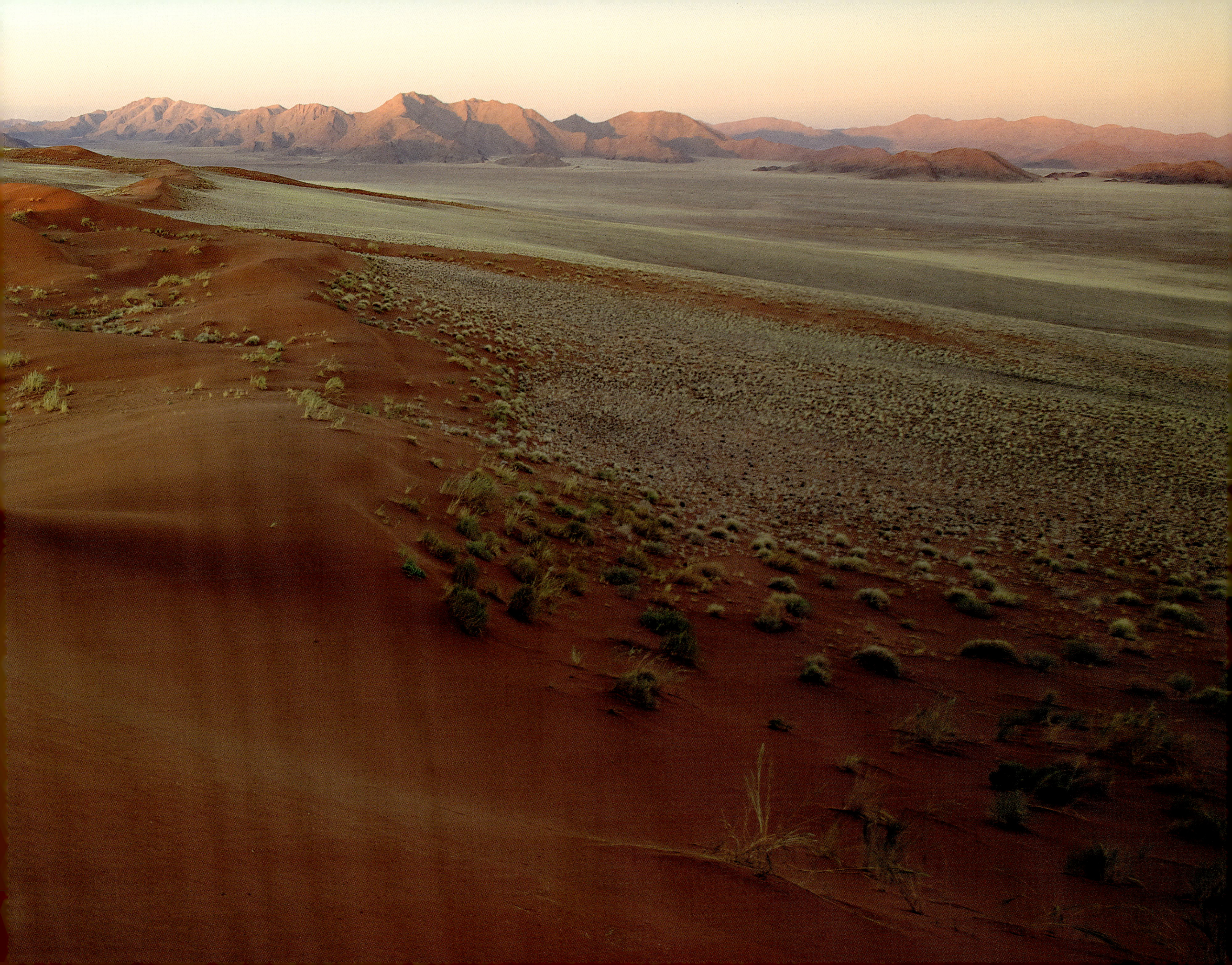

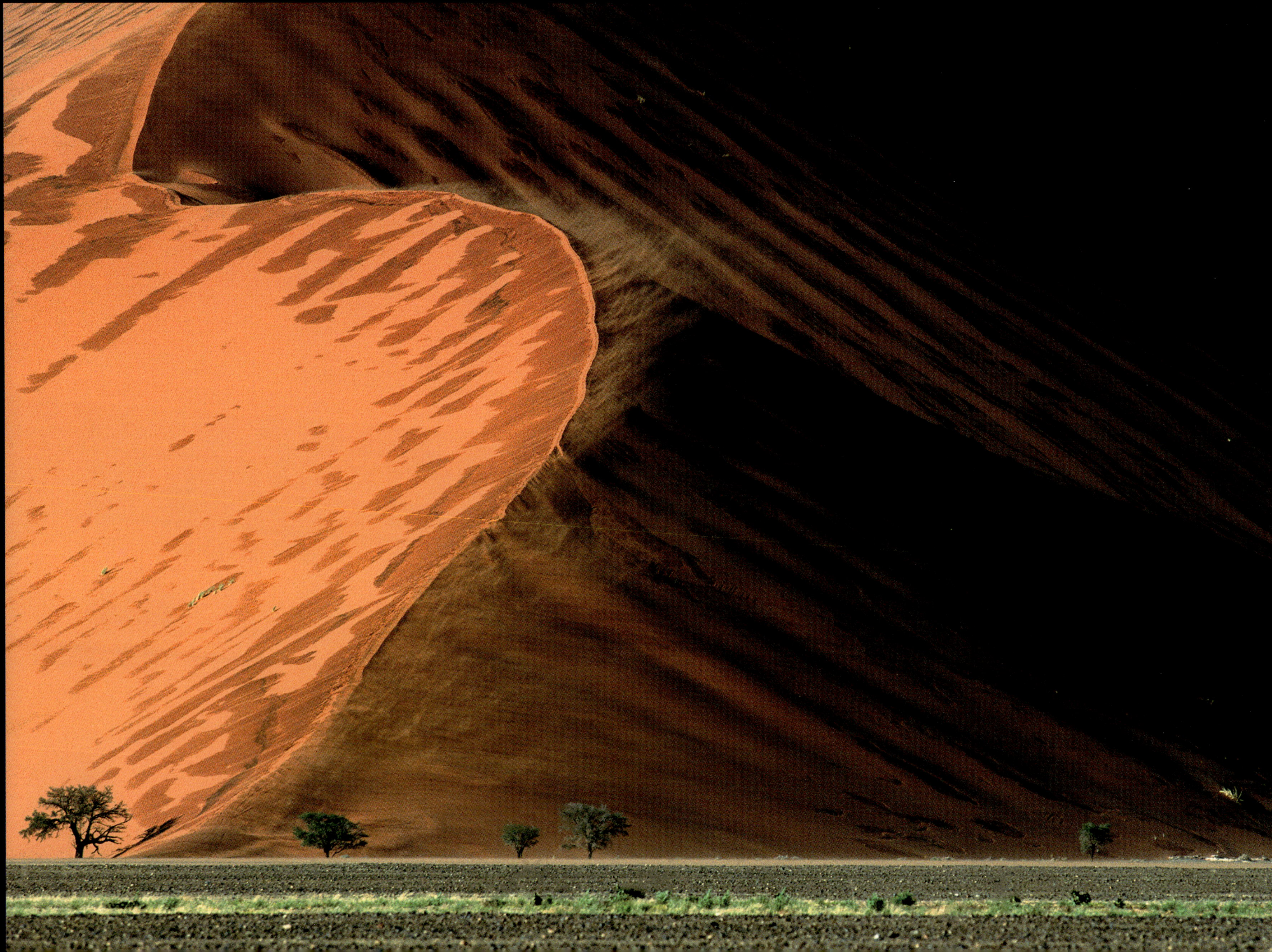

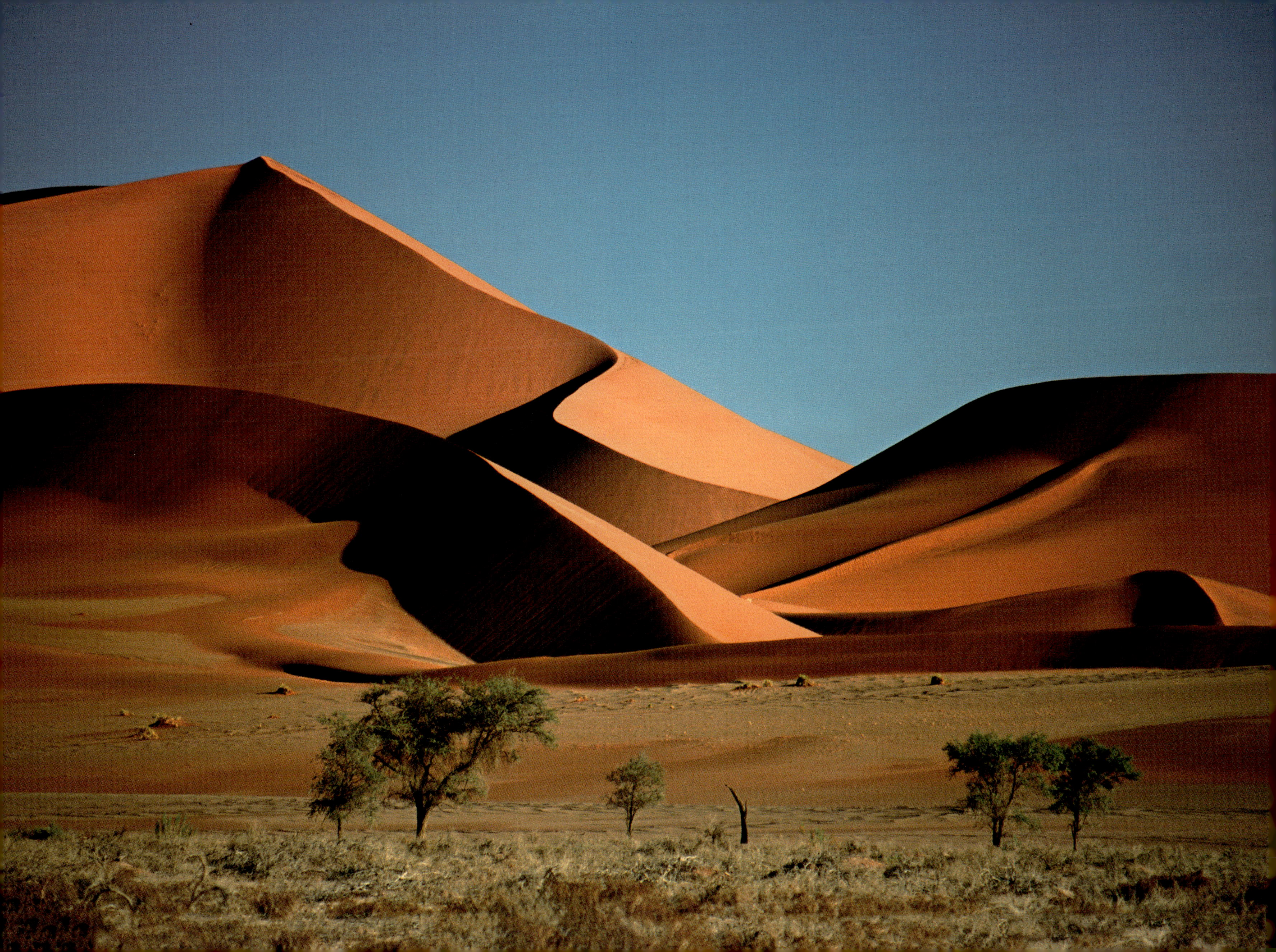

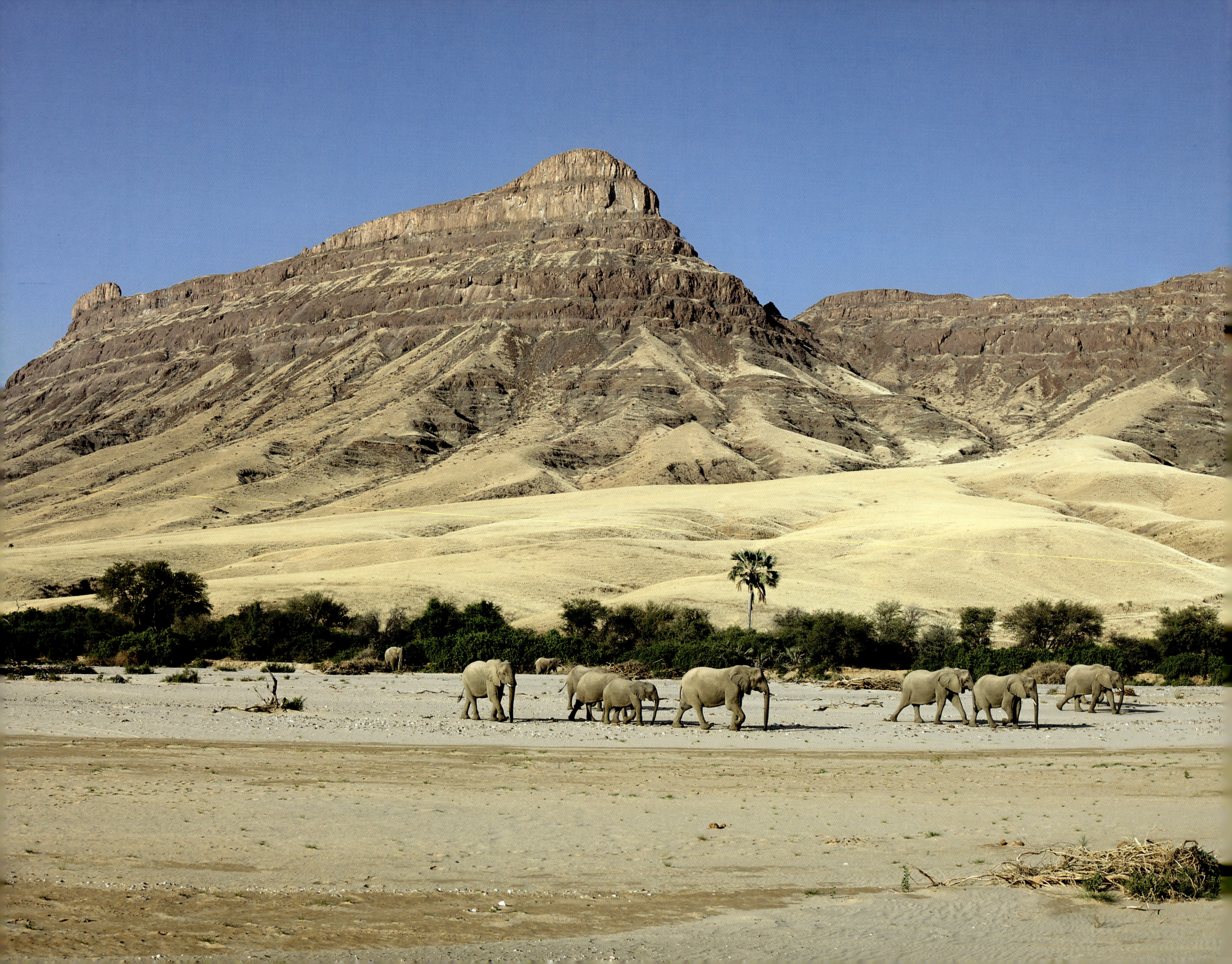

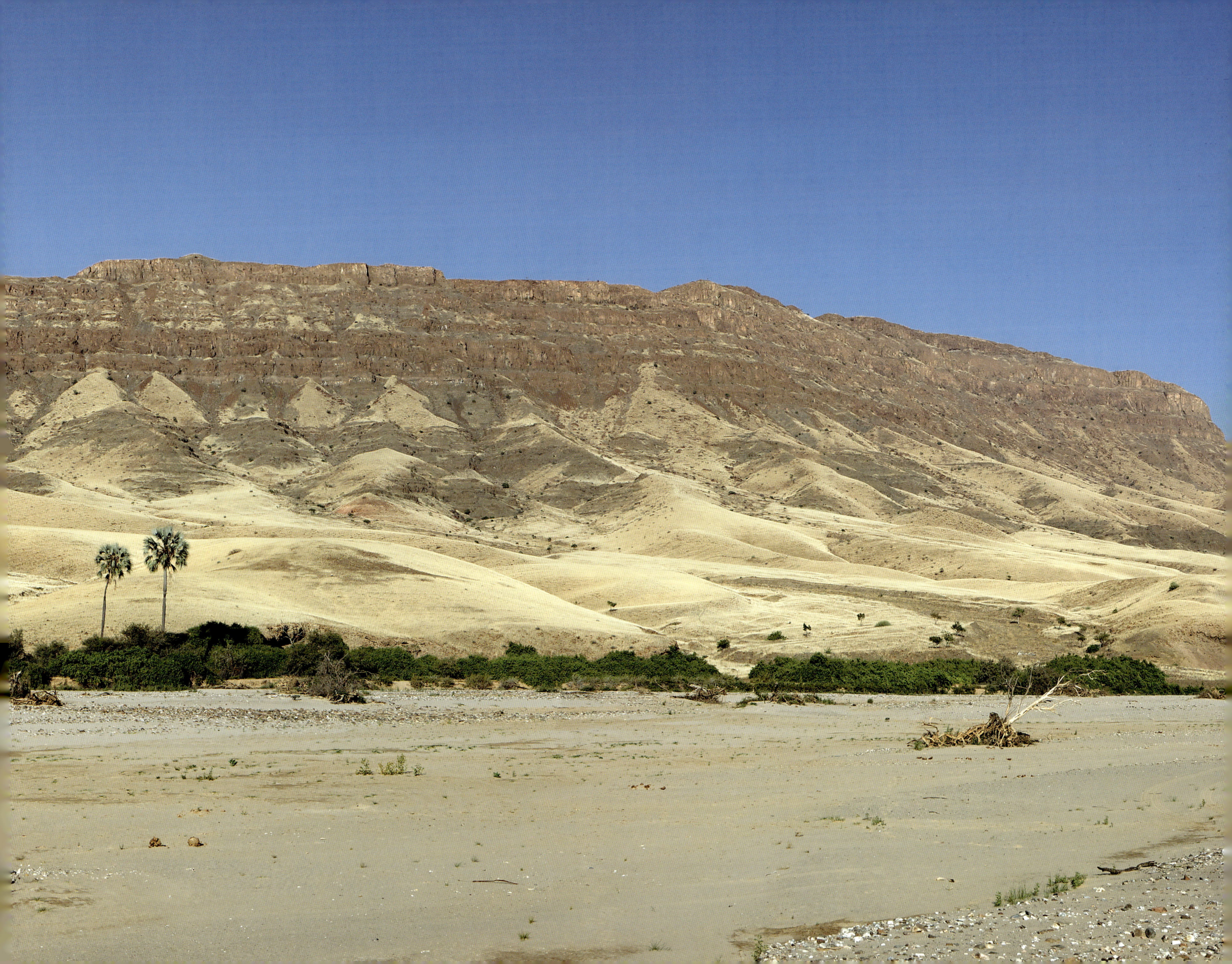

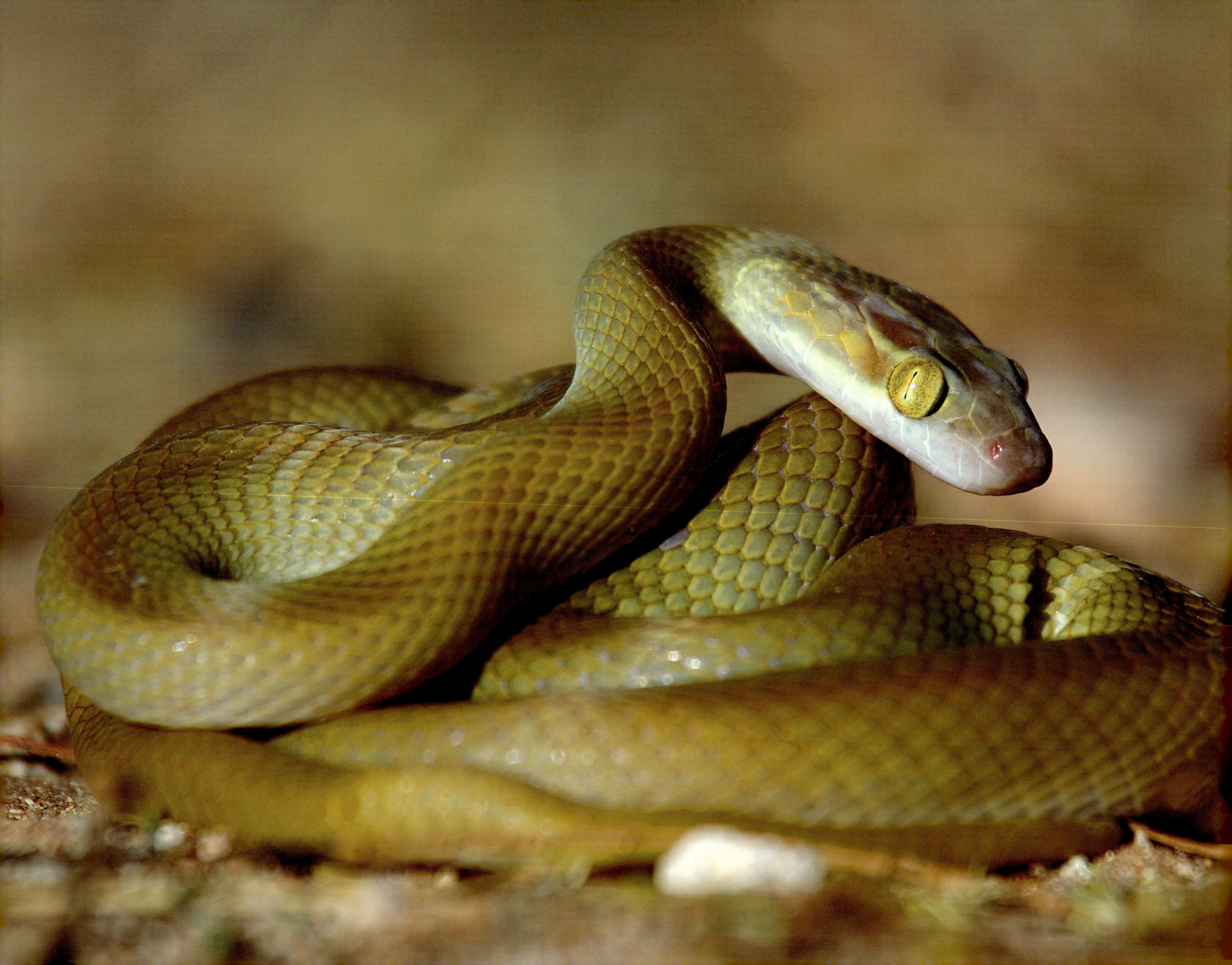

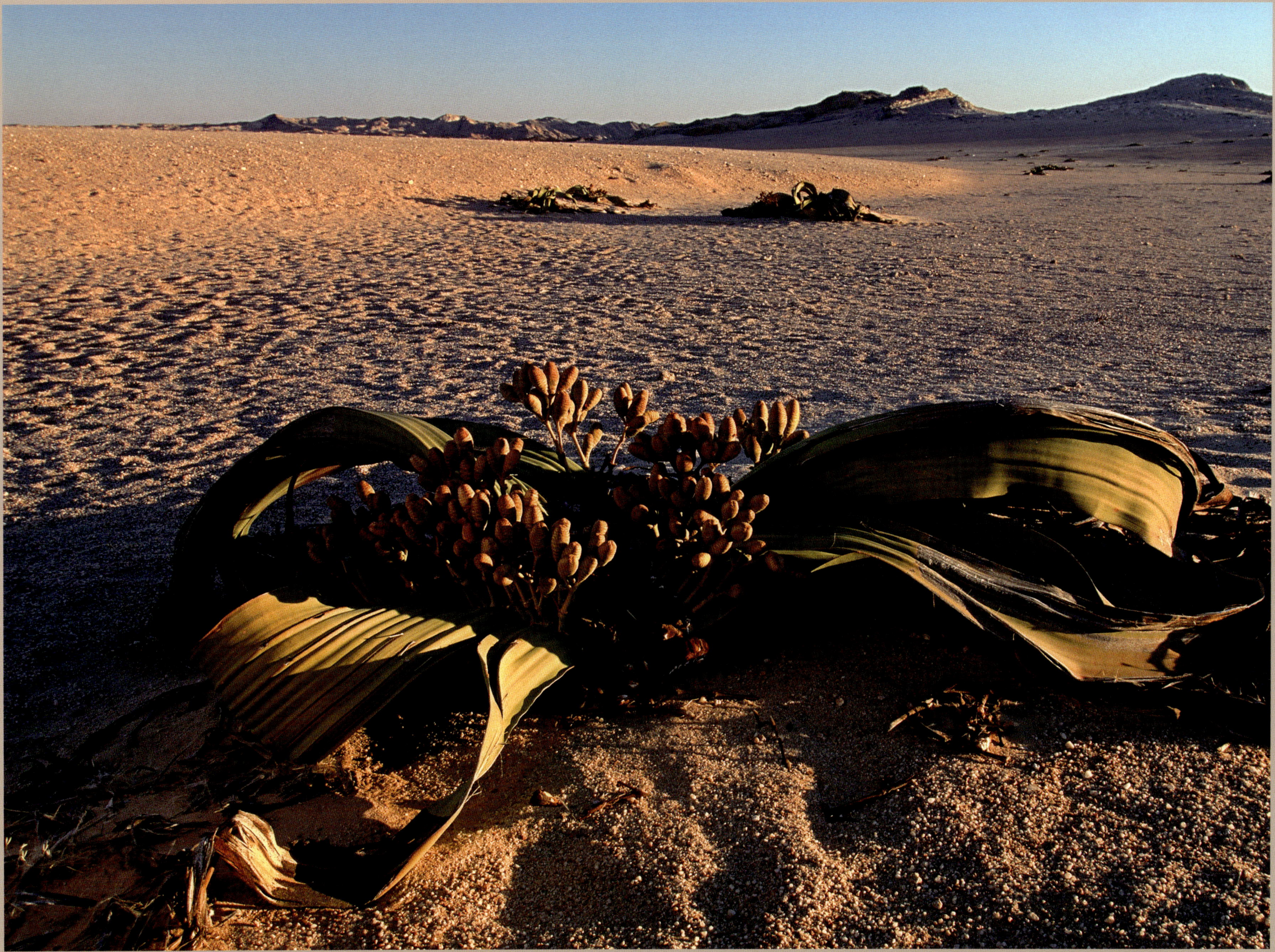

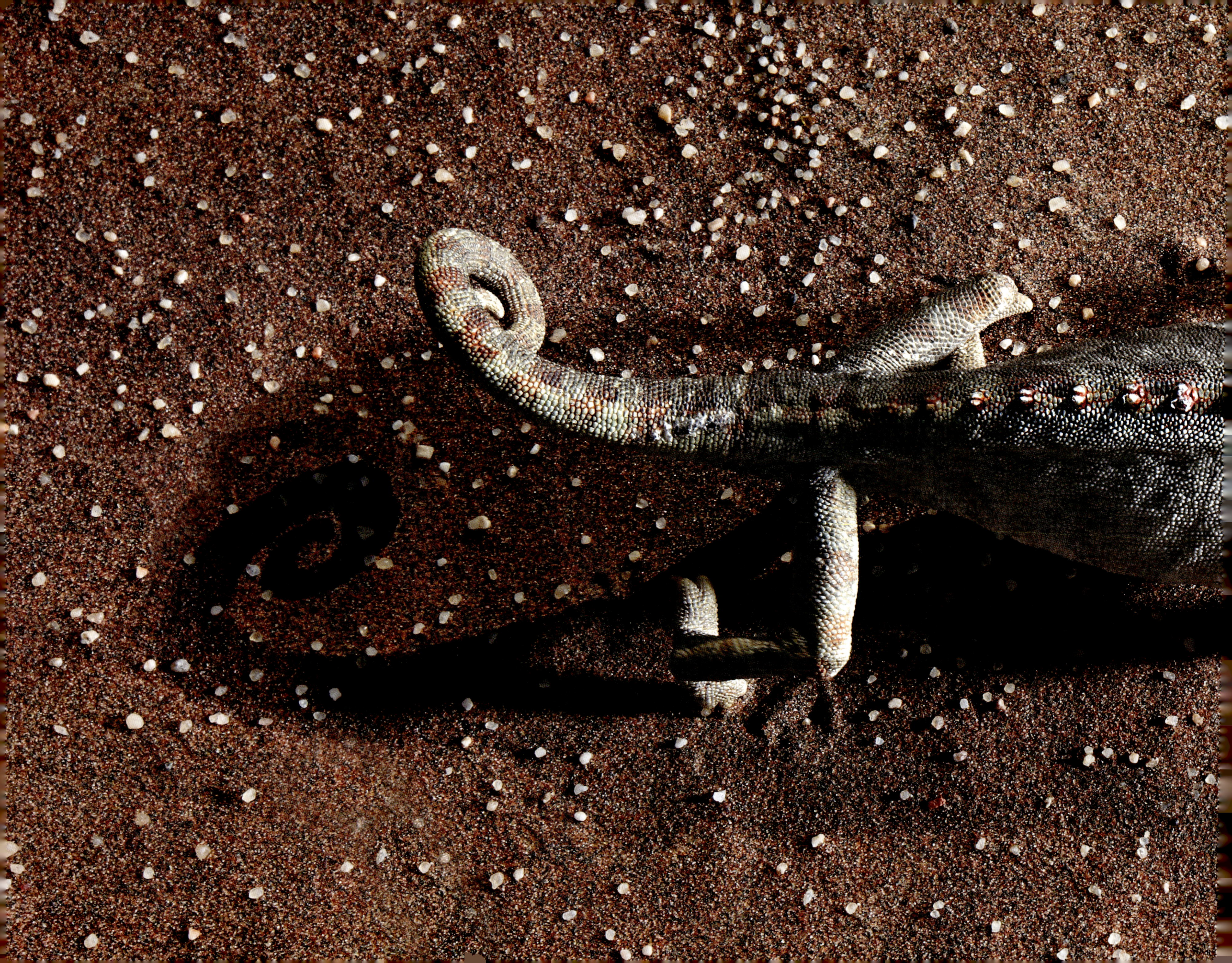

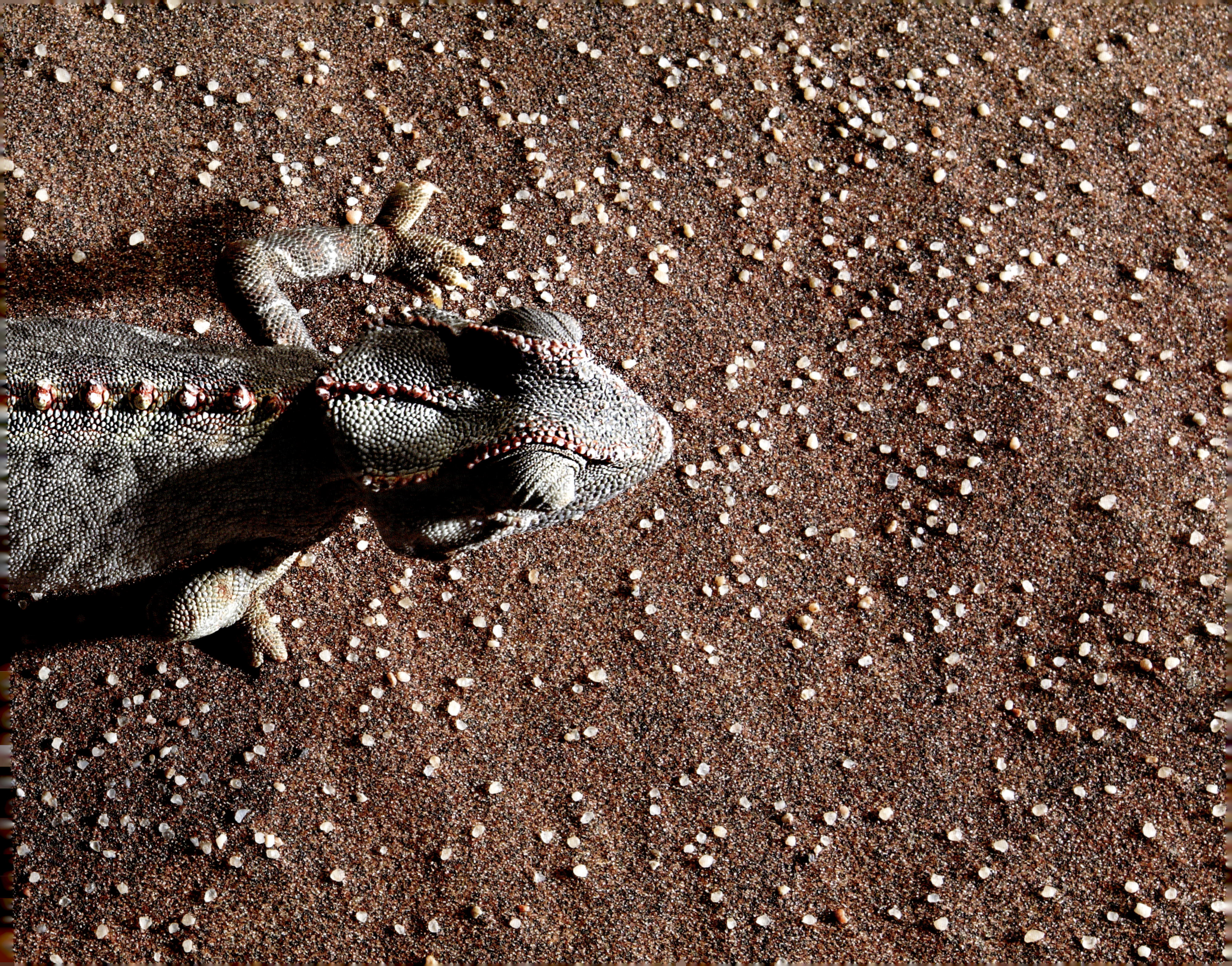

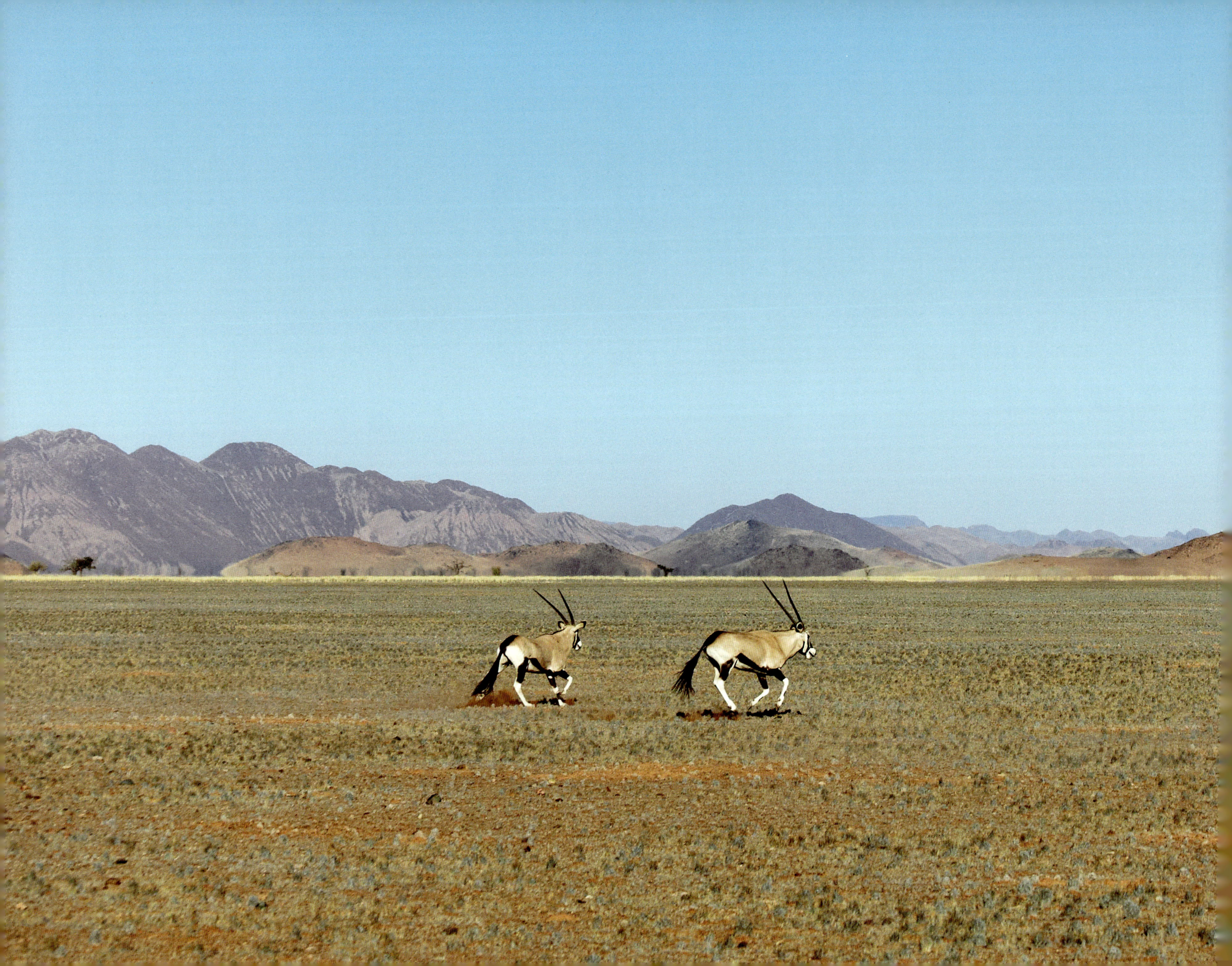

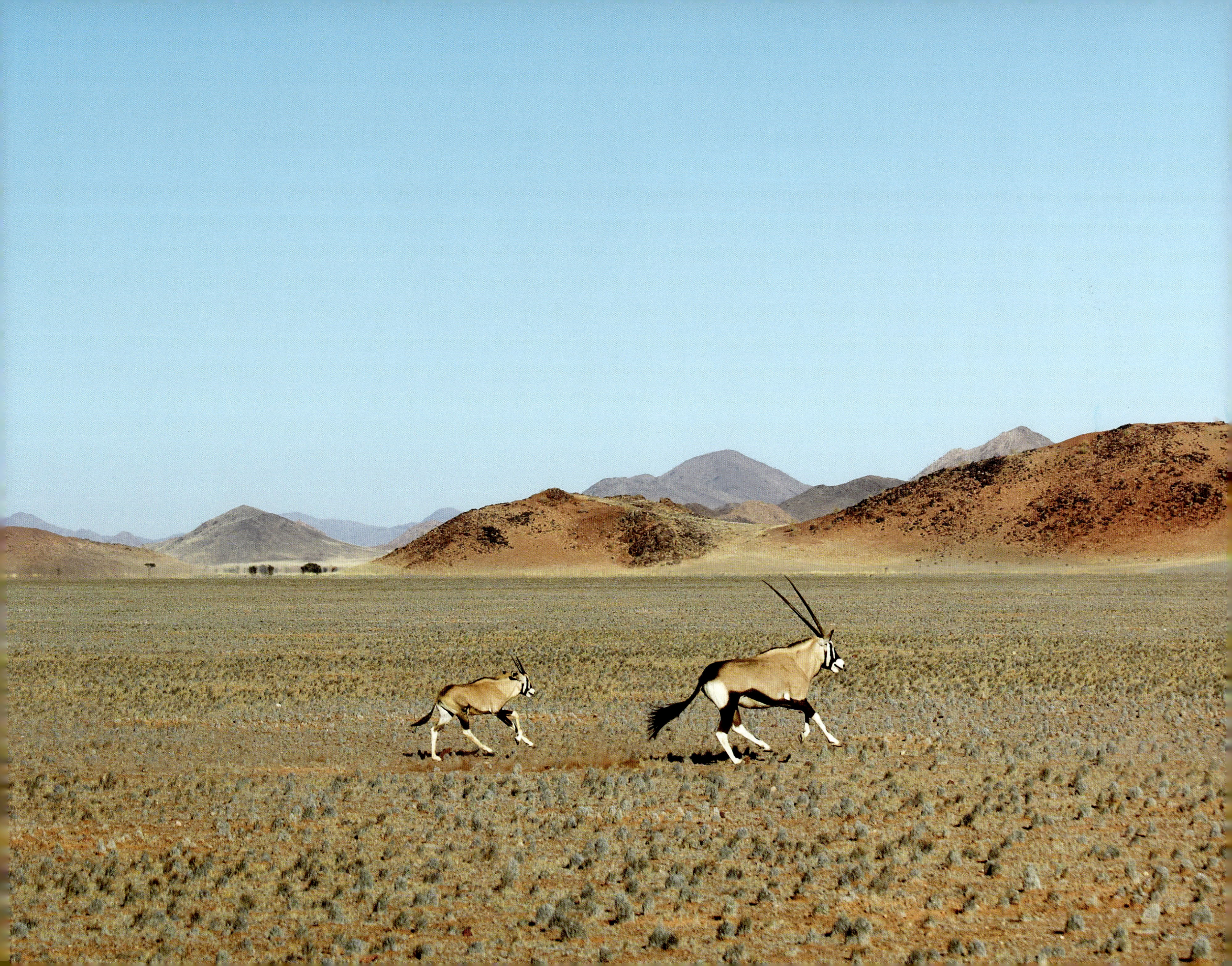

More about the Desert Biome

Like any other desert, the Namib presents different faces. In the south it merges with the Succulent Karoo with its winter rainfall and its own special life forms. In the north the meagre rain falls mainly in summer and its plants and animals differ accordingly. Along its western edge, the Namib receives an almost nightly invasion of fog from the Atlantic Ocean, extending 60km or more towards its eastern edge along the escarpment formed by the Naukluft and other mountain ranges.

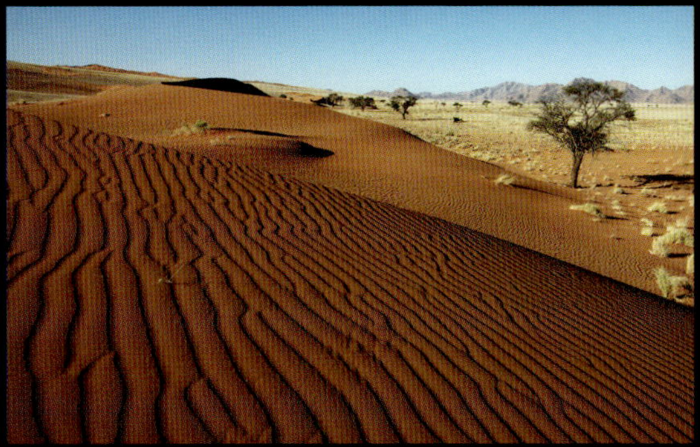

Where biomes meet – red dunes merge into the Nama Karoo

The warm air above the fog acts as a lid or inversion layer and prevents the formation of rain clouds, thereby increasing the aridity of the desert. At Sesriem near Sossusvlei (Namib Naukluft Park), the sand dunes end abruptly as the Desert Biome merges into the Nama Karoo.

Clouds are not common and when they occur they do not necessarily bring any rain. The reflection of the red, white and ochre sand may, however, transform the desert into a wonderland of earthen colours at sunset.

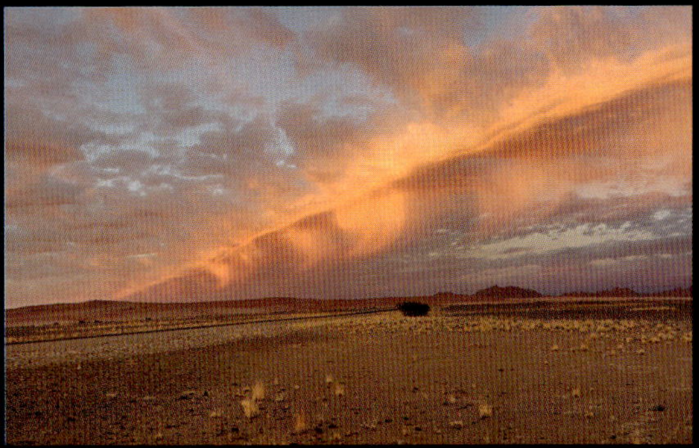

The fragile green blanket after light rain

Occasionally a 'good' rain year may follow years of no rains. When this occurs, even the lightest precipitation causes the desert grasses to sprout, covering the vast expanses with a fragile green blanket.

The Namib shows another face in the extensive gravel plains and the gravelly troughs between the dunes. Some of these troughs or 'valleys' are a kilometre or more wide. The gravel flats further inland form a somewhat stonier desert pavement. Even further inland you will find rough stony ground that merges with the brown rocky foothills of the escarpment.

The gravel plains have very different fauna and flora from those in the dune sea. The gemsbok is one of the few desert-adapted large mammals that may be found both on the gravel plains and even far into the dune sea. It is a powerfully-built antelope, easily recognisable by its magnificent v-shaped horns and distinctive facial markings. Living in areas where there is always a shortage of water, the gemsbok's resourcefulness in obtaining moisture from melons, succulent roots and bulbs allows it to survive in one of the hottest places on earth.

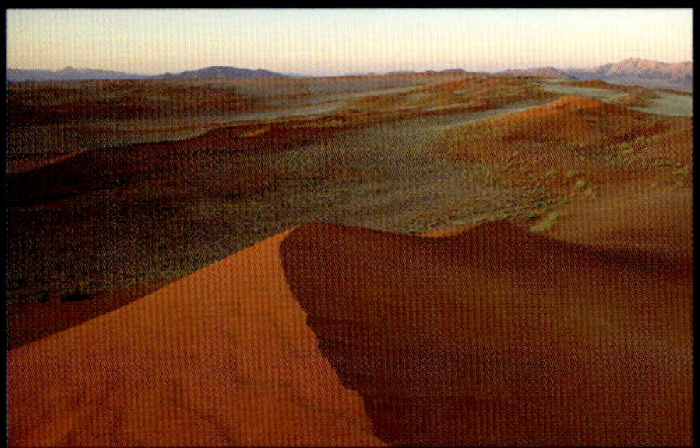

Clouds that do not spill their moisture

The gravel plains tend to be pale pinkish or greyish mauve in colour. Animals that inhabit these plains are camouflaged in similar pale colours – for example, the terrestrial Namaqua chameleon that may be found even in the harshest desert environments. This is a remarkable desert inhabitant because it exploits the normal colour-changing response of chameleons to other ends – it turns almost black in the early morning and late afternoon to absorb the maximum amount of heat from the sun, and during the hottest part of the day it turns pale to reflect the sun's rays.

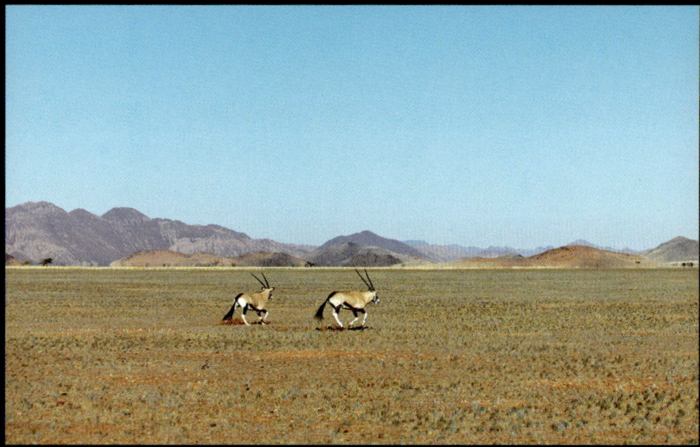

Gemsbok on a gravel plain

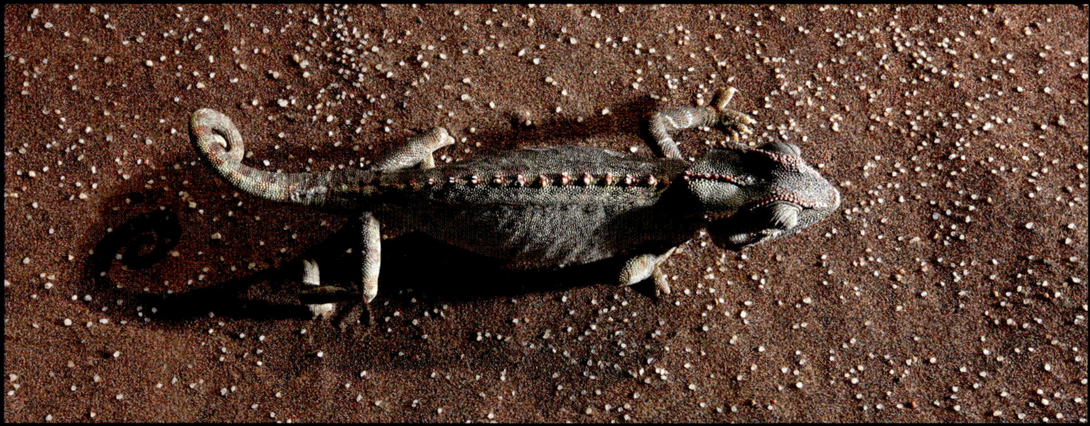

Namaqua chameleon

Snakes face similar temperature problems and they deal with these according to the type of ground surface where they live. Those that inhabit gravel plains will emerge sluggish from their night shelters and move to places where the sun can quickly warm them up.

The weird and wonderful *Welwitschia mirabilis*, simply called welwitschia, is often described as a living fossil. It is related

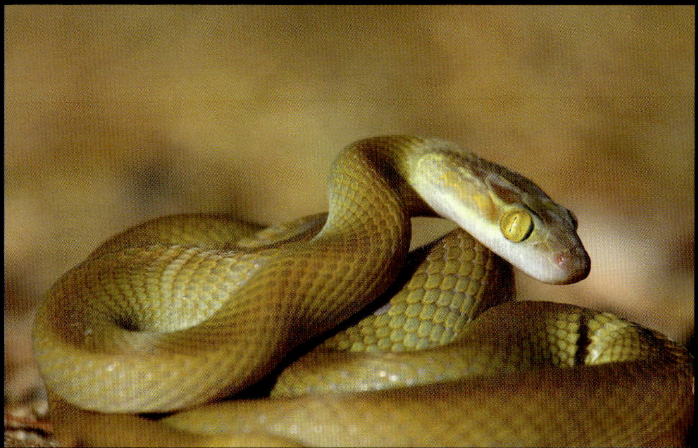

Desert brown house snake

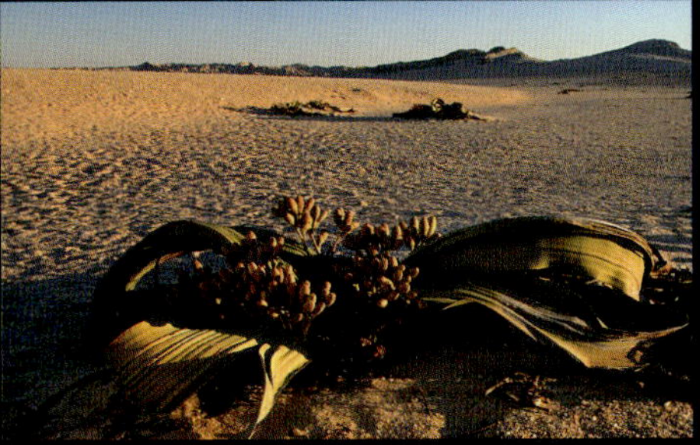

Welwitschia mirabilis

to the group of cone-bearing plants that include cycads and pines. Although they produce seeds, they bear no flowers – only cones. Furthermore, male and female cones are borne on separate plants.

What makes it unusual is that it is a dwarf tree that has only two leaves growing from opposite sides of an underground stem, but it looks as if there are more because wind shreds those two into many strips. These leaves are the only two the plant will ever have and continue to grow throughout its lifetime, sometimes reaching a width of 2m at their base. Some of these welwitschias in the Namib are estimated to be at least 2 000 years old. They are among the rarest plants on Earth, numbering only a few thousand, sparsely scattered over the extreme western strip of the Namib Desert from the Kuiseb River in Namibia to south-western Angola.

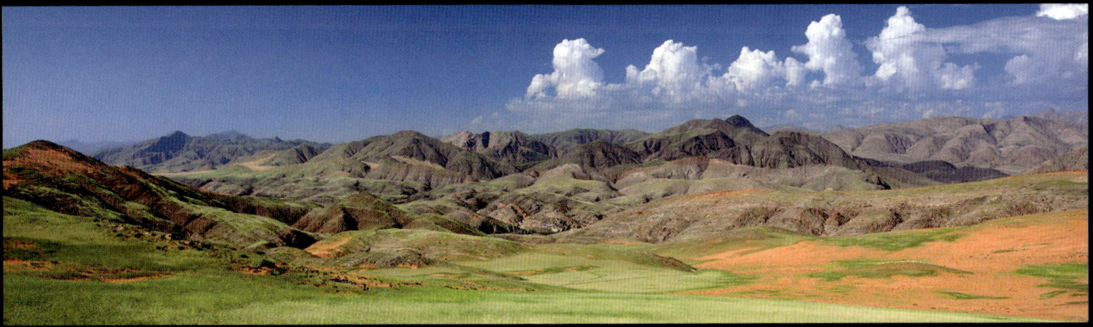

Kaokoland after good summer rains

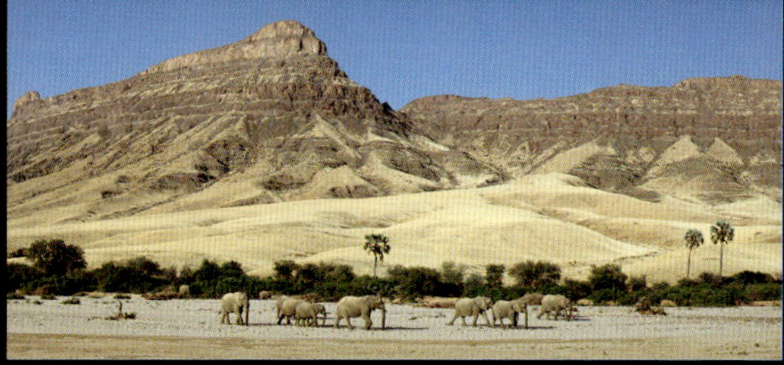

Desert elephant in Kaokoland

In the far north-western parts of Namibia, called Kaokoland, the desert responds well to occasional summer rains, changing the landscape from dull grey-brown to a mind-blowing wonderland. The large expanse of sand and gravel is covered by various species of *Stypagrostis*, commonly known as bushman grass. In the distance the Kunene River bisects the desert on its way to the ocean and is fringed by the mountainous Angolan hinterland.

A special sighting in Kaokoland is the desert elephant. They are not a separate species but a type of African elephant that has adapted to survive in the desert. They move great distances between their feeding grounds and waterholes, drinking only every third or fourth day, unlike elephant in the moist savannah that drink every day. Several dry river courses wend their way across the Namib, some of which have an almost permanent flow of water under the sand that forms their beds. Such watercourses provide special conditions, often including occasional surface water, and are inhabited by species of plants and animals that are not strictly desert-adapted. They may also have some fairly impressive gallery forests of ana, Cape ebony and mustard trees, even deep into the desert proper. However, trees are scarce and desert elephant are careful not to debark or uproot them. Elephant browse in the dry season, but during the rains they eat more grass, which gives the trees time to recover.

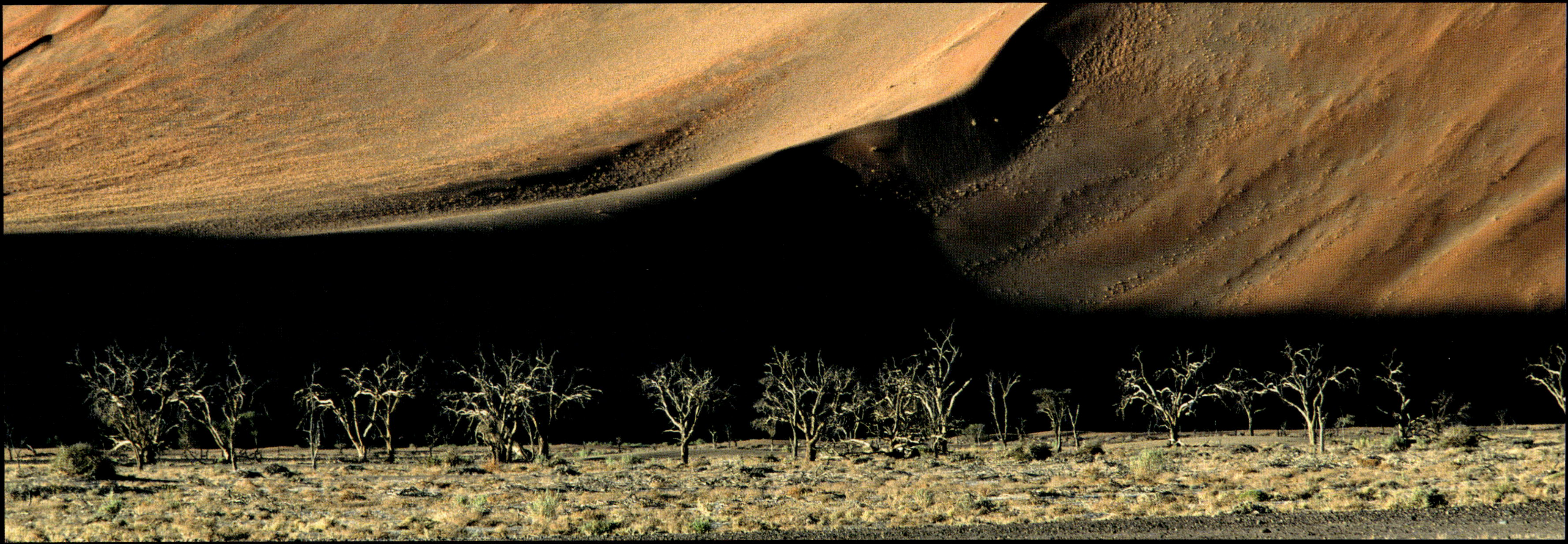

Tree skeletons against the dunes on the way to Sossusvlei

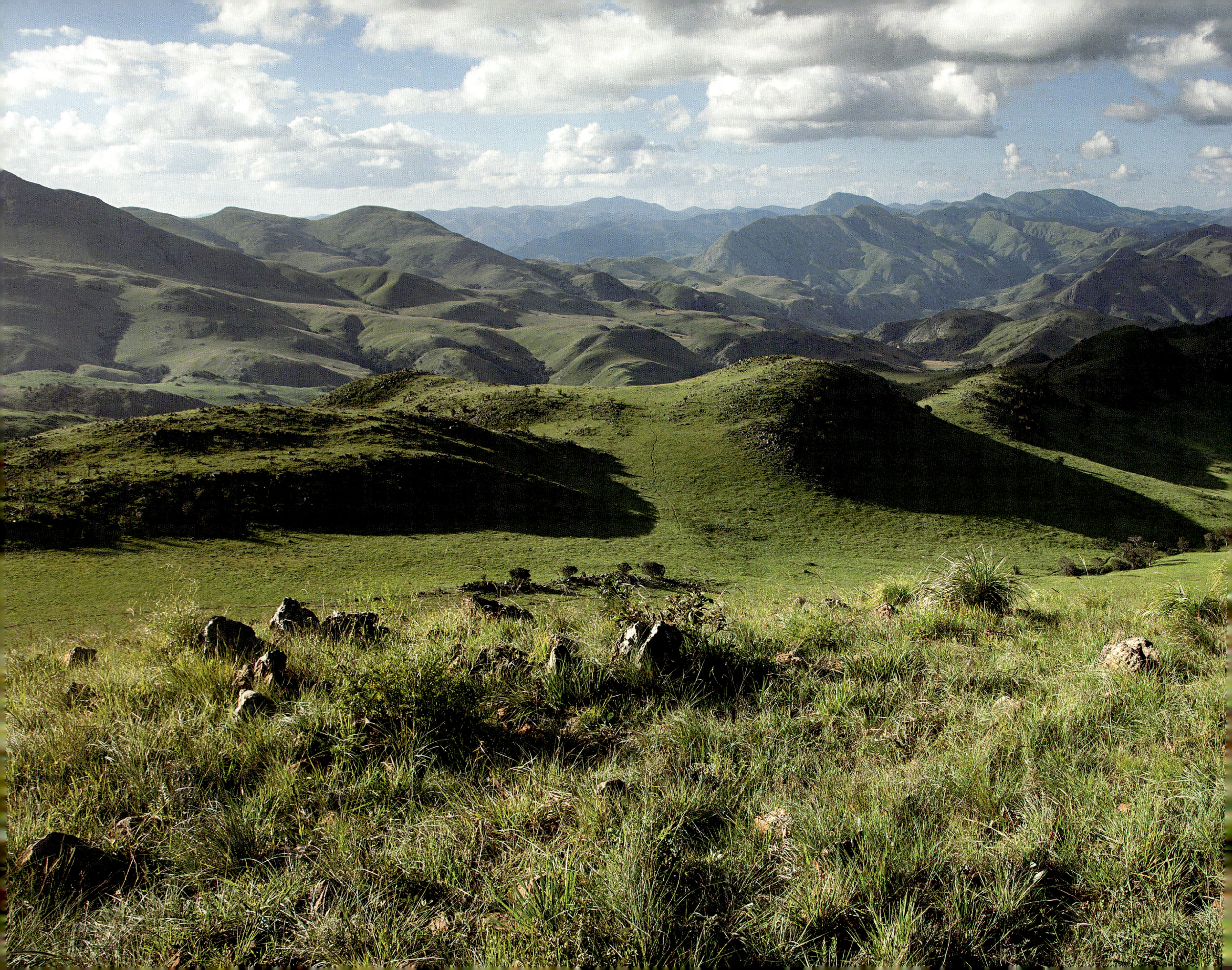

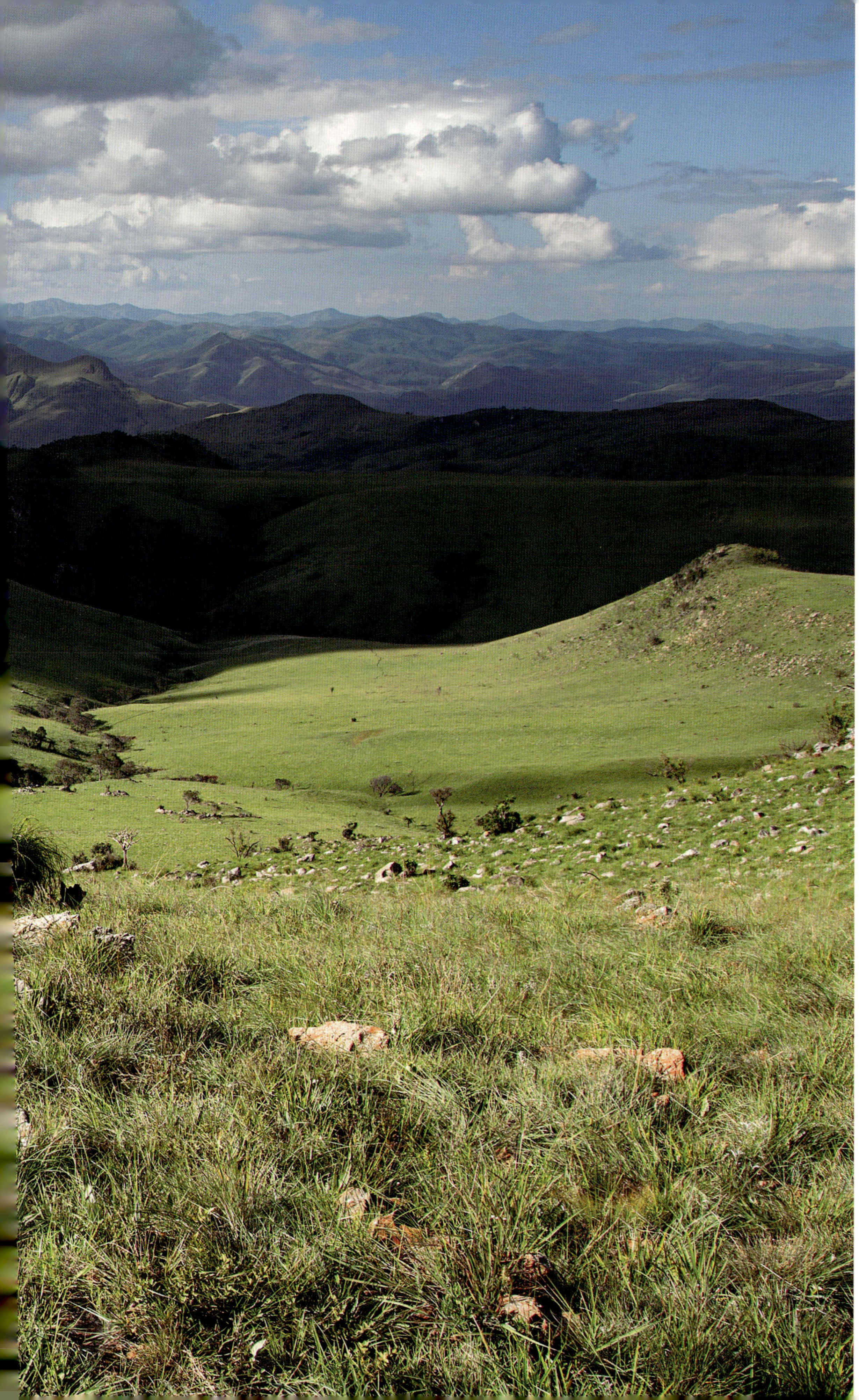

Grassland

Rich in biodiversity

The huge central plateau of southern Africa is often referred to as the 'Highveld' – open, treeless, big-sky country with vistas into tomorrow. The area is mainly flat or slightly undulating and was originally covered by waving grasslands with huge herds of game. Today this area forms the agricultural and economic heartland of the country and much of the original vegetation has vanished. And so have the herds of game.

The Grassland Biome, however, extends further than the Highveld plateau and includes the escarpment itself – the mountainous areas and valleys of KwaZulu-Natal, Mpumalanga and the Eastern Cape. Here fairly large pockets of pristine grassland have survived the onslaught of human intervention and can still be found in the wilderness areas of the uKhahlamba-Drakensberg Park and on the mountain ranges along the Mpumalanga escarpment.

Grassland is typical of high country where the altitude may vary from a few hundred to almost 3 000m above sea level and is dependent on good summer rains – varying between 400mm and 2 000mm annually. In such areas, thunderstorms and hail are common in summer and frost occurs in the dry, cold winter. Fire, frost and grazing are important factors in the maintenance of a healthy grass cover and preventing the establishment of trees.

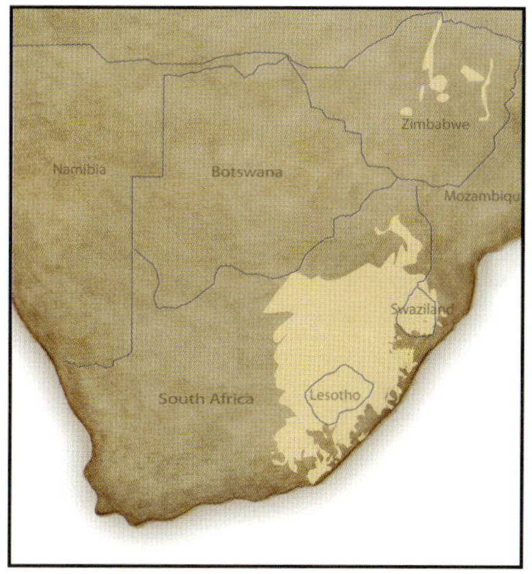

Surprisingly, the Grassland Biome ranks with Fynbos and the Succulent Karoo Biomes as one of the best vegetation types in which to see wild flowers. As far as richness of species is concerned, it ranks second to Fynbos. Apart from grasses, other plants are mainly bulbous species adapted to co-exist with the various kinds of perennial grasses.

In higher rainfall areas receiving 625mm per annum and more, soils are more acidic and sour grasses prevail. Sour grasses are unpalatable to grazers since they have a higher fibre content and tend to withdraw their nutrients from the leaves during winter. In lower rainfall areas, grasses have a lower fibre content and nutrients remain in the leaf blades in winter. These are the sweet grasses since they are more palatable.

Key grassland mammals are the blesbok, black wildebeest, grey rhebuck and mountain reedbuck. While there is a variety of grass species, woody plants are rare – localised within rocky outcrops and ravines, or totally absent.

Good places to visit grasslands

The grassland region is heavily utilised for agriculture, but fortunately there are still areas set aside for conservation. These include the Suikerbosrand Nature Reserve near Heidelberg, the Rietvlei Nature Reserve in Gauteng, the Willem Pretorius Game Reserve in the Free State, Verlorenvlei Nature Reserve in Mpumalanga, Nooitgedacht Dam, Bronkhorstspruit Dam, Vaal Dam, Spioenkop and Koppies Nature Reserve. Highland grassland occurs in the Golden Gate National Park and the adjacent uKhahlamba-Drakensberg Park and along the eastern escarpment in the Mpumalanga and the Wolkberg Wilderness Area in the Limpopo province.

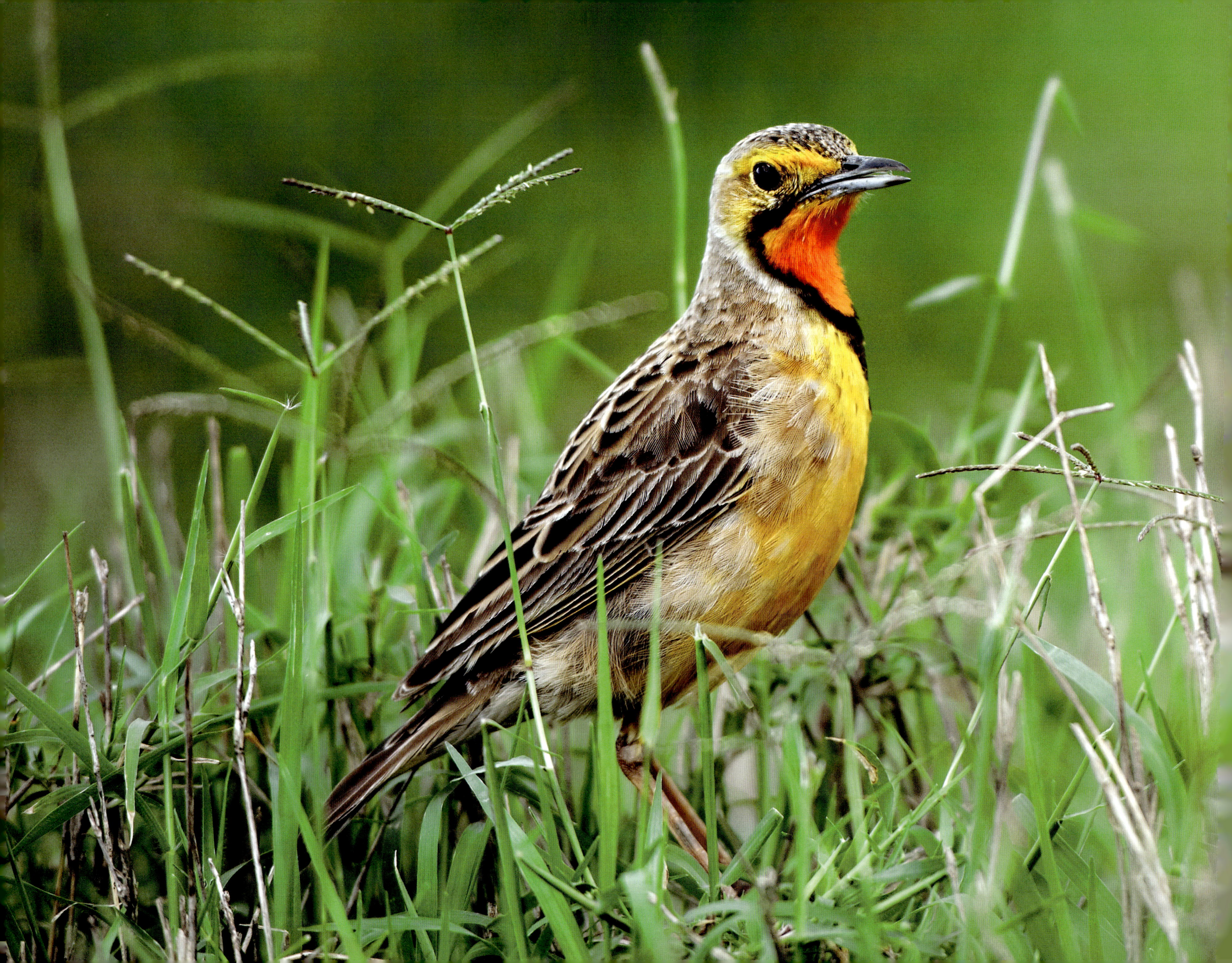

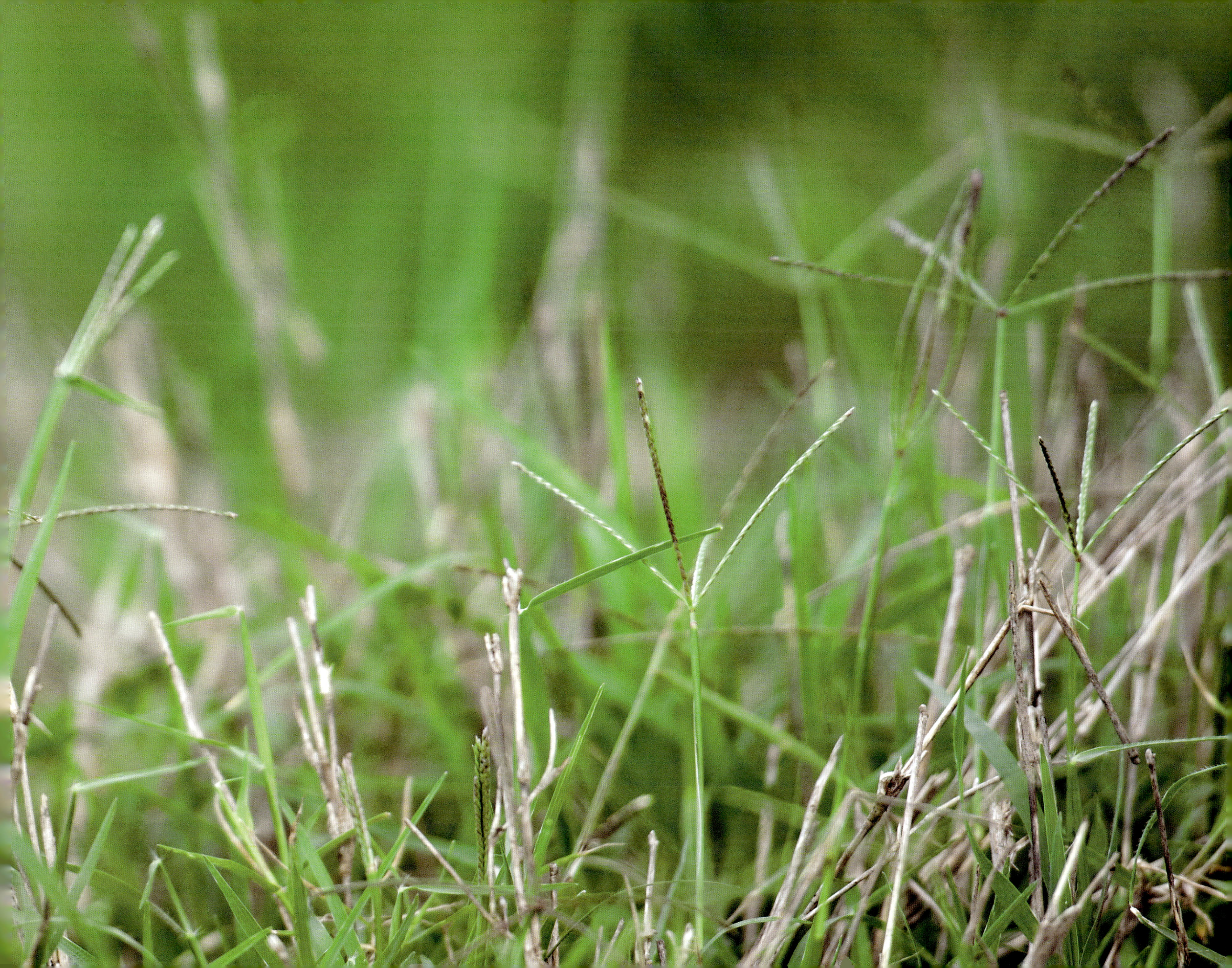

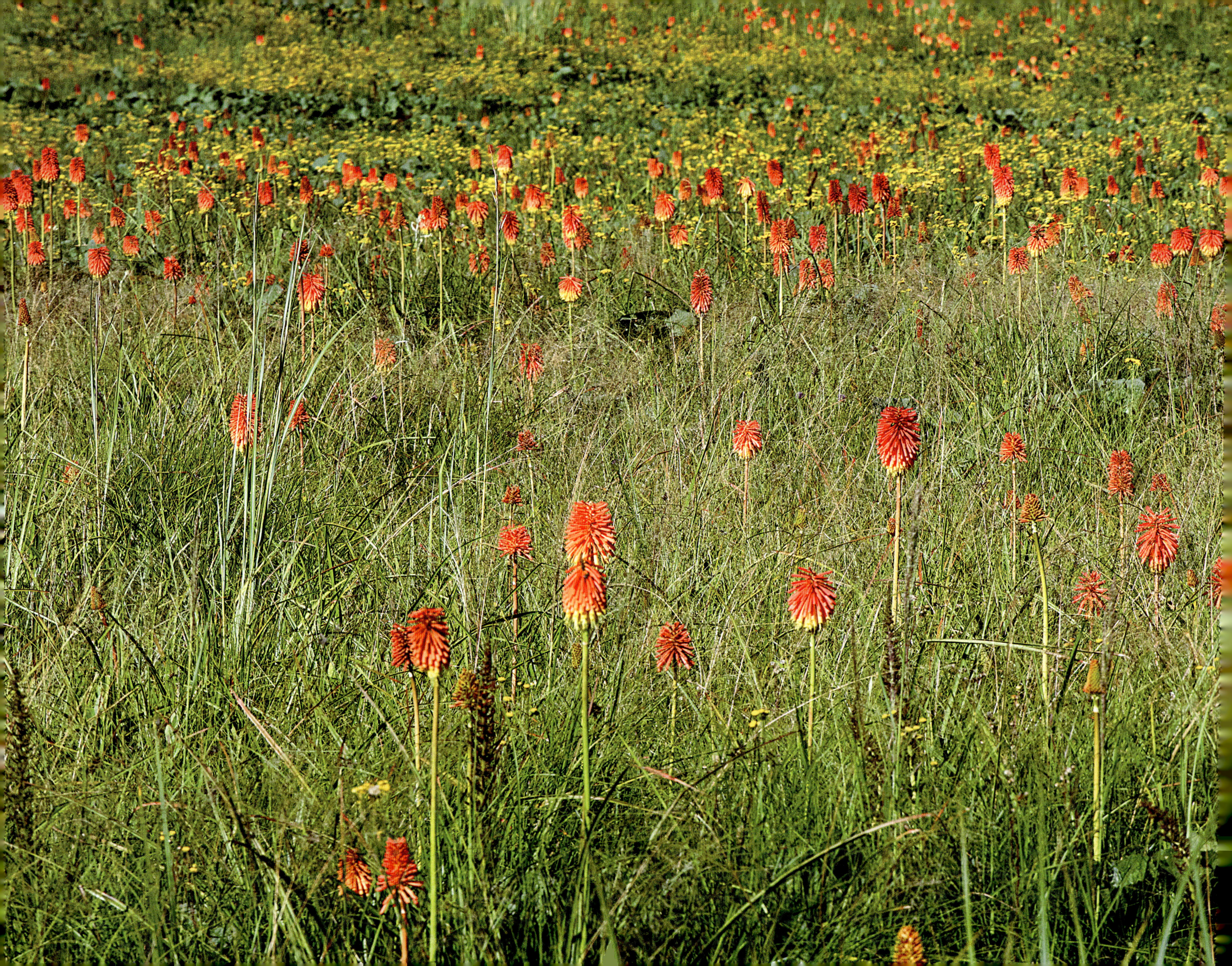

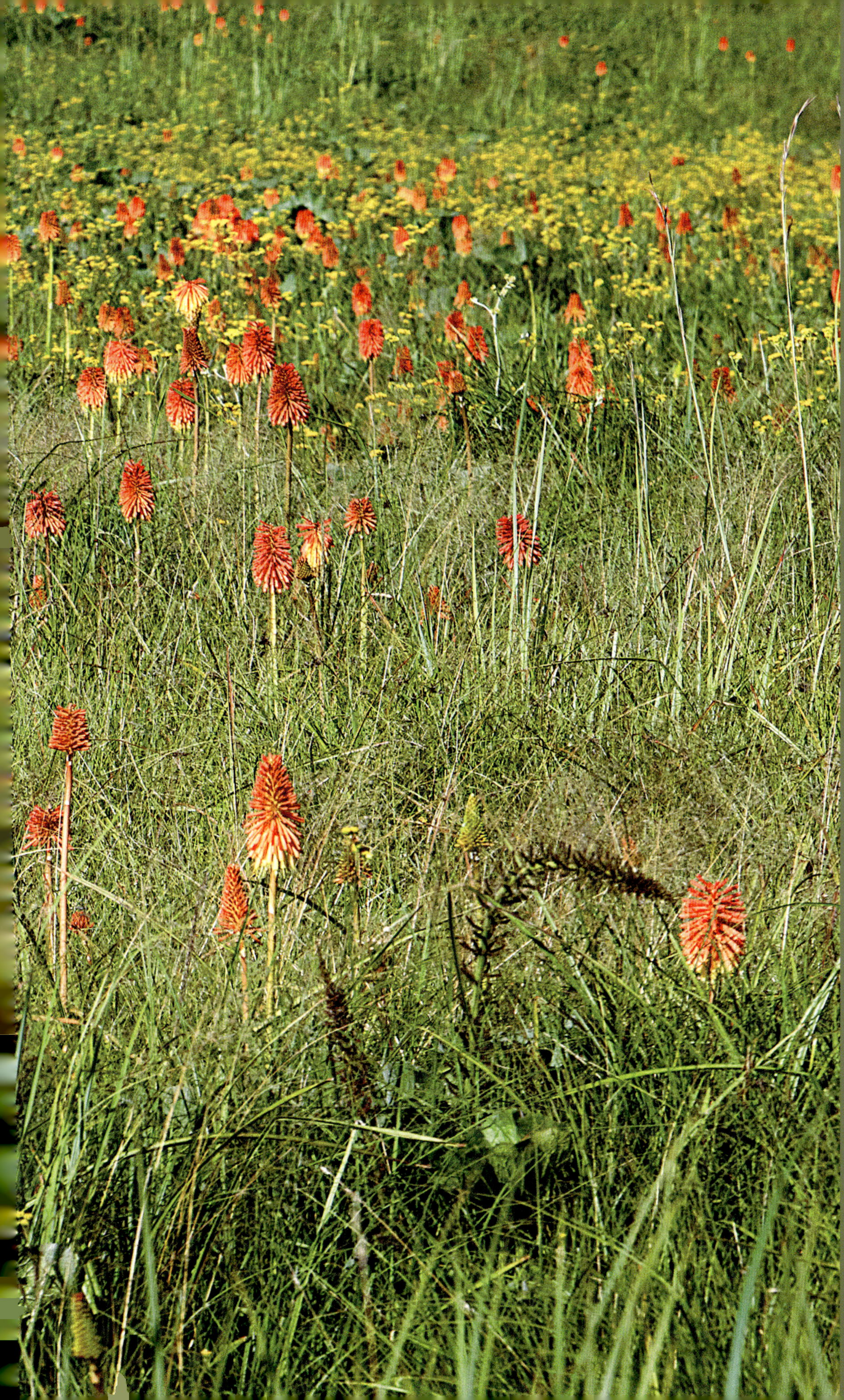

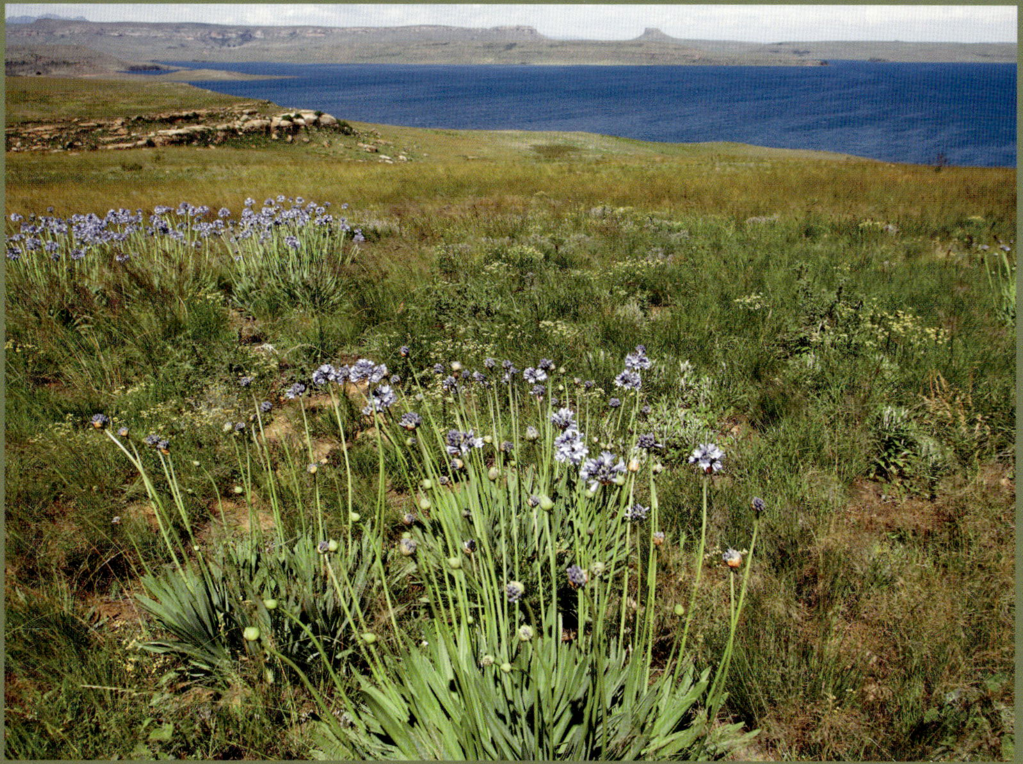

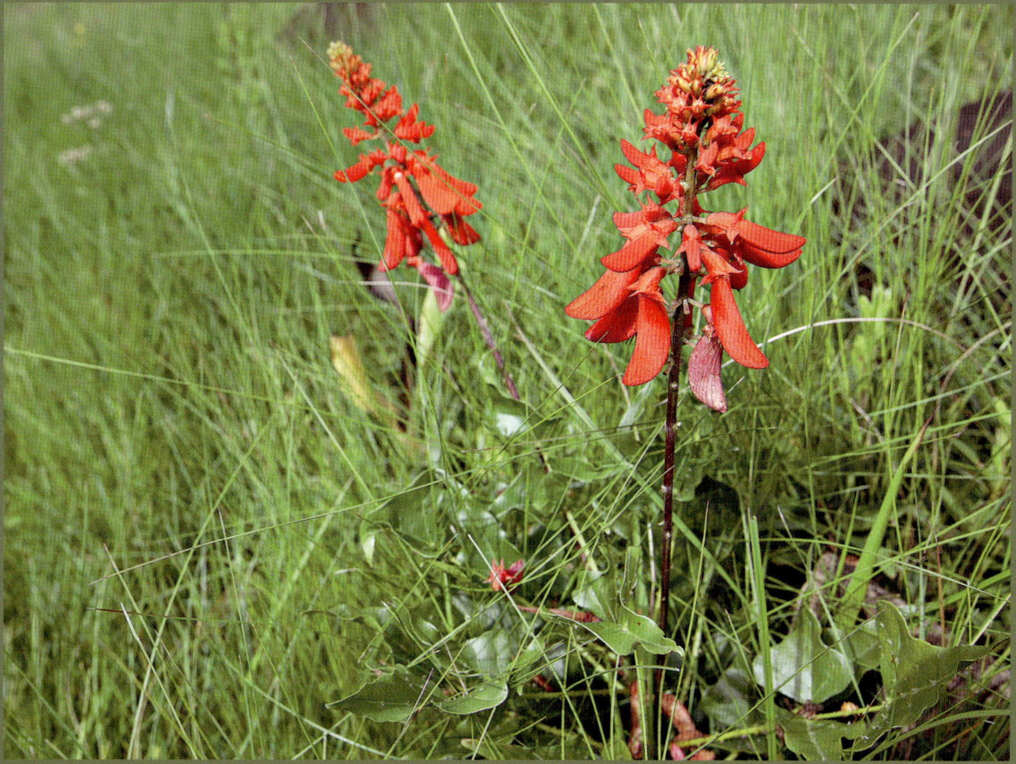

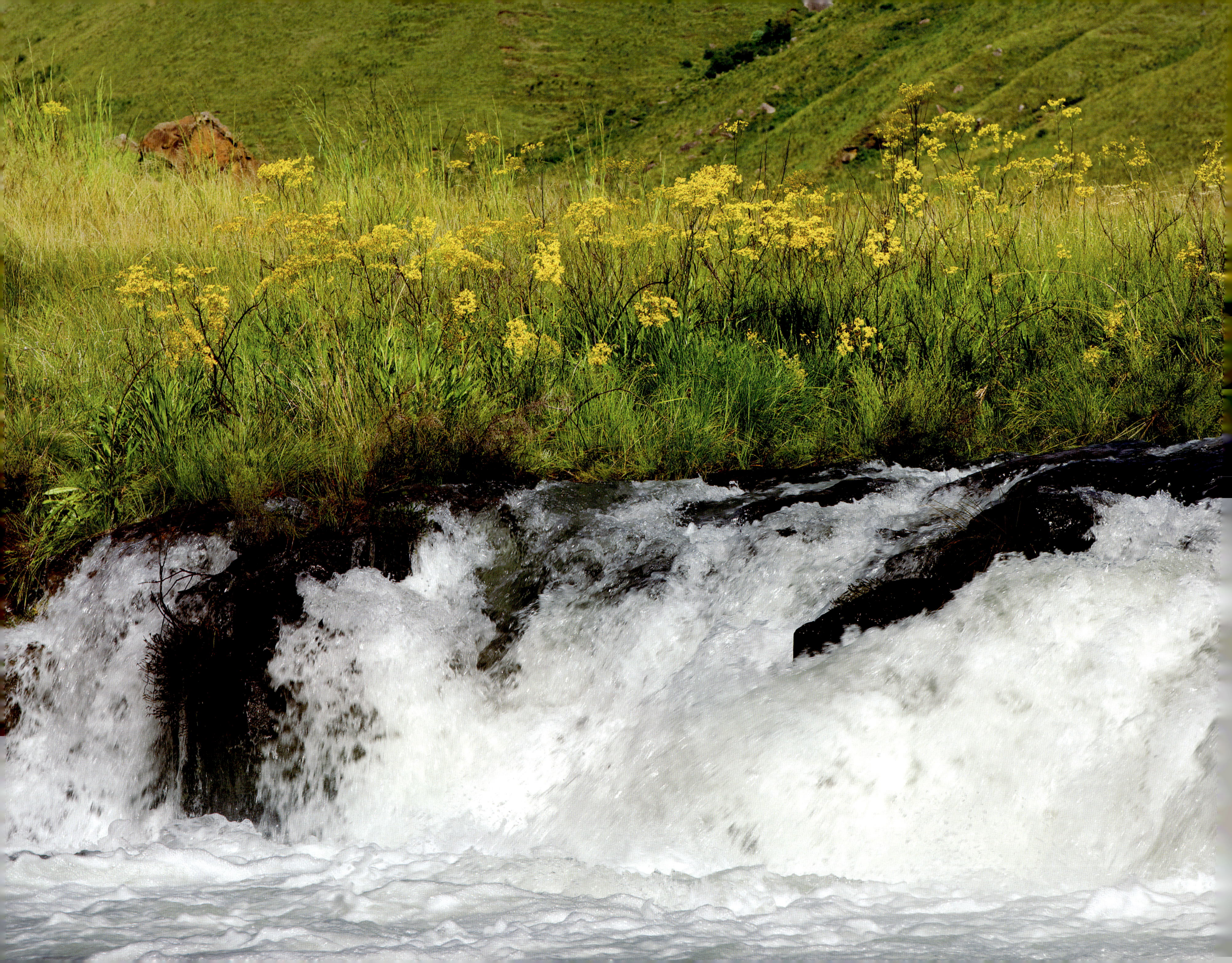

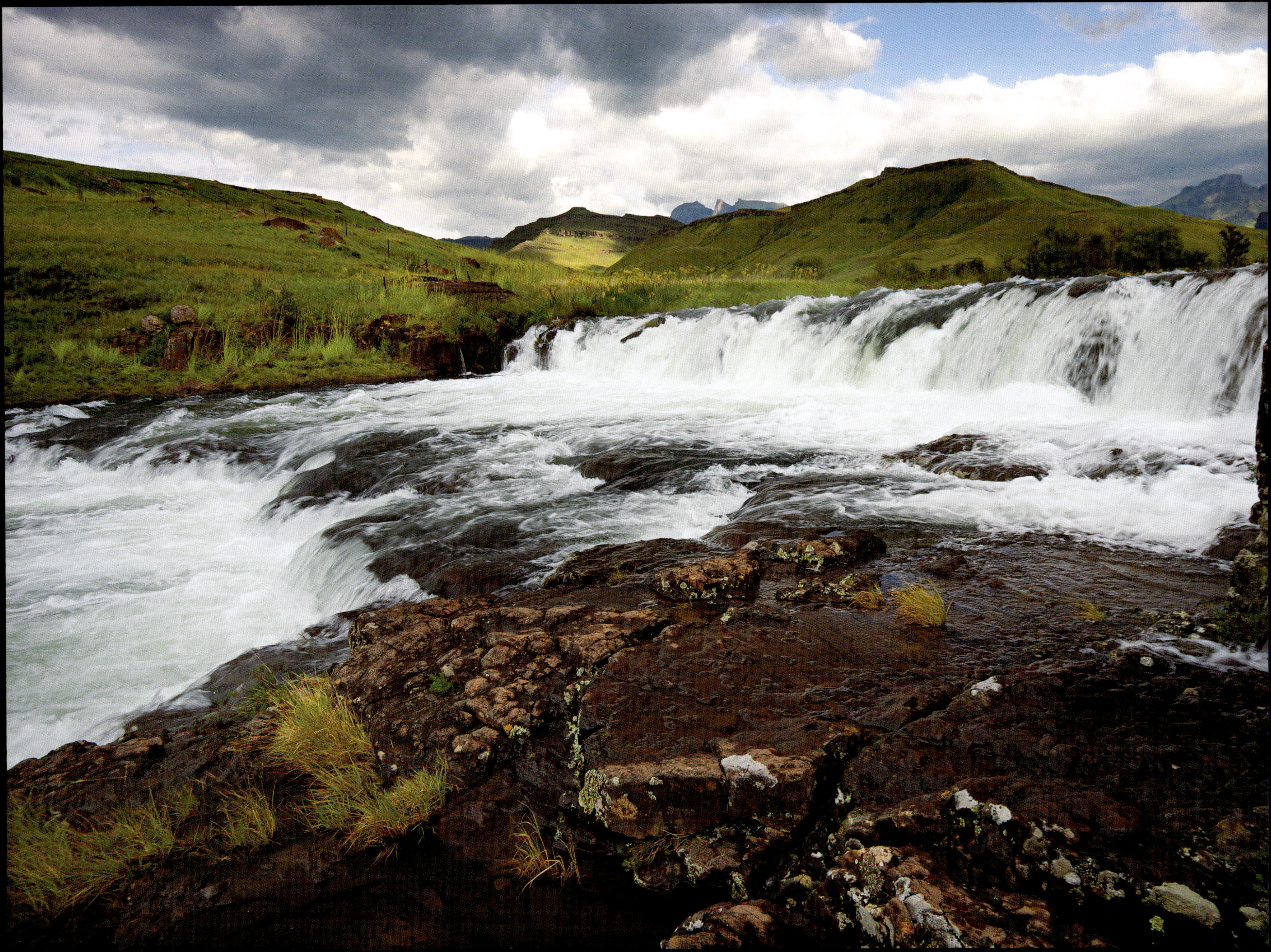

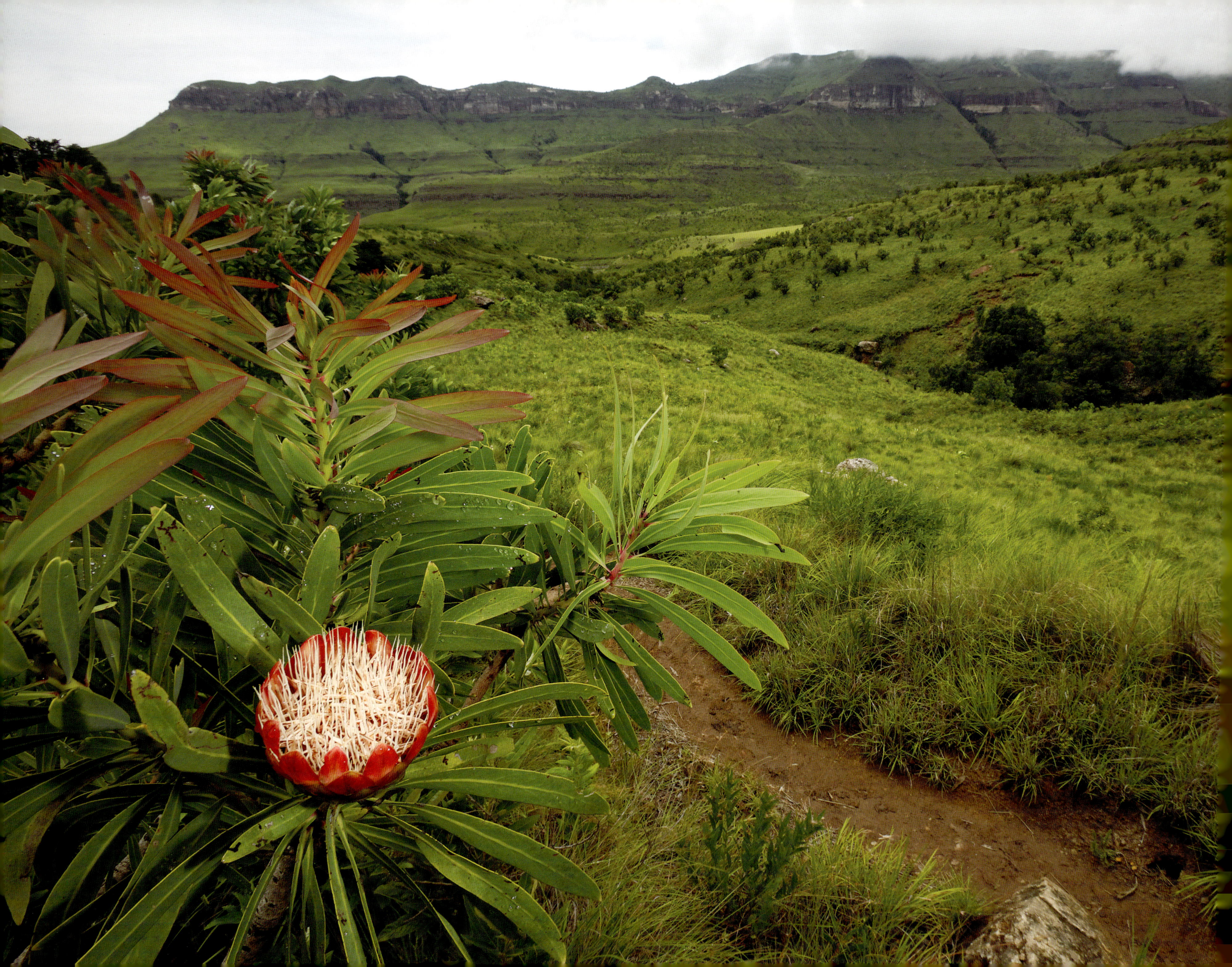

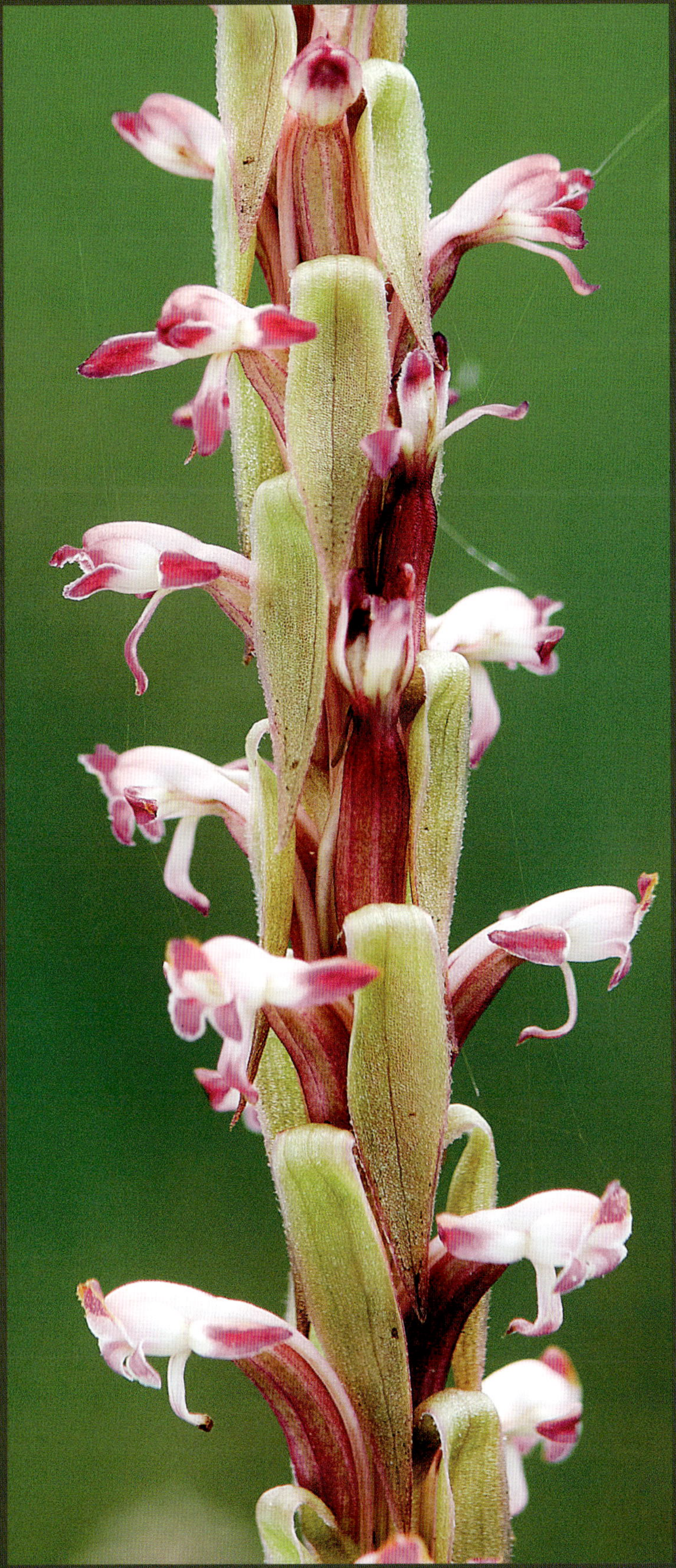
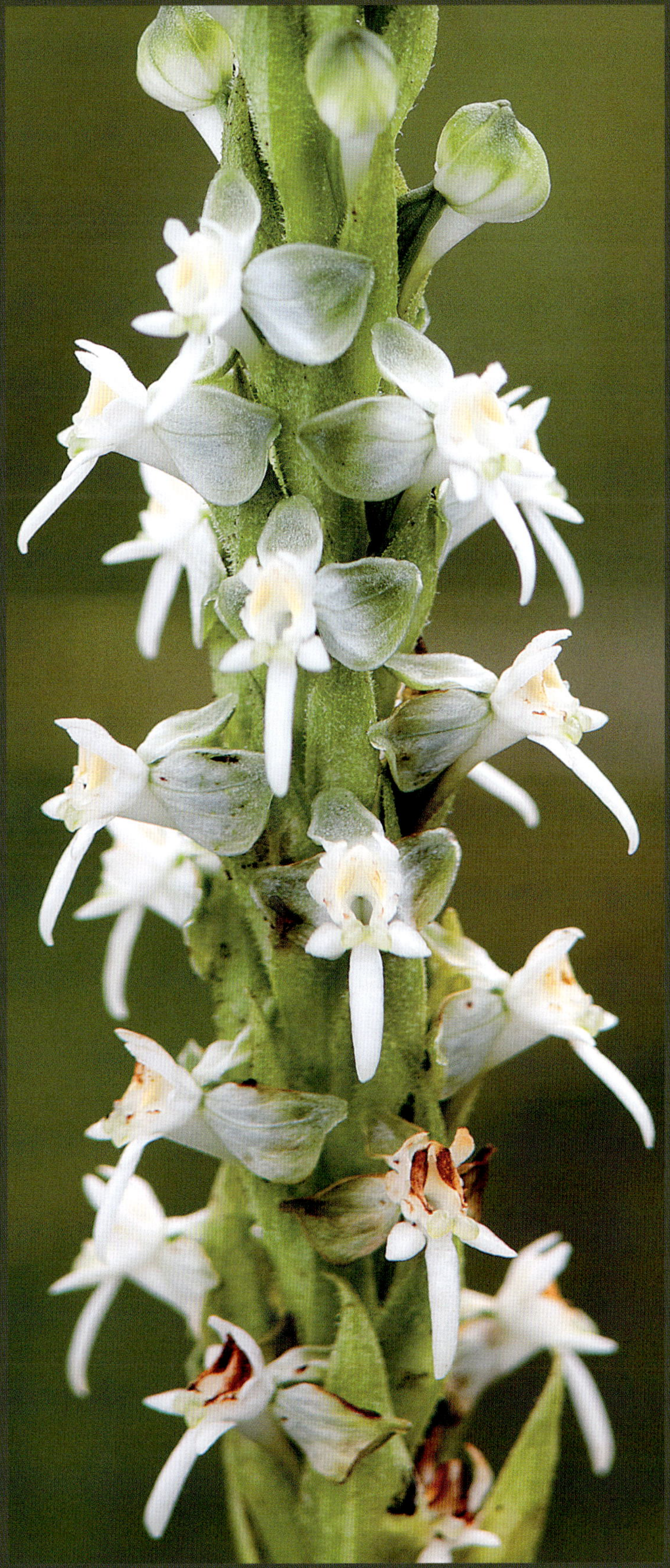
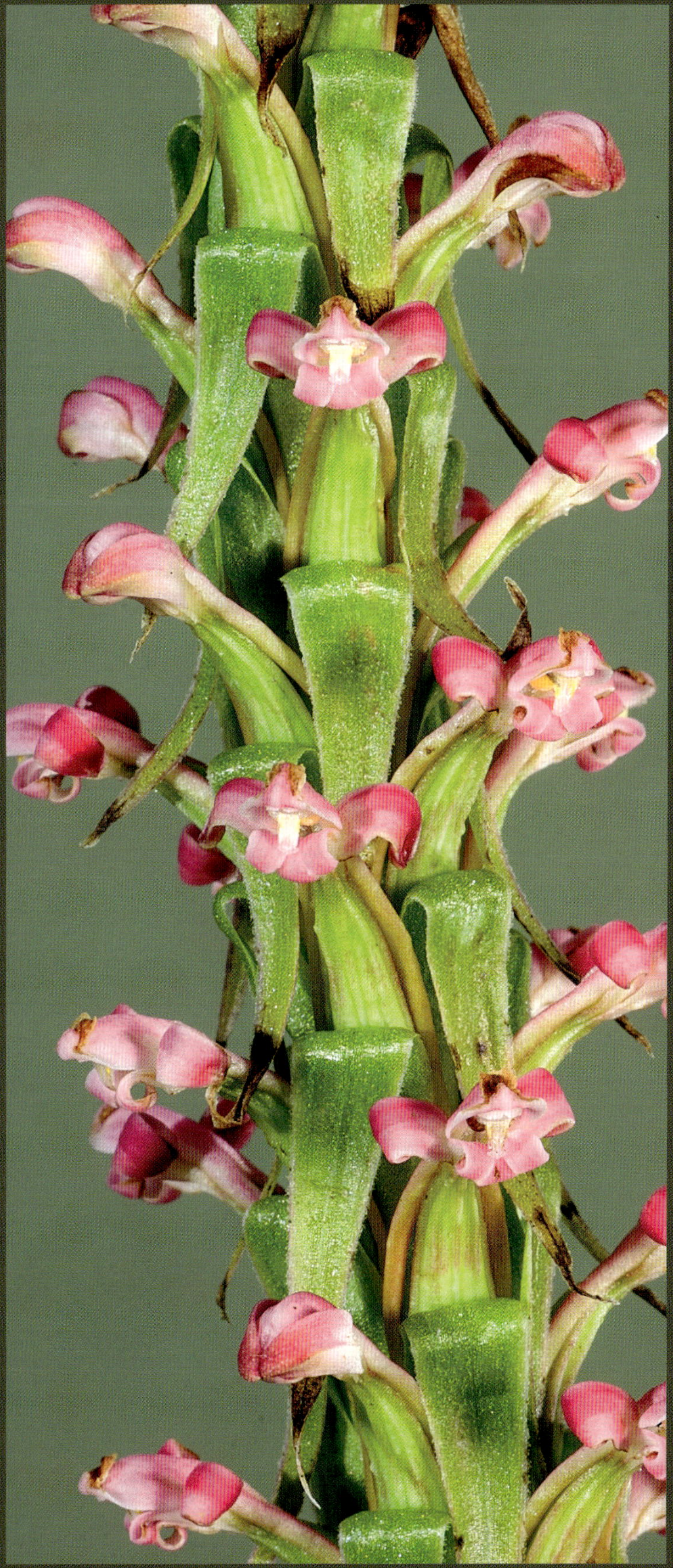

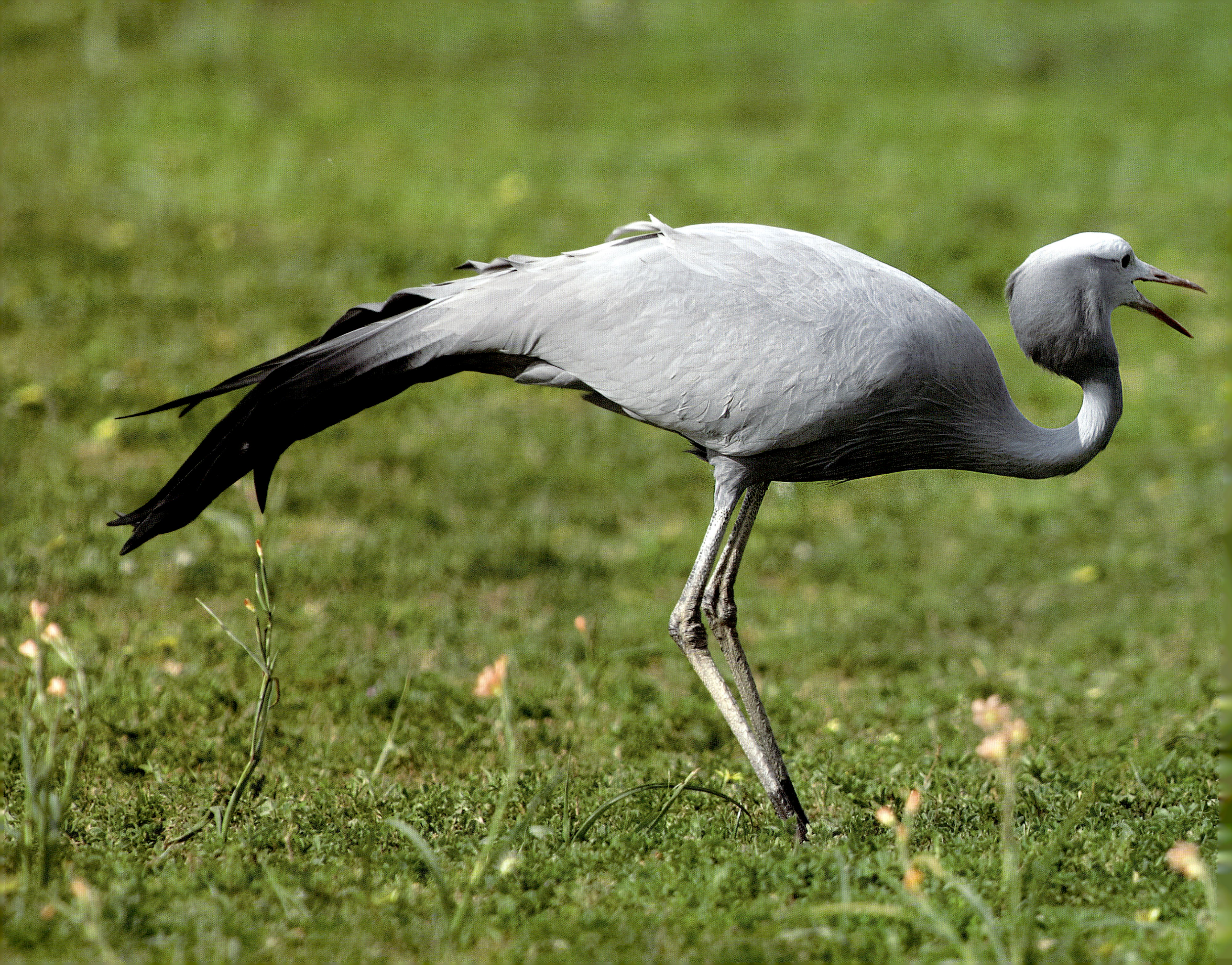

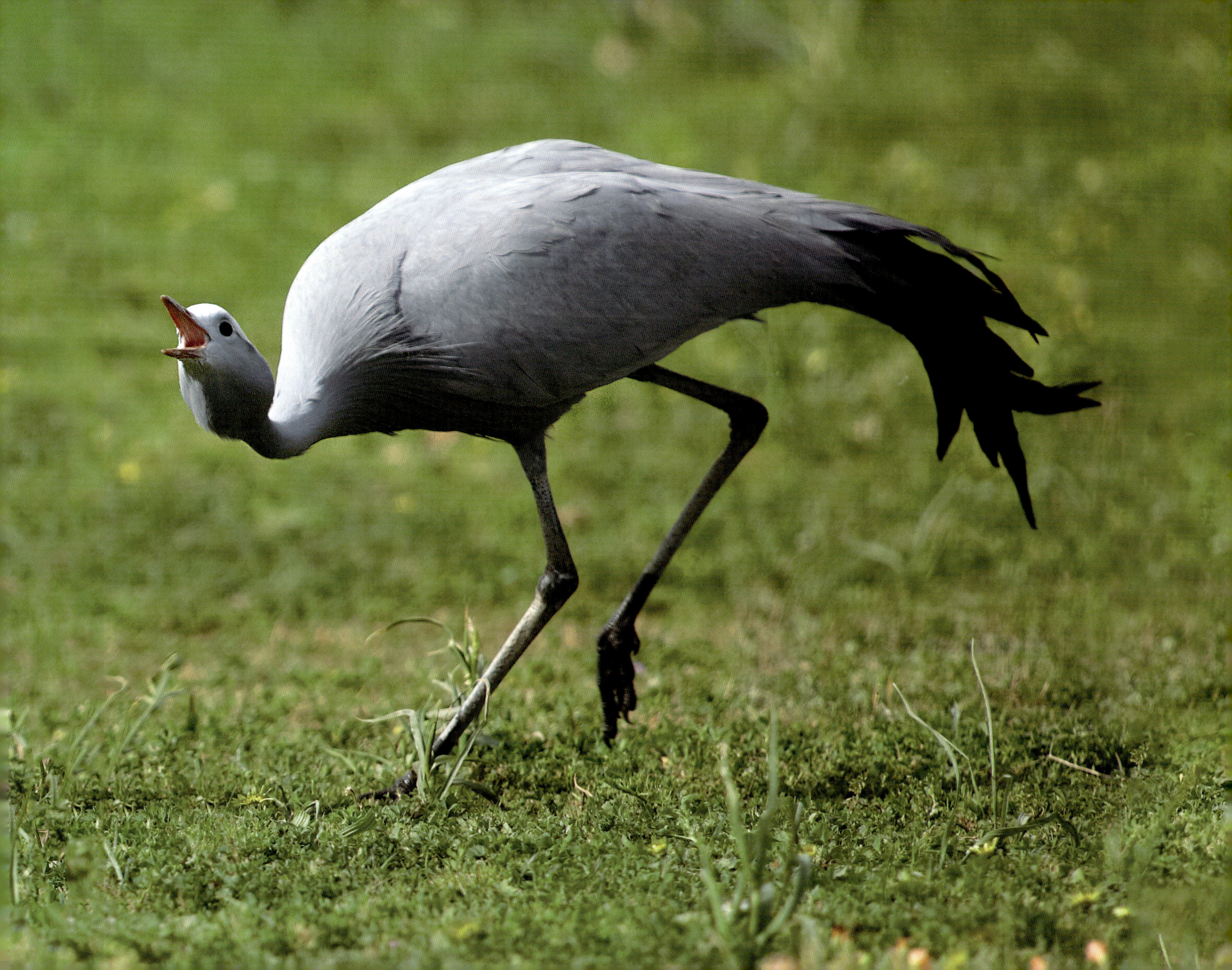

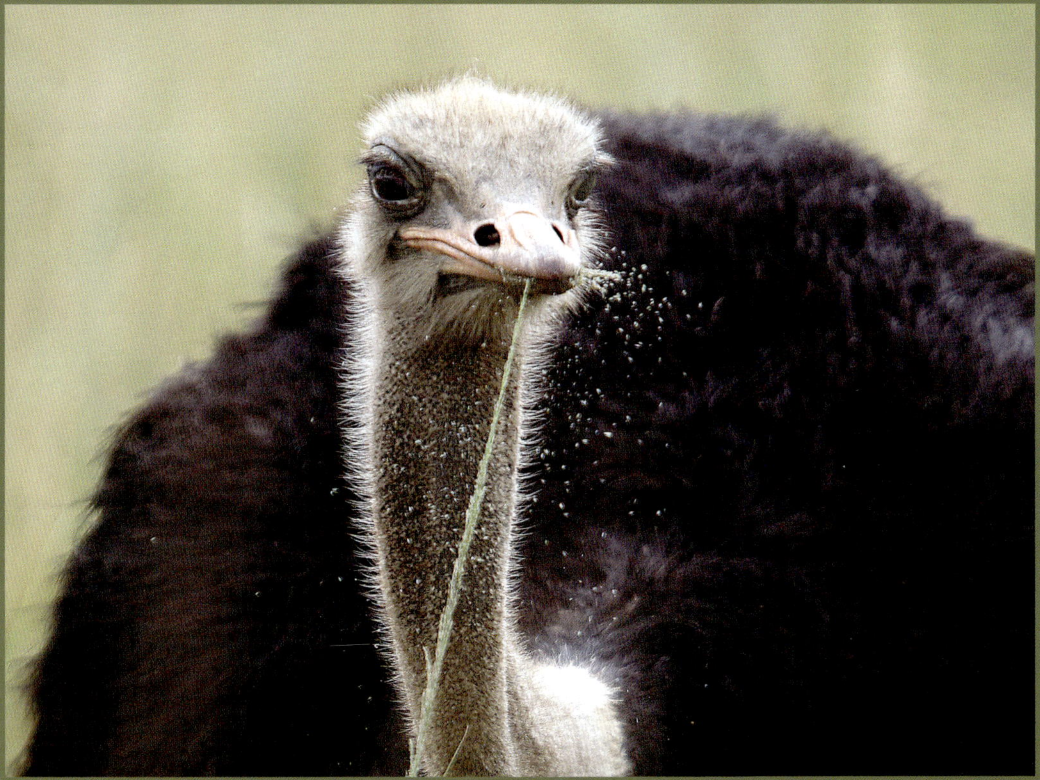

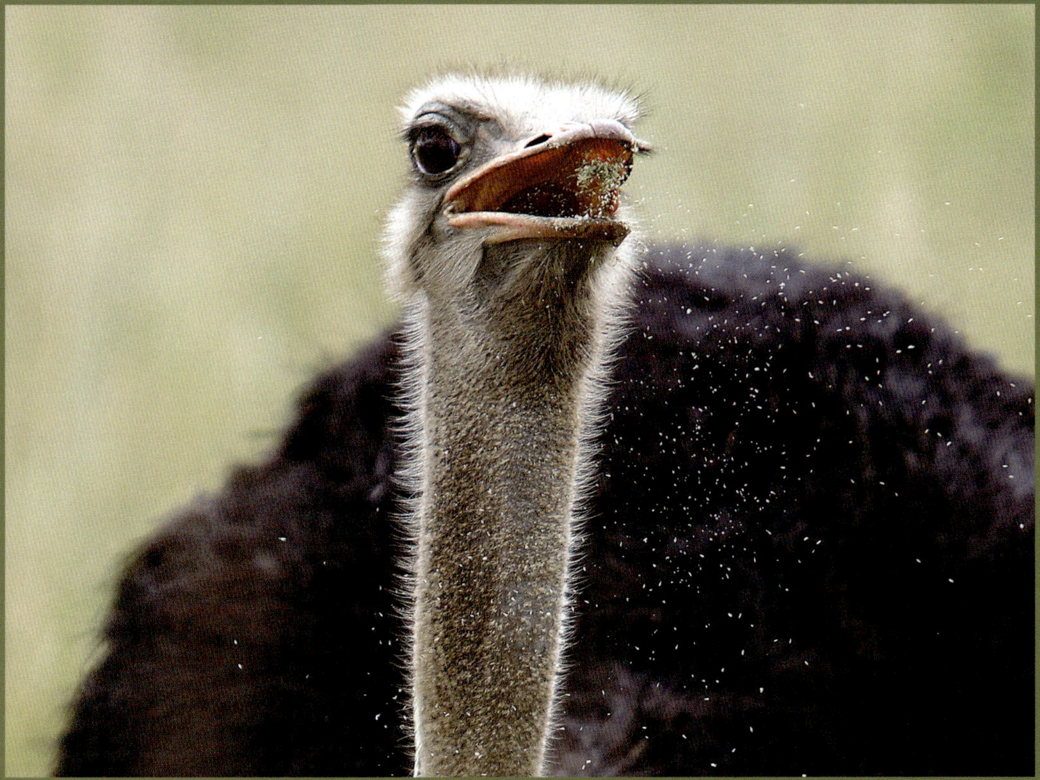

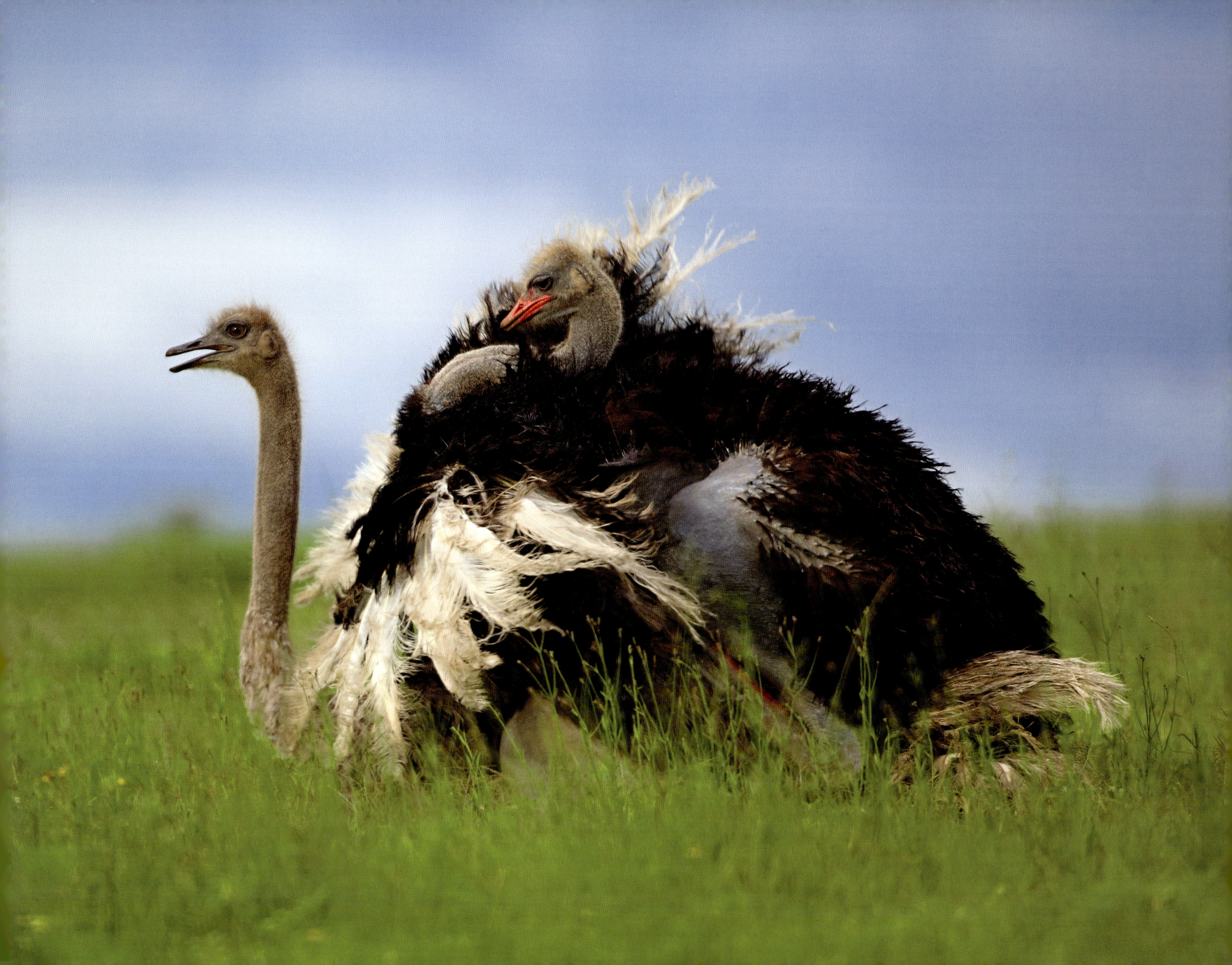

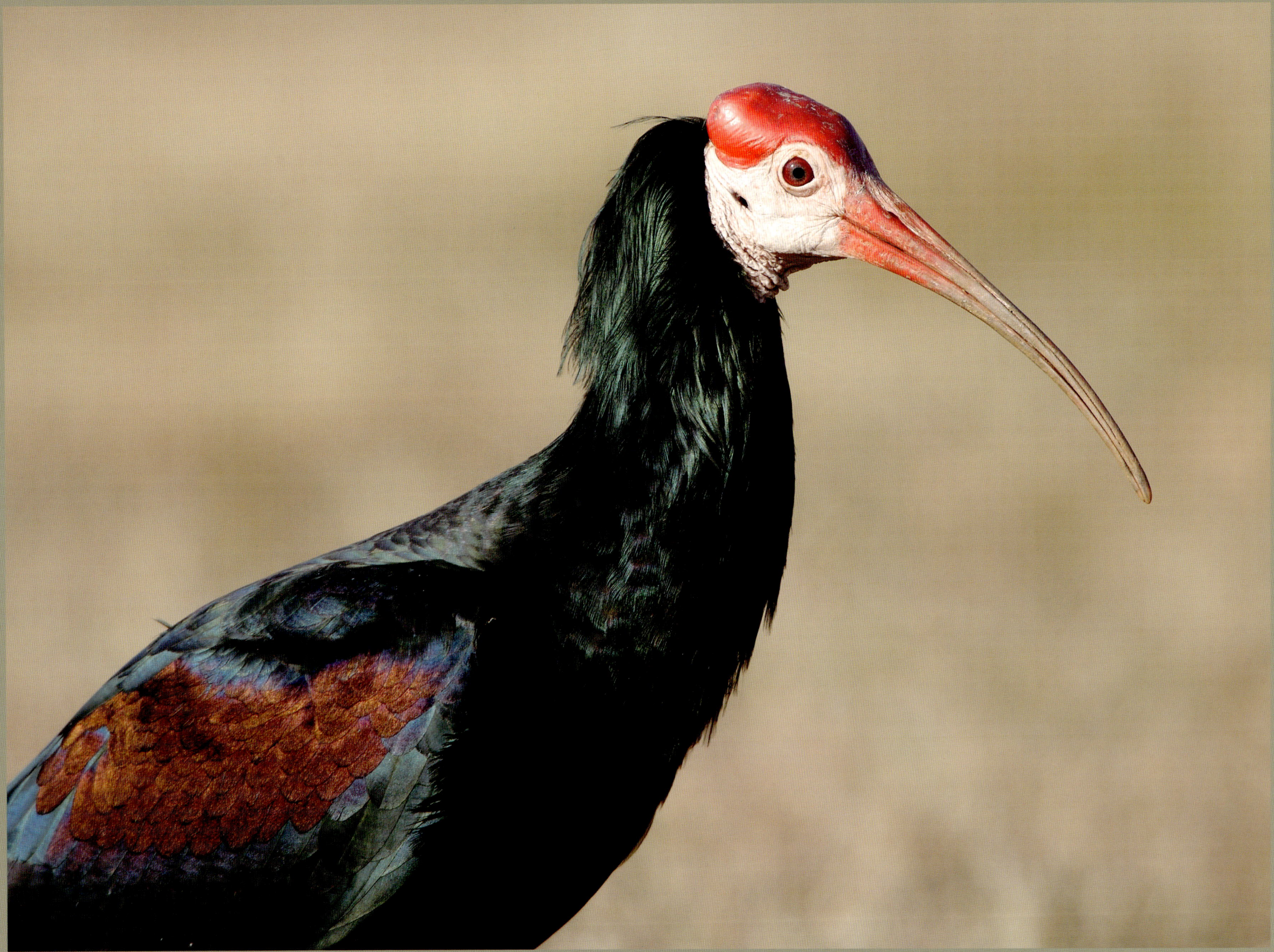

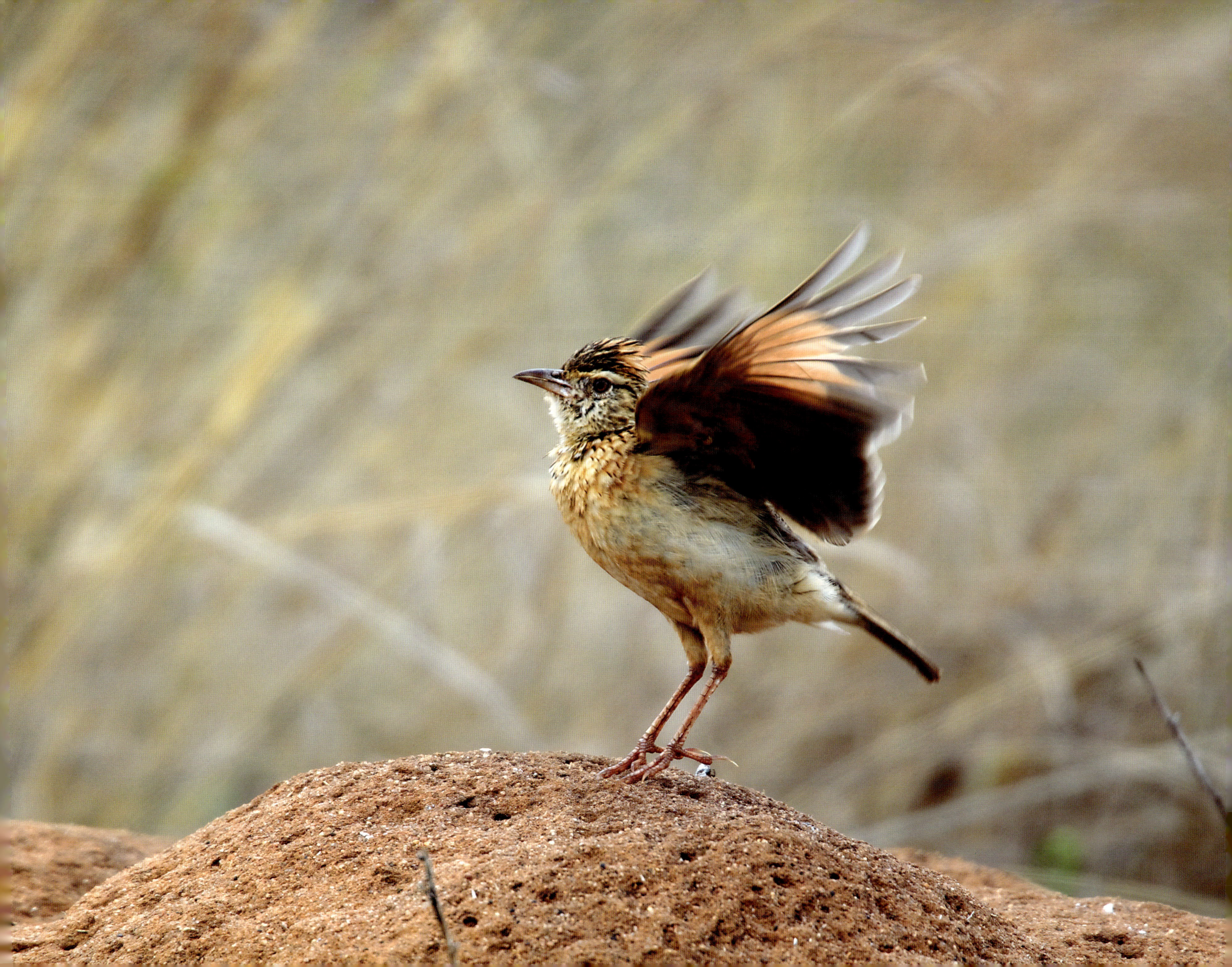

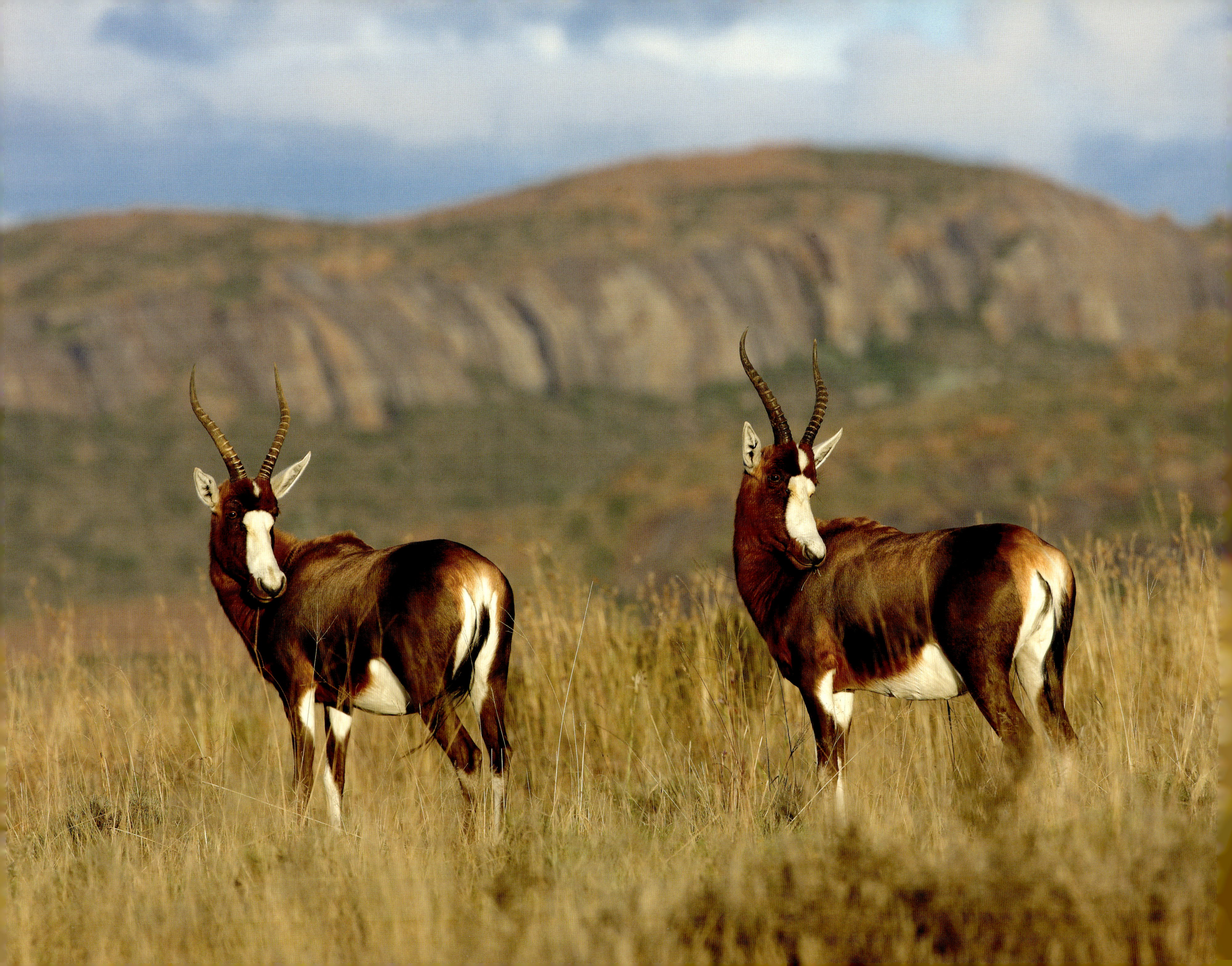

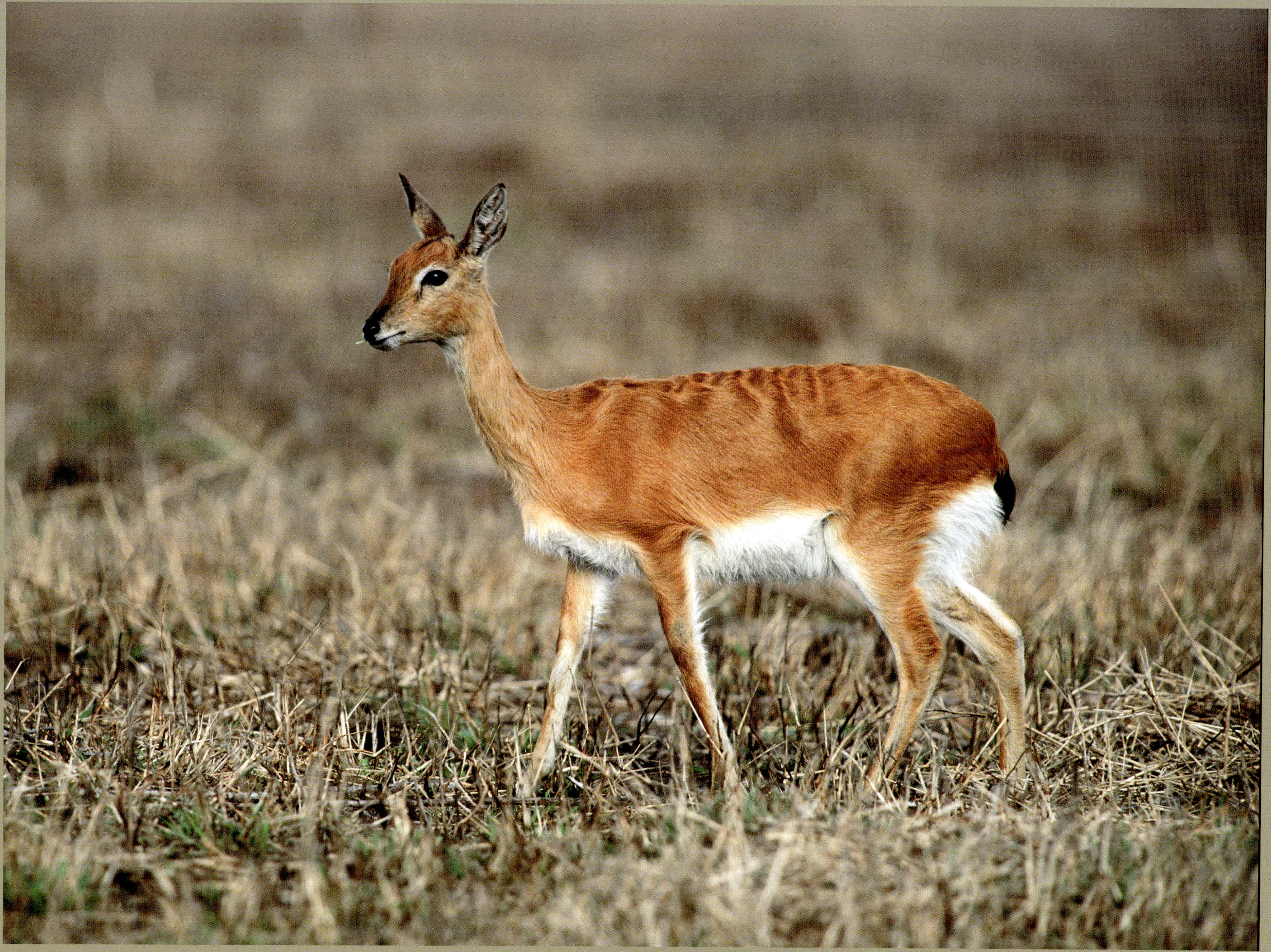

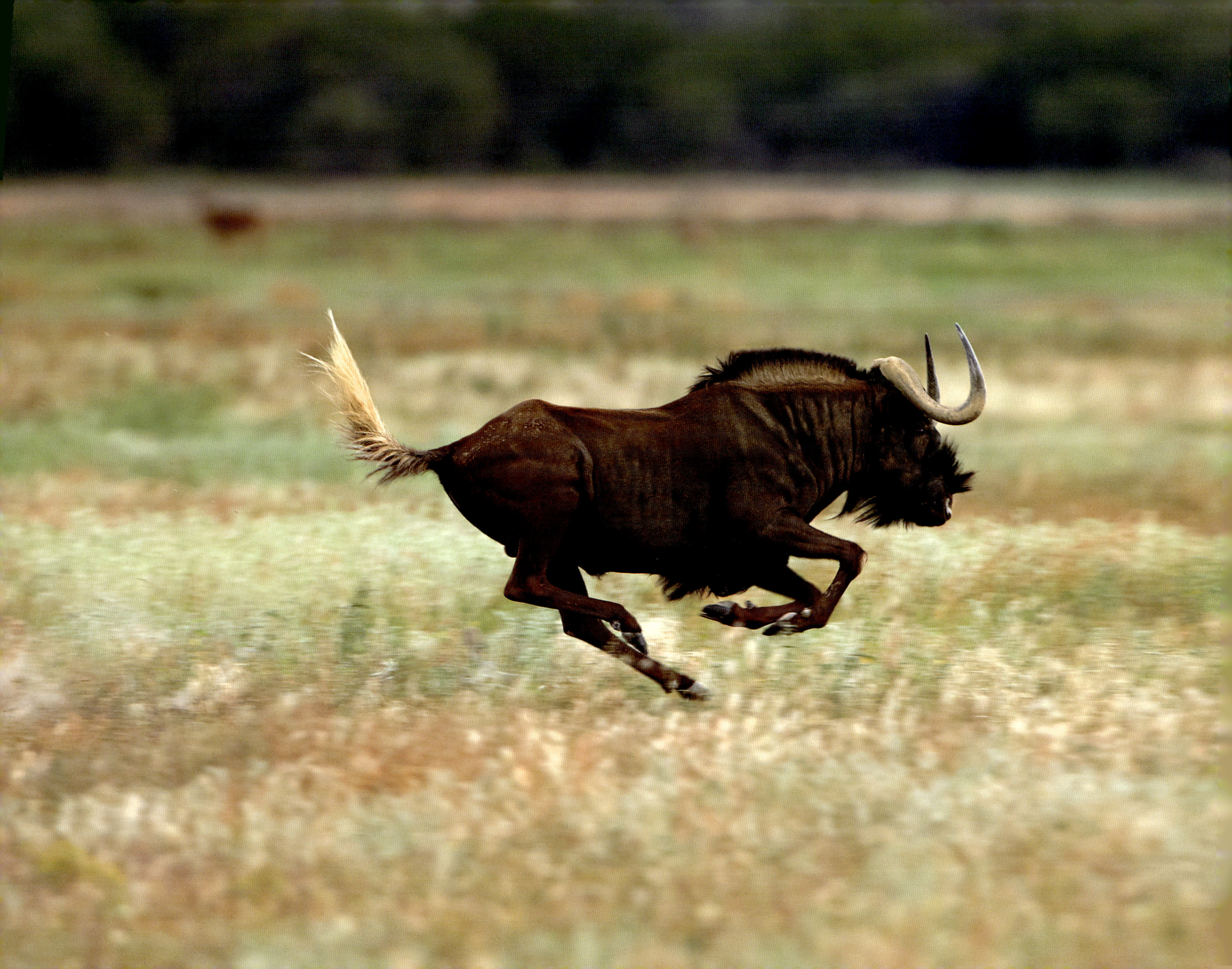

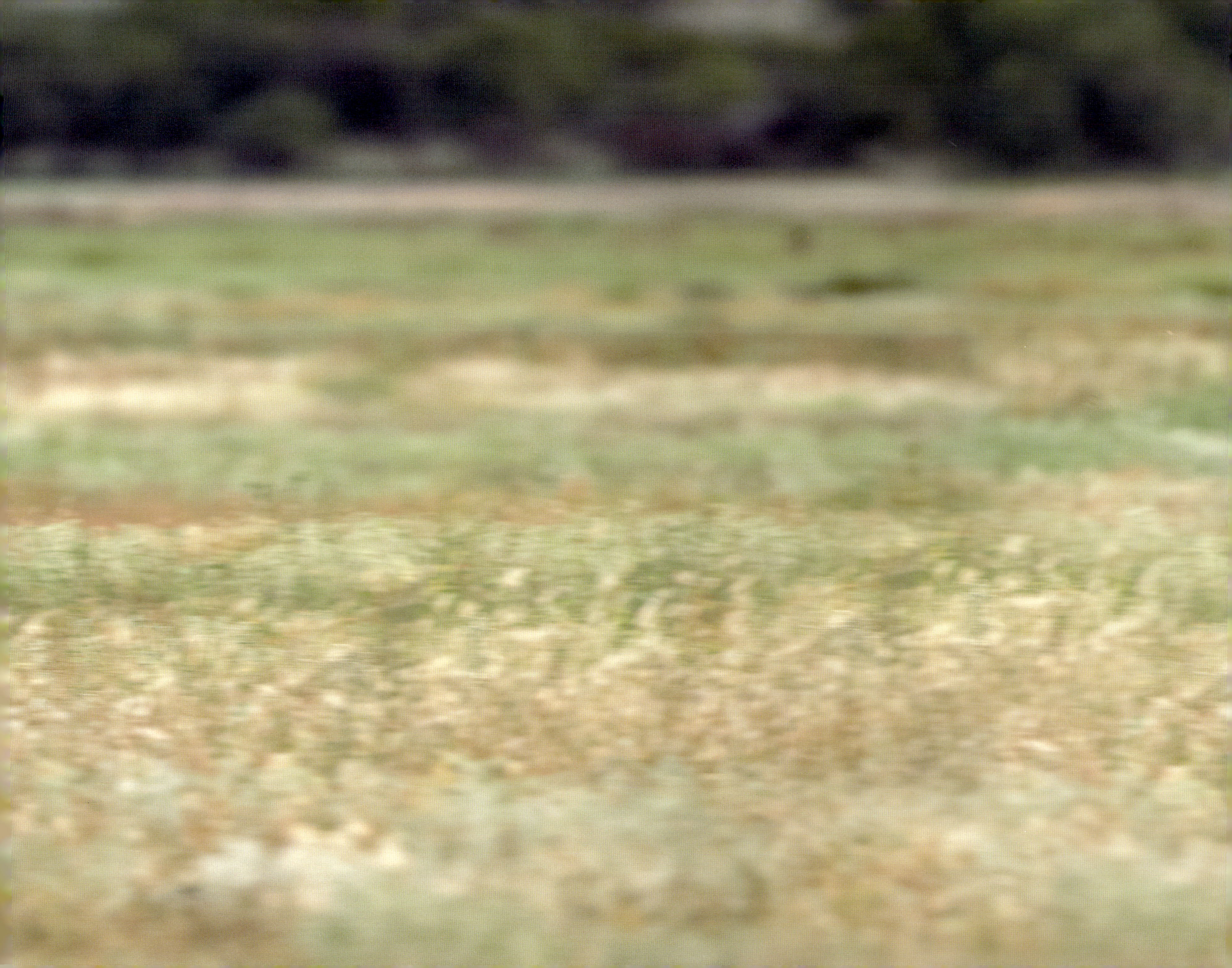

More about the Grassland Biome

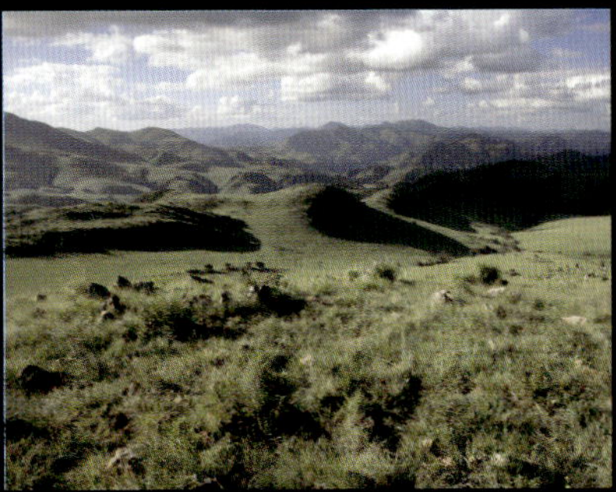

Ukhahlamba-Drakensberg

Although most of southern Africa's Grassland Biome suffers heavily from over-utilisation and destruction in favour of agriculture, mining and extensive development of urban areas, there are still pristine wilderness places where grasslands dominate. Few people are aware that only recently a number of important 'hot spots' in southern Africa's grasslands have been identified and acknowledged by the World's Centre of Plant Diversity as areas of global botanical importance.

One of those 'hot spots' recently declared as a World Heritage Site is the uKhahlamba-Drakensberg Park that straddles the South Africa/Lesotho border. The largest section of this park is still pristine and can only be reached on foot. The slopes are mainly covered by temperate grassland, in which sweet redgrass, *Themeda triandra*, is the dominant grass species. The floral splendour and profusion at high altitudes in early summer, when the high mountain slopes have changed into multicoloured floral fields, is unexpected and breathtakingly beautiful. Here annual rainfall is as high as 2 000mm.

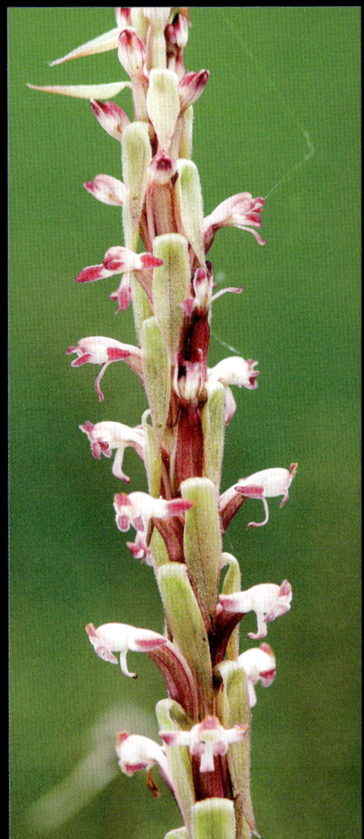

Grass orchids

Many of the other plants found in grasslands have underground corms, bulbs or other woody rootstocks. In this way, they store nutrients and protect their renewal buds from fire and freezing. This enables them to sprout rapidly in spring before the grass cover can replenish itself. Common families are the daisies, legumes, snapdragons, orchids, irises, crassulas and senecios. Everlastings of the *Helichrysum sp* are a special delight.

The well-known grassland bulbous lily, the agapanthus, has a name rich in legend – combining the Greek words 'agape' and 'anthos' and so giving us 'love flower'. Although this plant was traditionally used as a protective charm, it now has become a popular garden plant and is widely used as a good cut-flower. Its tuberous root stock protects it from drought and frost and it therefore survives the cold, frost and fire so typical of grasslands in high altitudes.

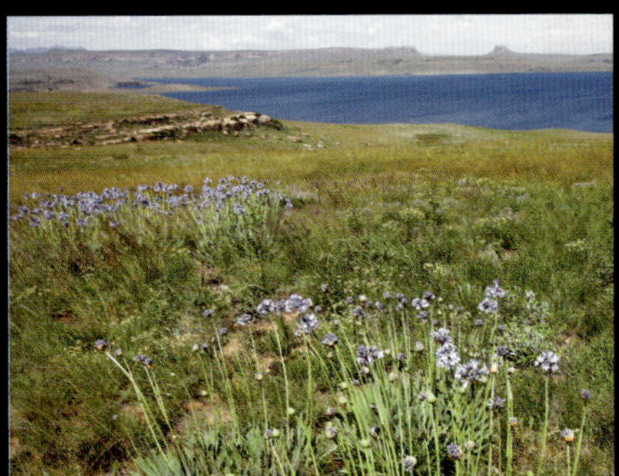

Agapanthus plants

Although grasslands are devoid of shrubs and trees, one of the coral tree species has made a special adaptation – it has gone underground. With its large subterranean stem, *Erythrina zeyheri*, colloquially known as the ploughbreaker, can withstand extreme cold, frost and fires. It produces its large red inflorescences in late spring and early summer.

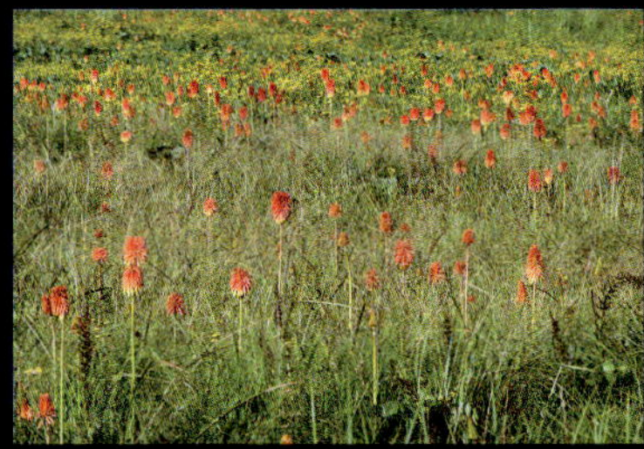

Red-hot poker or Kniphofia caulescence

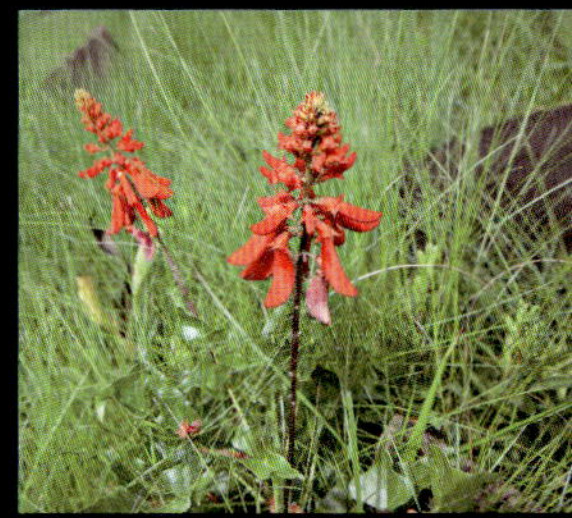

Erythrina zeyheri

In grassland marshes on damp high-altitude mountain slopes, the Lesotho red-hot poker, *Kniphofia caulescence*, grows in large colonies amongst rocks and grasses. Such a large stand can be seen close to the winding road up the Sani Pass near Himeville in KwaZulu-Natal from January to March.

Where there are plants, there are associated animals of one kind or another. Many of these mammals, reptiles, frogs and insects are endemic to southern Africa.

Like most other true grassland birds, the Orange-throated Longclaw is well camouflaged from above or behind with its dull brownish colouration. Yet, when it turns to face you, the bright orange chin, throat and neck, emphasised by a black necklace, surprises and delights. It occurs in short dense and moist grassland and is yet another endemic.

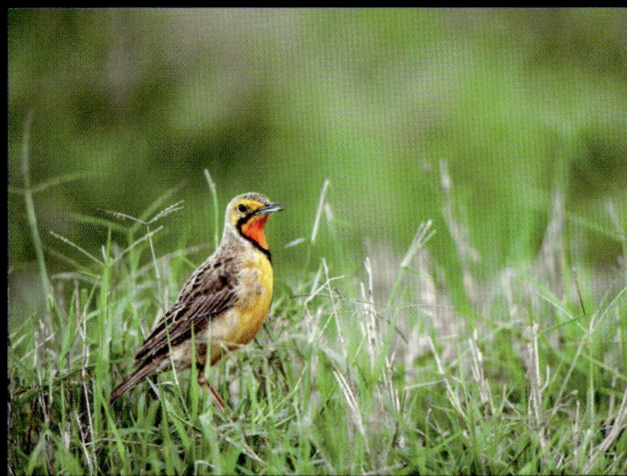

Orange-throated Longclaw

The graceful Blue Crane is South Africa's national bird and endemic to southern Africa. Once numerous, these birds are now classed as 'vulnerable' and are no longer the well-known icons of the grassy plains they used to be. This is probably due to poisoning, loss and degradation of their habitat, destruction of eggs and chicks during harvesting, and predation by dogs. They feed on vegetation in grasslands and also in agricultural fields, and are therefore blamed for damaging crops. However, they do also feed on insects and crop pests such as locusts, boll worm and others.

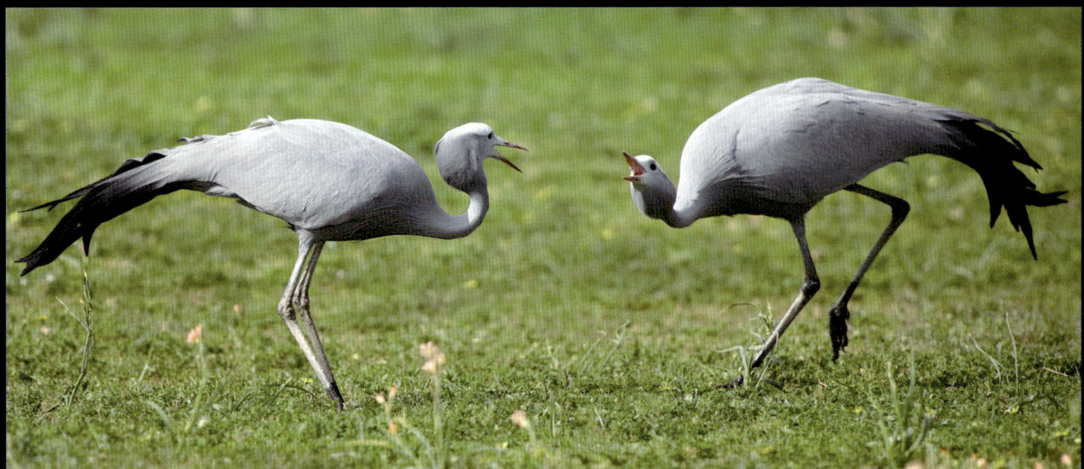

Blue Crane

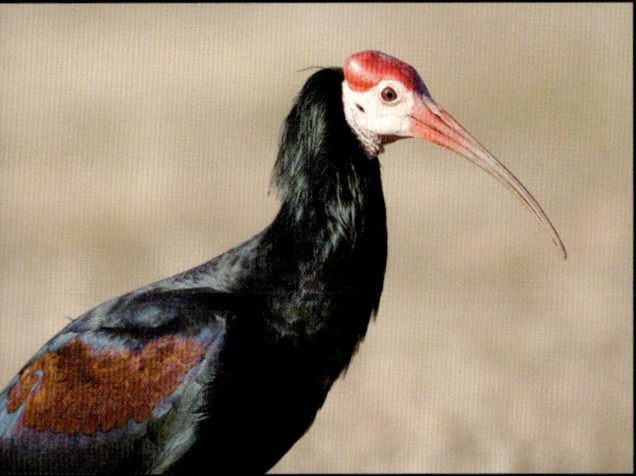

Bald Ibis

Generally uncommon and endemic to the grasslands of the Great Escarpment, the Bald Ibis is gregarious and roosts in groups with other ibises and herons. It occurs at high altitudes where rainfall is high and favours sour grasslands where the grasses are dense and short. They are often seen in recently burnt, ploughed, mown or heavily grazed fields and cultivated lands.

The Rufous-naped Lark is probably the most common of all the lark species, since it can survive in a wide range of habitats and is encountered in both grassland and savannah. It usually perches fairly close to the ground, raising the feathers on its crown while uttering its characteristic 'hope-you-see-meee' call. There is not much variation in the sound, which is repeated for long periods at intervals of

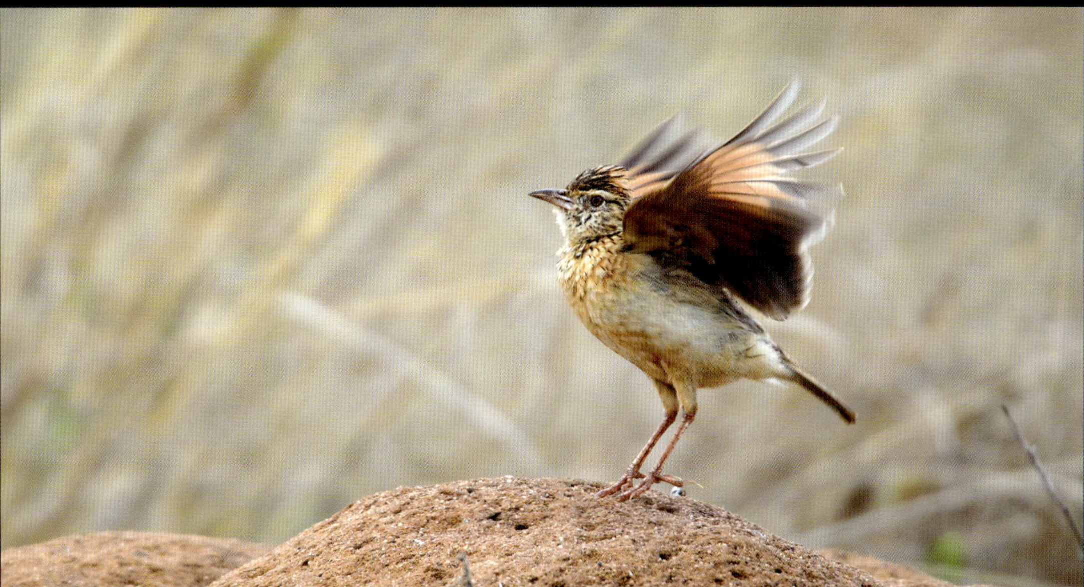

Rufous-naped Lark

a few seconds and is surprisingly loud.

Ostriches are most abundant on short-grass plains and in arid savannah. This is an example of a species that has a wide range of tolerance for environmental factors and occurs in several biomes. It is amazing that a bird so well adapted for extremely arid, hot conditions, can just as

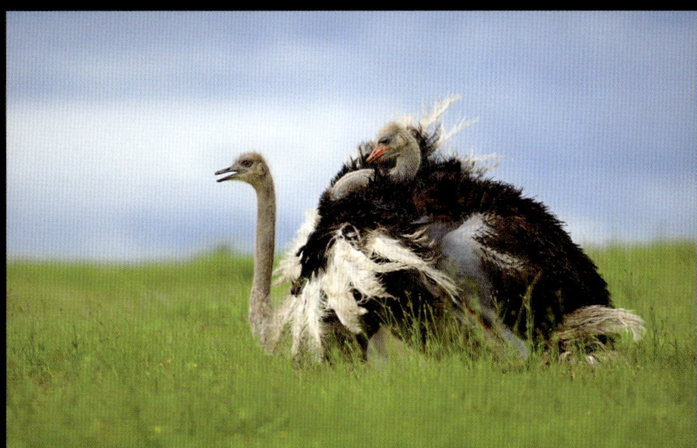

Mating Ostriches

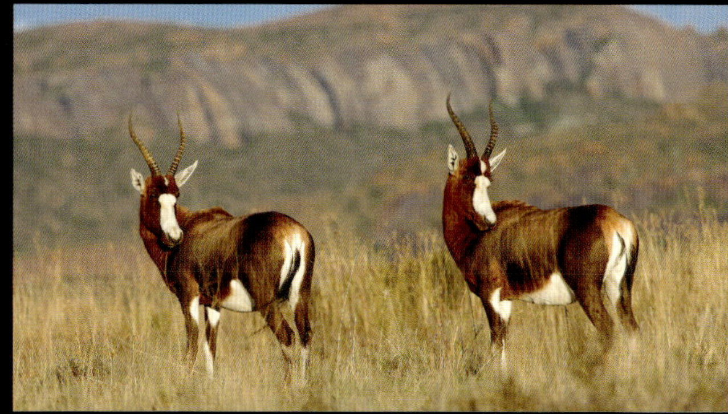

Blesbok

easily adapt to less arid ones.

Amongst the antelope, the blesbok is one of the key grassland species endemic to South Africa. The English name is derived from the Afrikaans word 'bles', which refers to the blaze on its forehead. Blesbok once occurred in their thousands, but excessive hunting in the 19th century for their meat and hides reduced their numbers to a mere 2 000. Fortunately their numbers recovered remarkably after numerous introductions and reintroductions from the remaining stock. They are now numerous and out of danger but are mainly restricted to fenced farms and reserves. As they feed purely on grass, they have an extensive rumen and intestinal capacity. They are partial to sprouting grasses that appear on burnt grassland after the first rain.

The dainty oribi has become one of the most threatened antelope in the grasslands. As a flagship grassland species, it is the first to disappear in badly-managed grassland and therefore serves as an indicator of declining quality. As a selective grazer, it requires a combination of high-quality short grass for food and long grass for resting and hiding its young. It prefers redgrass, which is an indicator species of climax grassland. Oribi are not dependent on water sources but need to feed on grasses with a high moisture content.

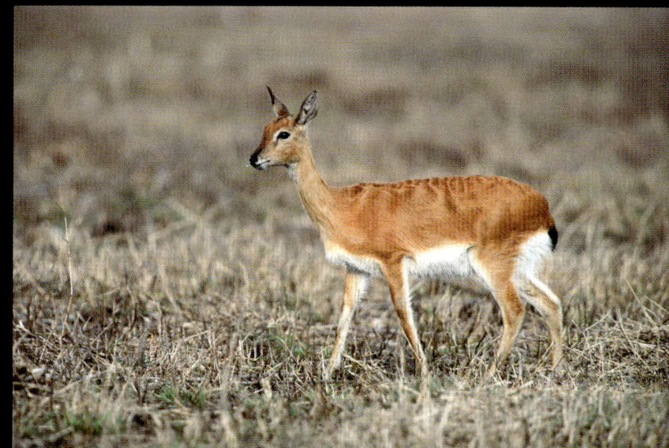

Oribi

The buffy-brown black wildebeest is another endemic of grasslands and is a species of the central plateau's open plains. They show a distinct preference for grazing where the grass is short. In a highly characteristic display, a territorial male will often gallop at high speed, lashing and swishing its tail and tossing its head, sometimes kicking out with its back legs.

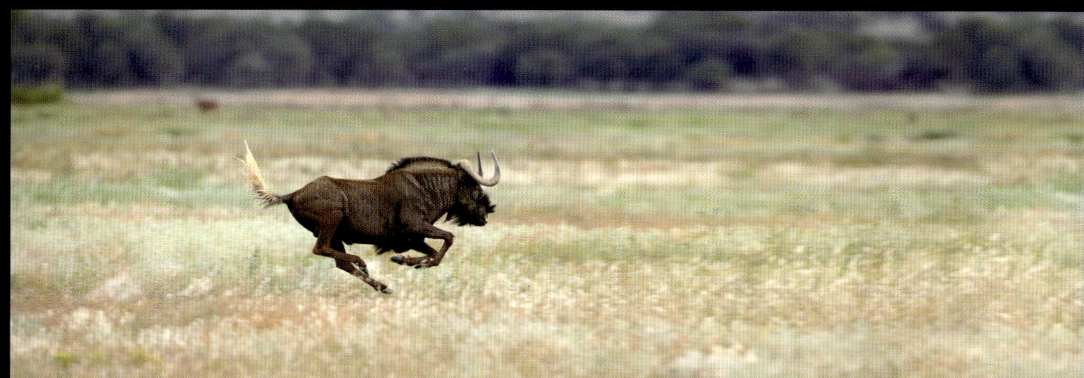

Black wildebeest

Savannah

The biome richest in wildlife

Bushveld is the colloquial term often used for savannah and is synonymous with African wildlife. Places termed savannah can be described as grassland with trees, where the trees are mostly deciduous and are sufficiently small or widely spaced so that the canopy does not close. The open canopy allows sufficient light to reach the ground to support a healthy grass and herb cover.

The largest part of southern Africa, about half the total land surface, is covered by savannah. Depending on rainfall, vegetation and other factors, more than 40 different kinds of savannah can be distinguished. Rainfall ranges from as little as 235mm in the western parts (Kalahari) to 1 000mm in parts of the eastern section.

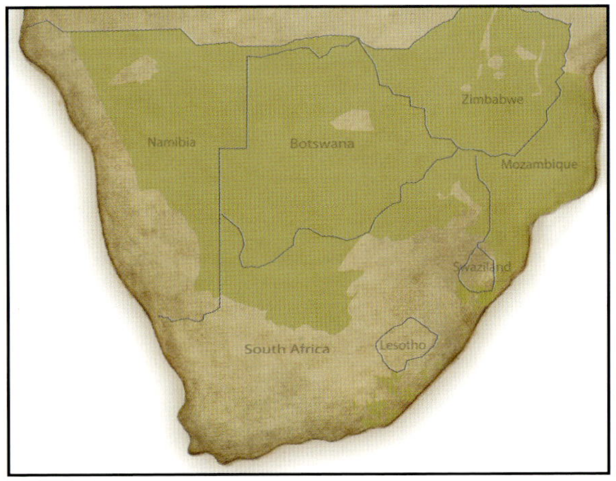

Savannah supports a large variety of game species including the so-called Big Five. The huge diversity of animals is associated with the rich plant diversity. Since large tracts of grassland are interspersed with true savannah, many typical grassland species may also occur in the savannah. Savannah also includes valley bushveld, the veld type where there is the greatest range of seasonal rainfall in southern Africa. As in grasslands, fire is a crucial factor in the ecology of all savannahs and is therefore a regular natural feature of this environment.

Since savannah occurs in areas where rainfall is confined to the summer, water availability is important. Savannahs are frequently in a transitional zone between forest, desert or grassland. Most game reserves in southern Africa are found in the savannah regions of the subcontinent and rank amongst the best destinations for wildlife viewing in Africa.

The savannah is home to over 200 larger mammals, more than 500 bird species and a wealth of reptiles, amphibians, fish and invertebrates.

Good places for wildlife viewing in bushveld areas

There is, indeed, a wealth of conservation areas within this region. Namibia has the world famous Etosha National Park. Moremi and Chobe National Parks in Botswana are within this biome and so is the Central Kalahari Game Reserve, Mashatu Game Reserve and several other smaller ones. Zimbabwe can boast the Victoria Falls and Zambezi, Hwange, Matobo, Kariba, Nyanga, Gonarezhou and Mana Pools national parks. Mozambique has several huge conservation areas, of which the Zinave, Banhine, Gorongosa and Limpopo national parks fall within the confines of southern Africa. The latter is part of the Greater Limpopo Transfrontier Park straddling three countries. South Africa also has many national and private game reserves and parks within the Savannah Biome. The most visited conservation region is the Kruger National Park and surrounding private reserves. Other popular destinations are the Kgalagadi Transfrontier Park (semi-arid savannah), Marakele, Mapungubwe and Pilanesberg national parks, and several parks in KwaZulu-Natal such as iSimangaliso and iMfolozi-Hluhluwe run by Ezemvelo KZN Wildlife. The Mlilwane, Hlane and a few smaller reserves are found in Swaziland.

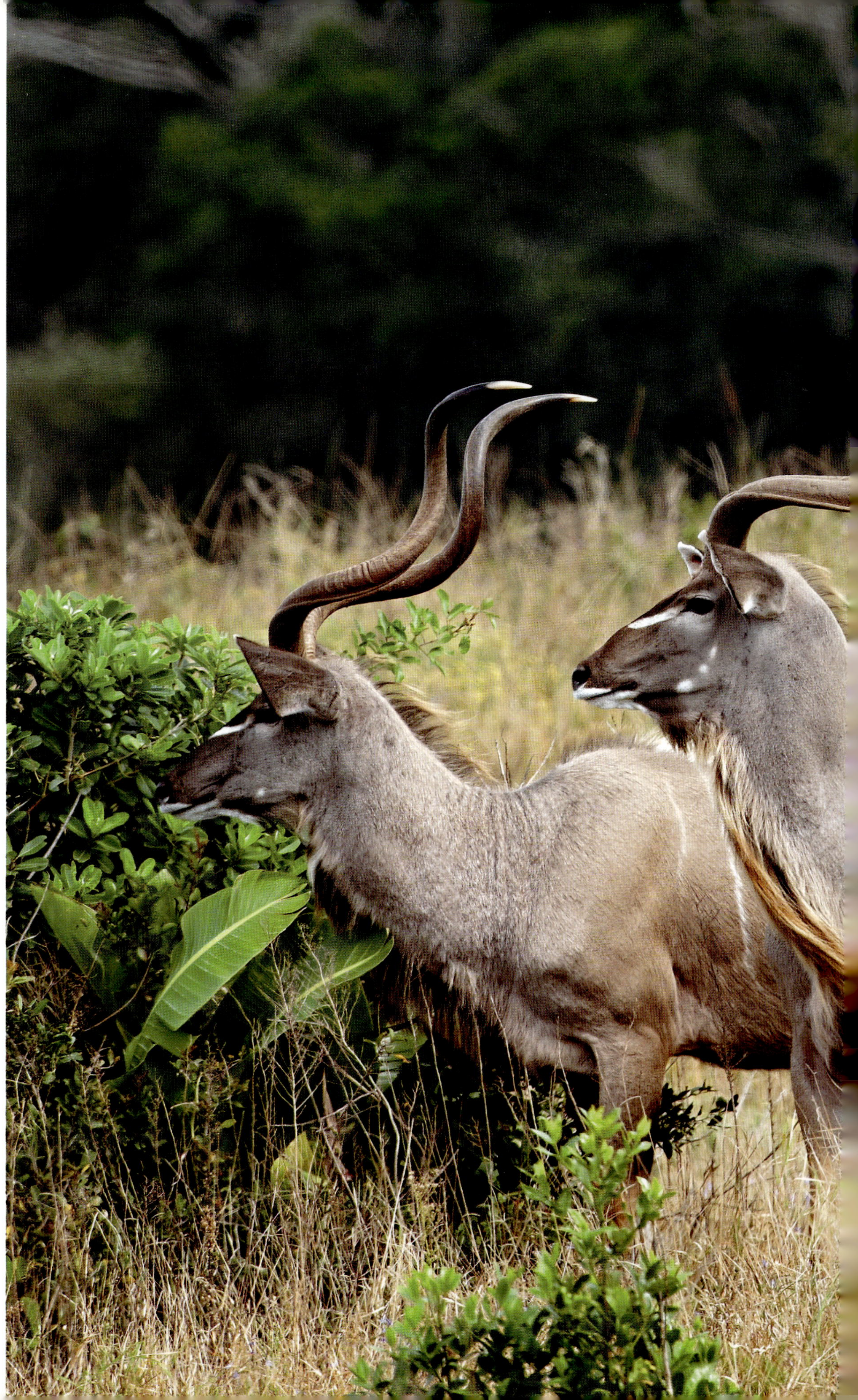

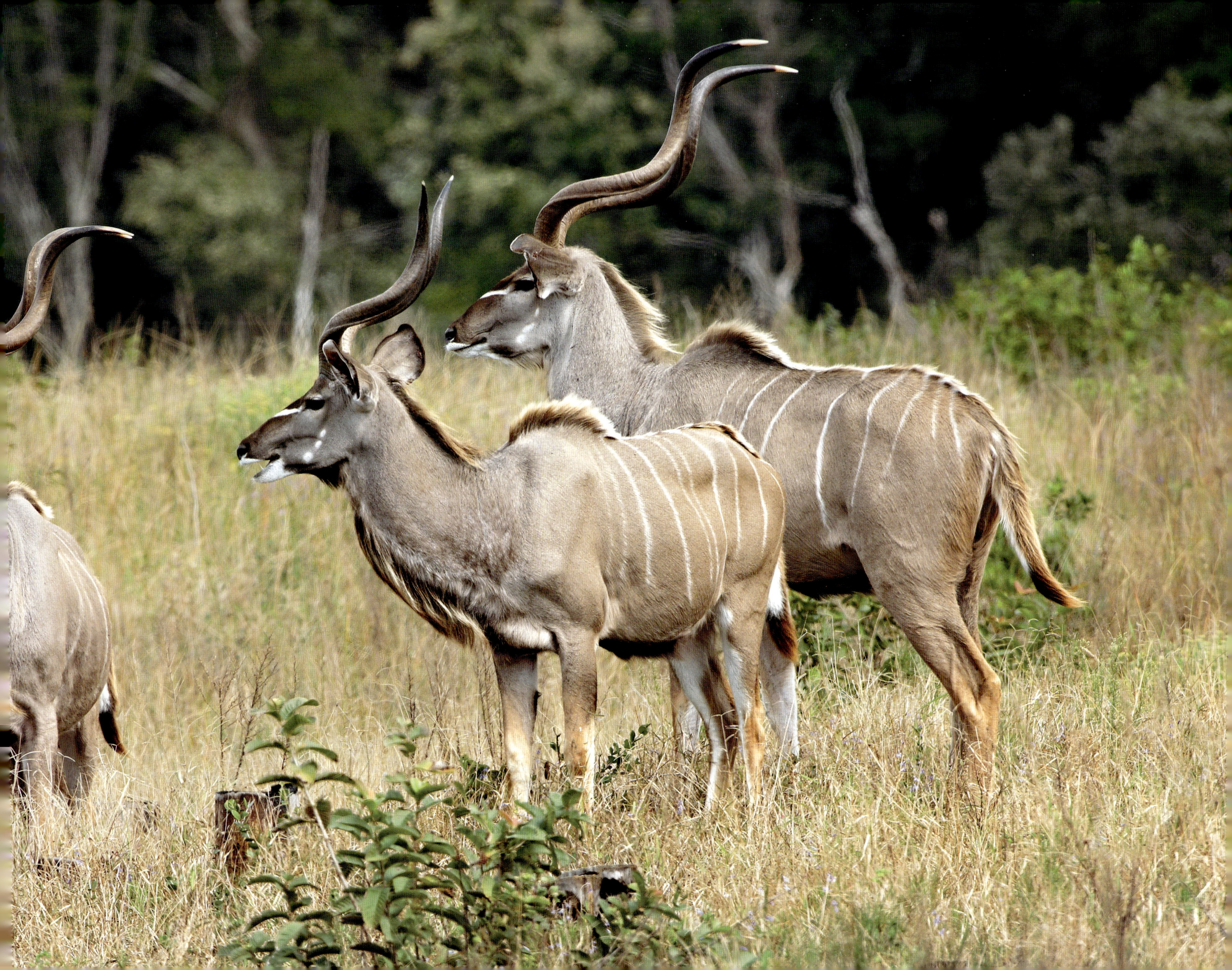

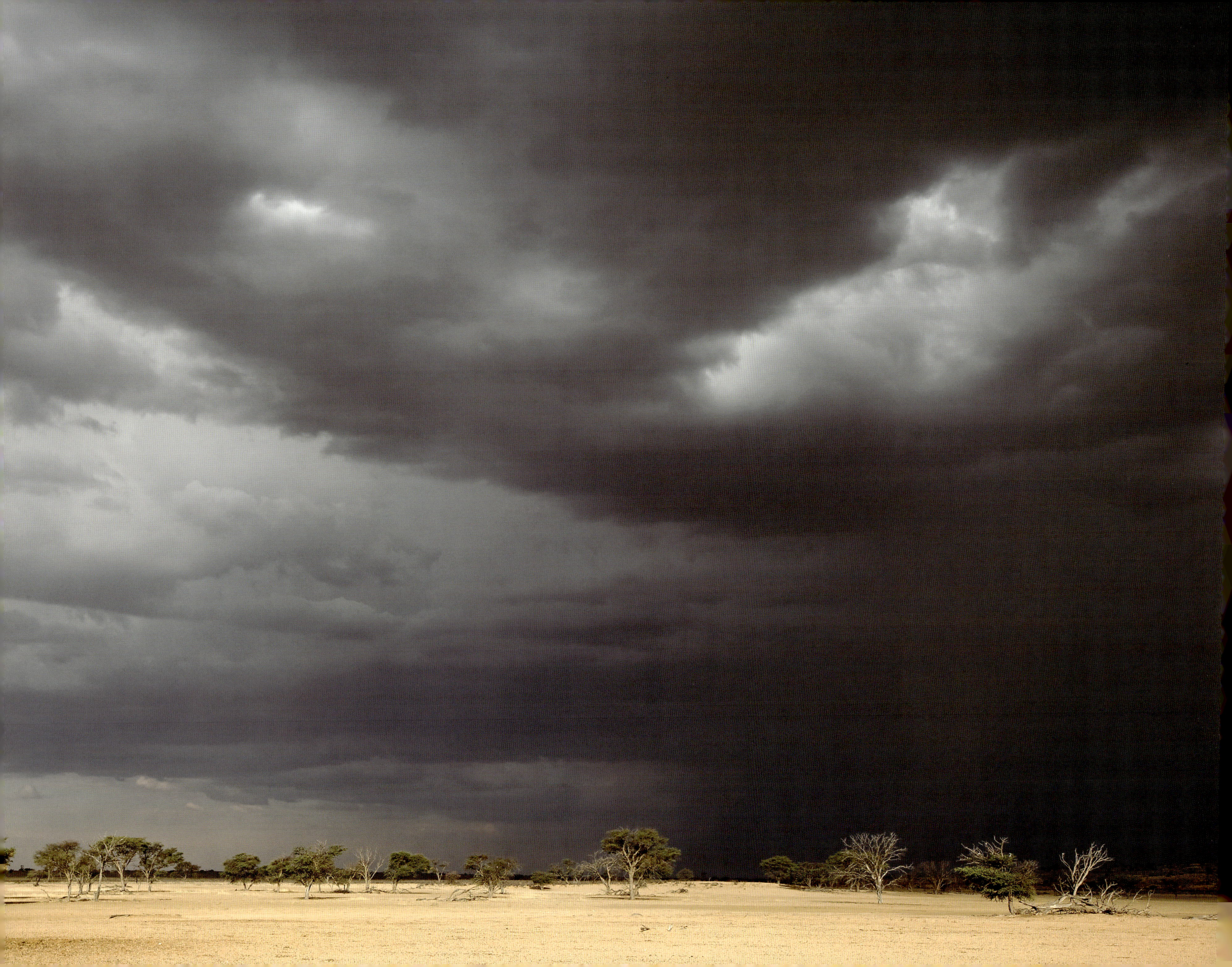

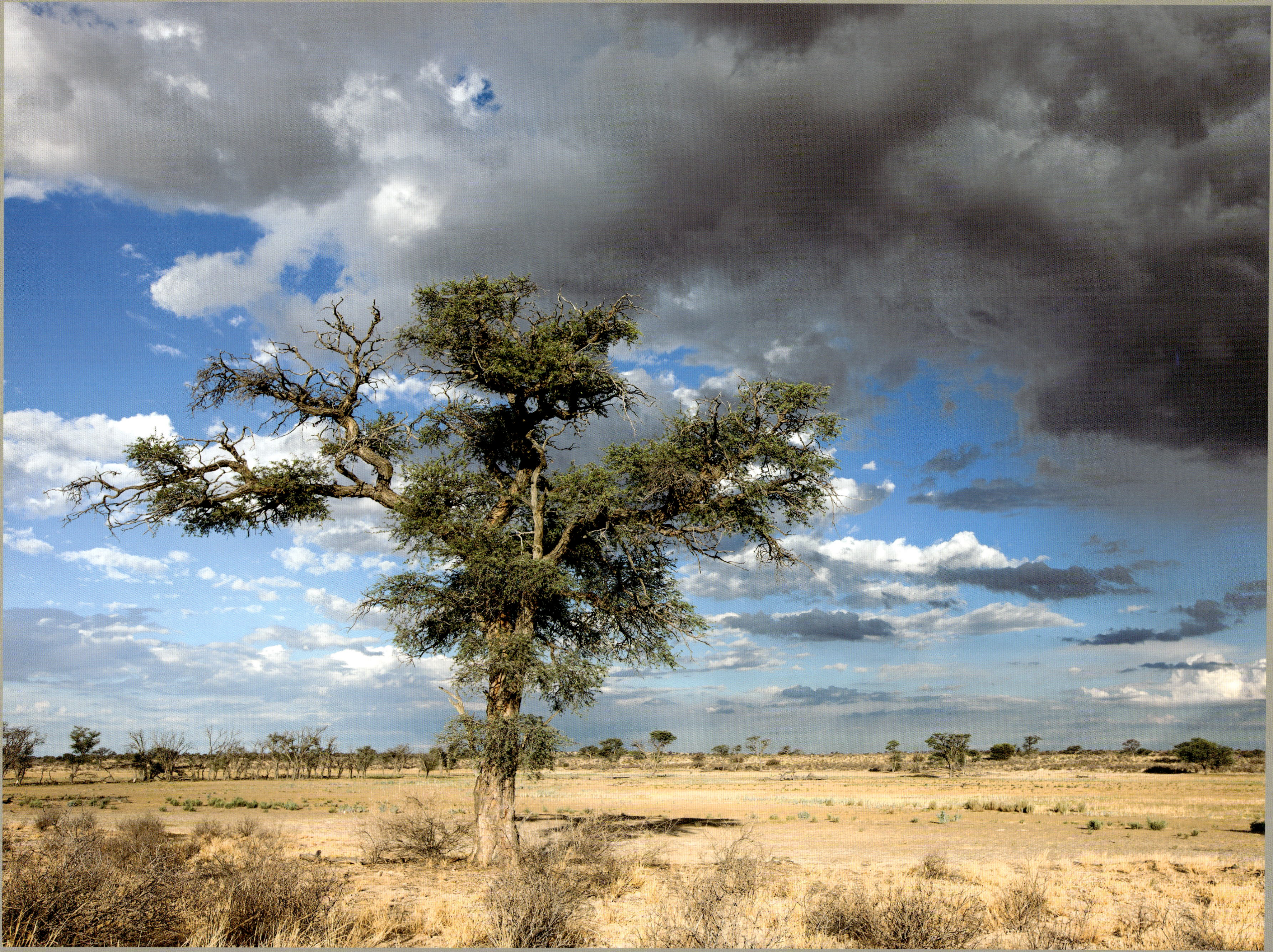

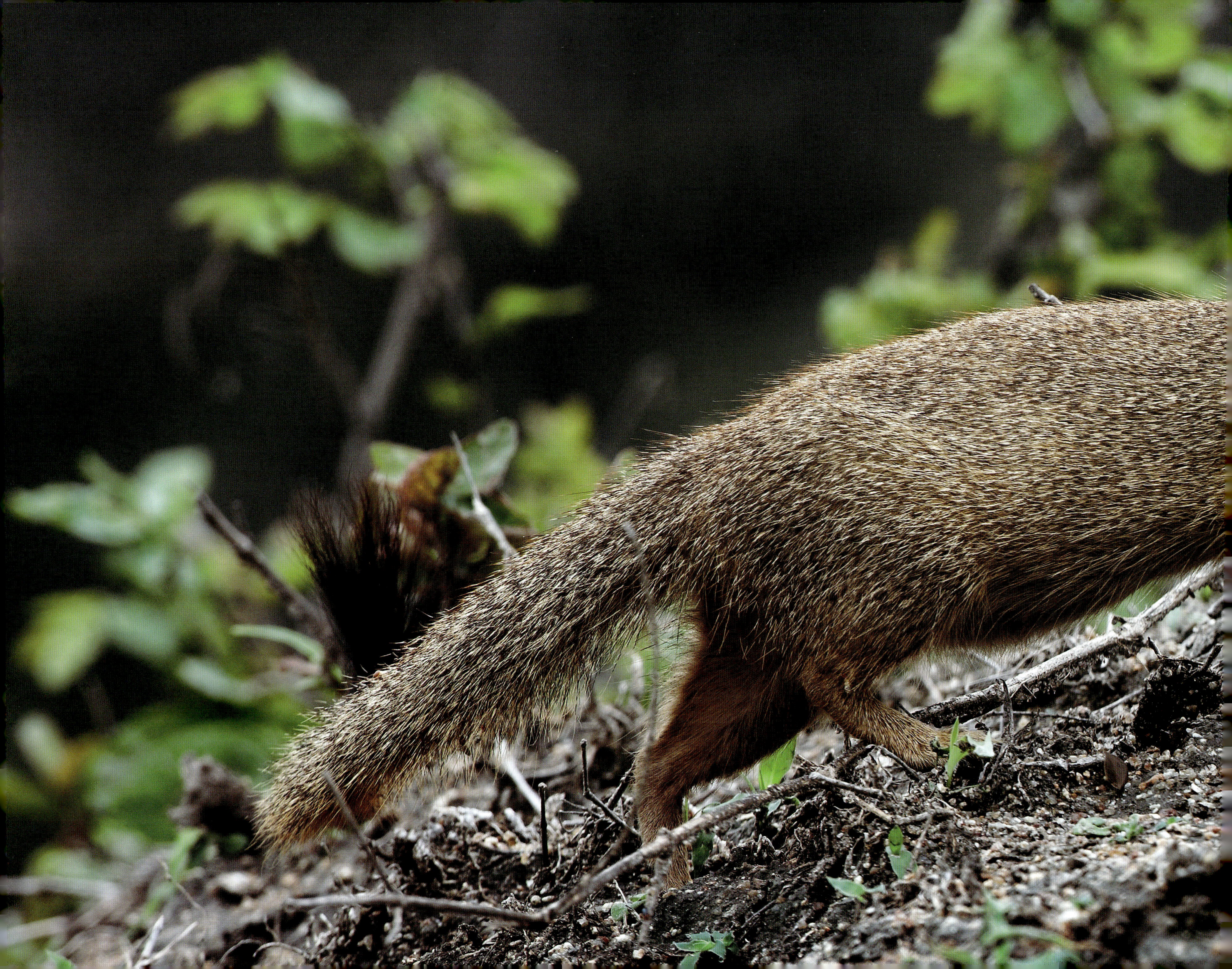

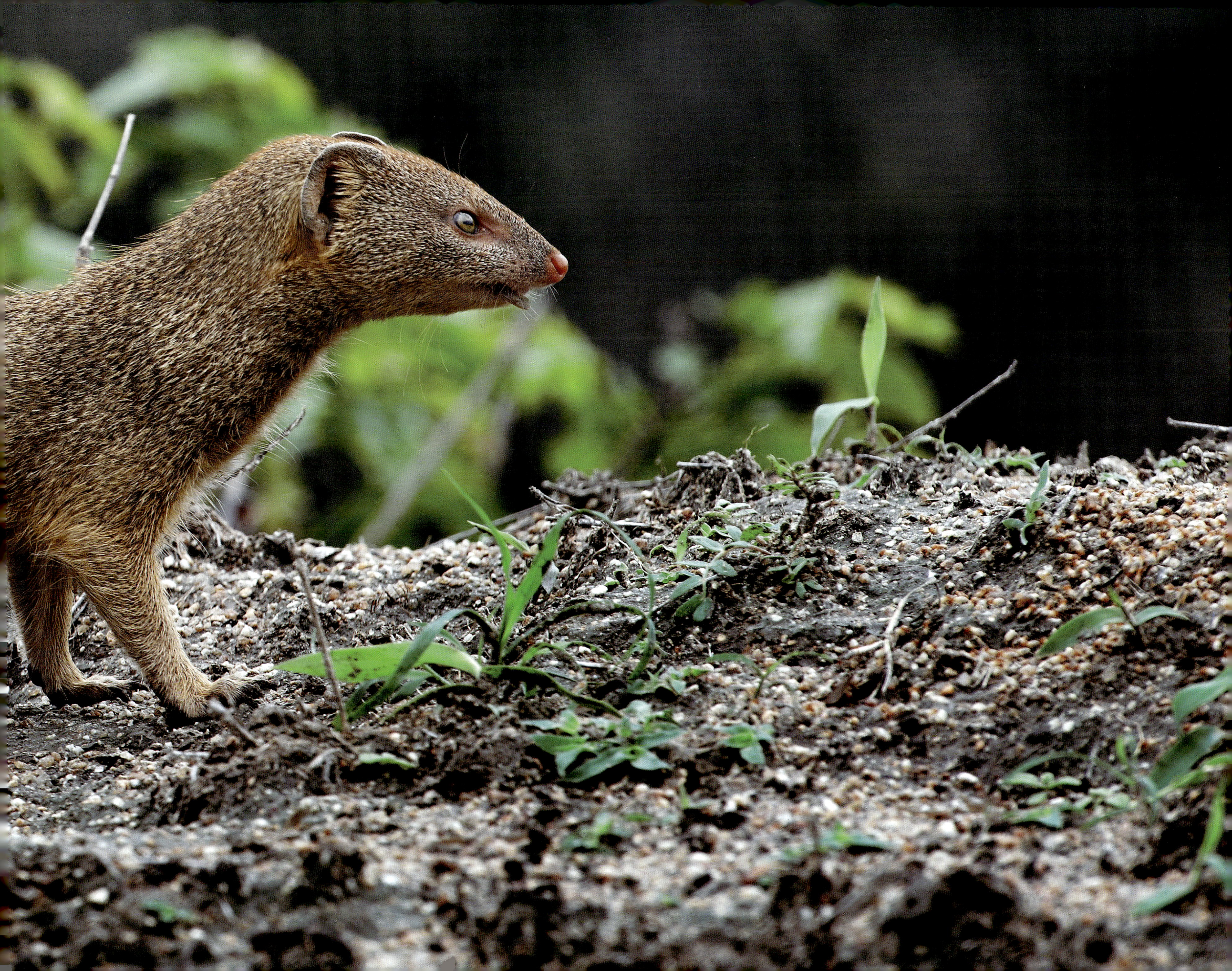

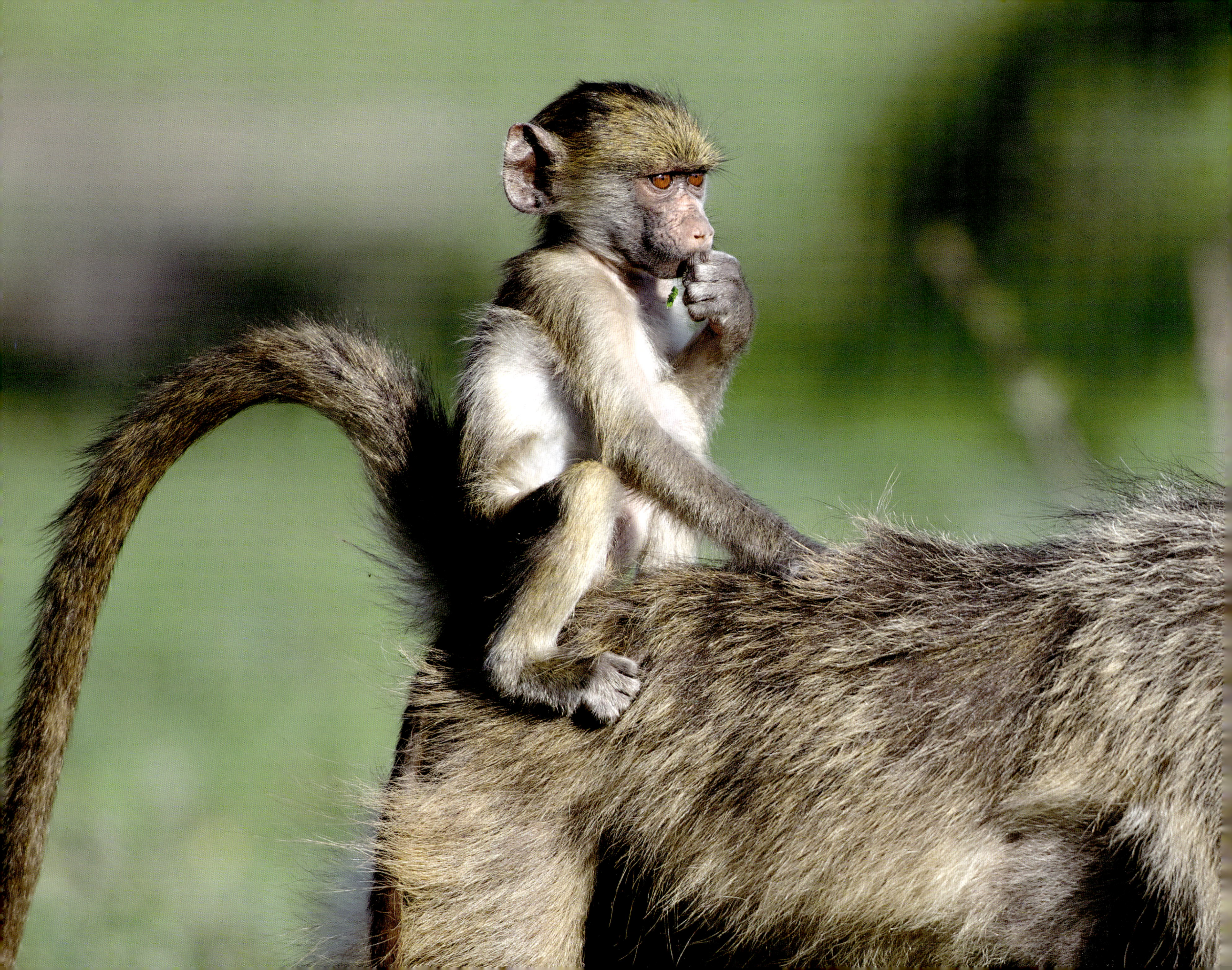

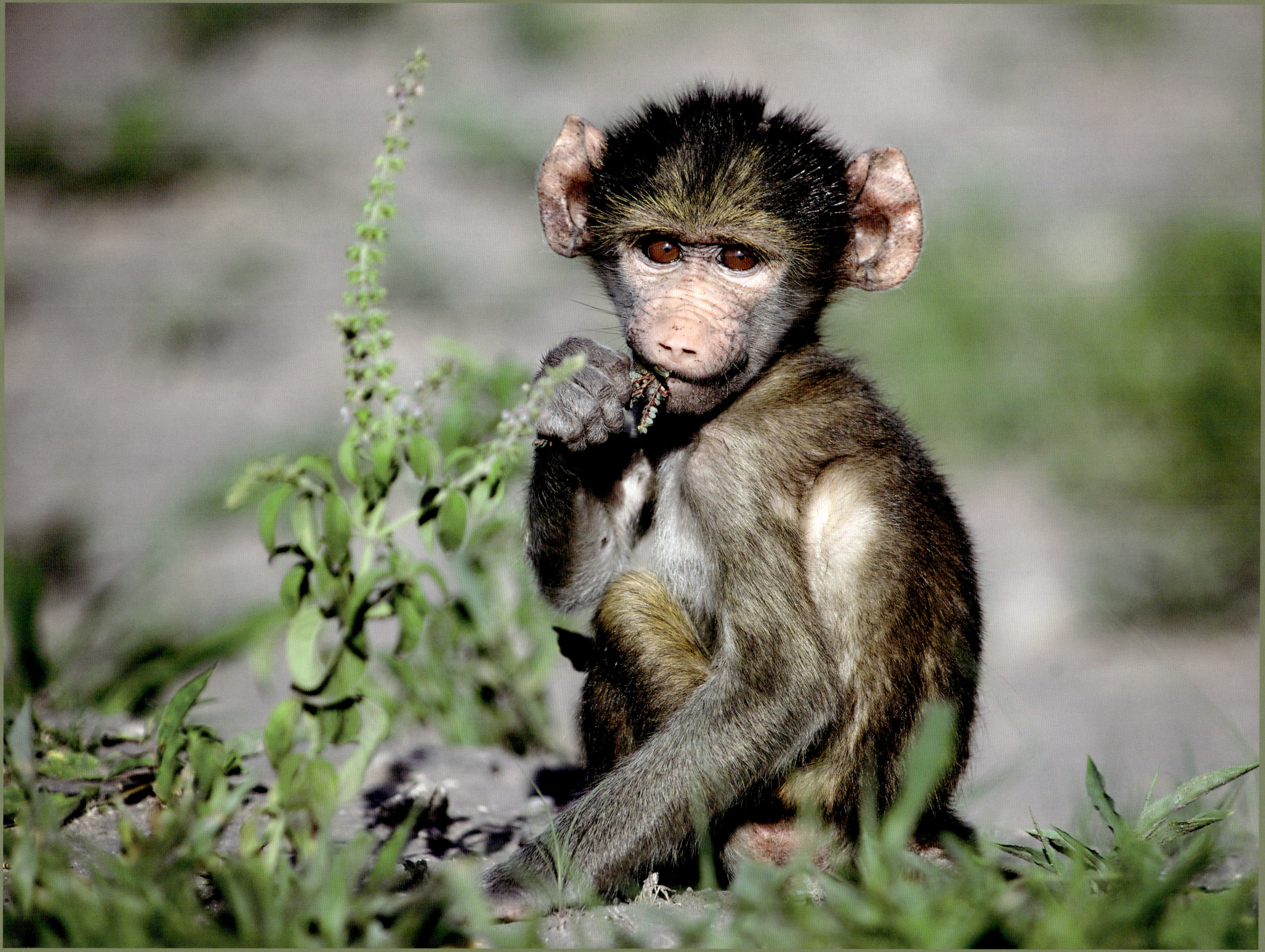

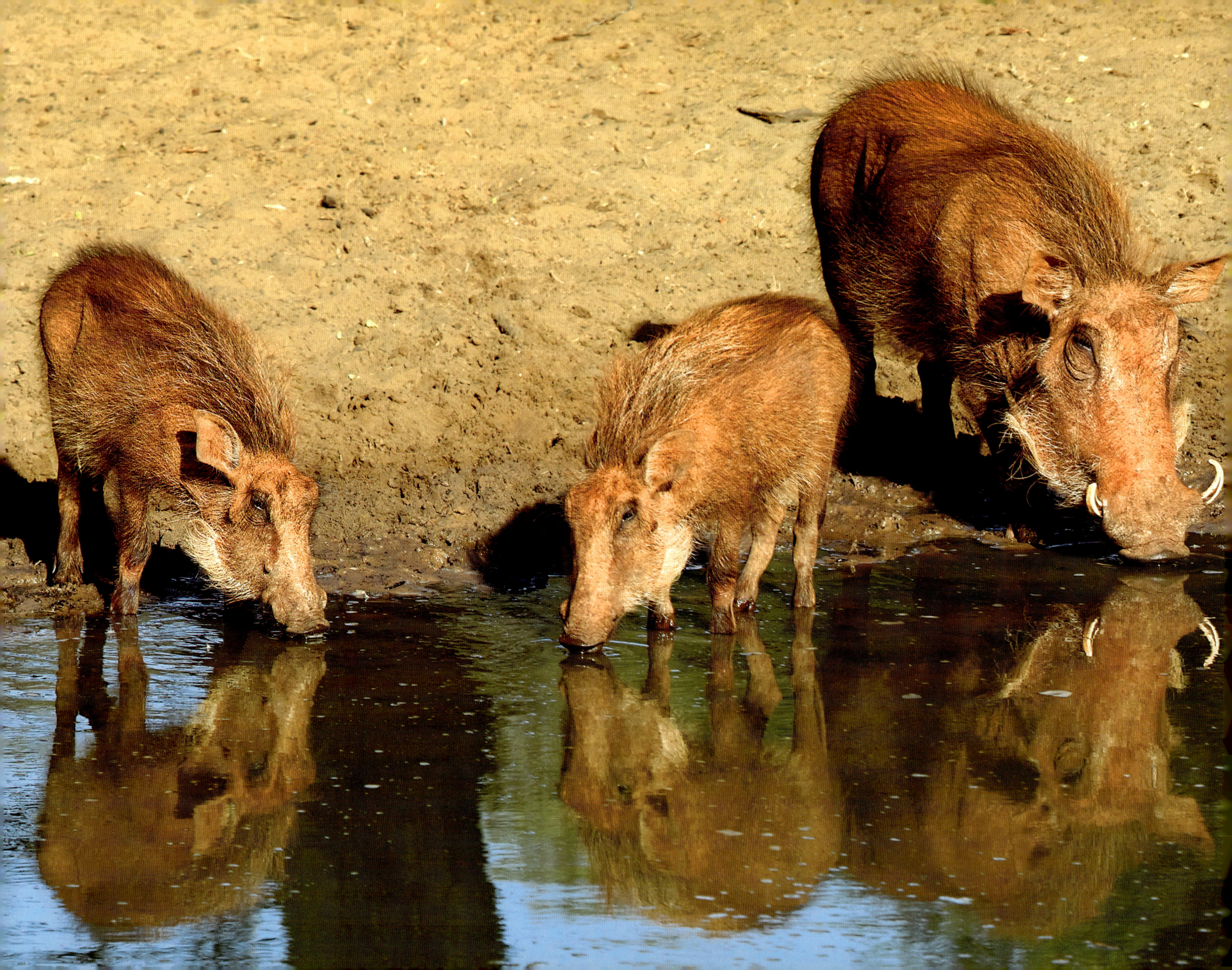

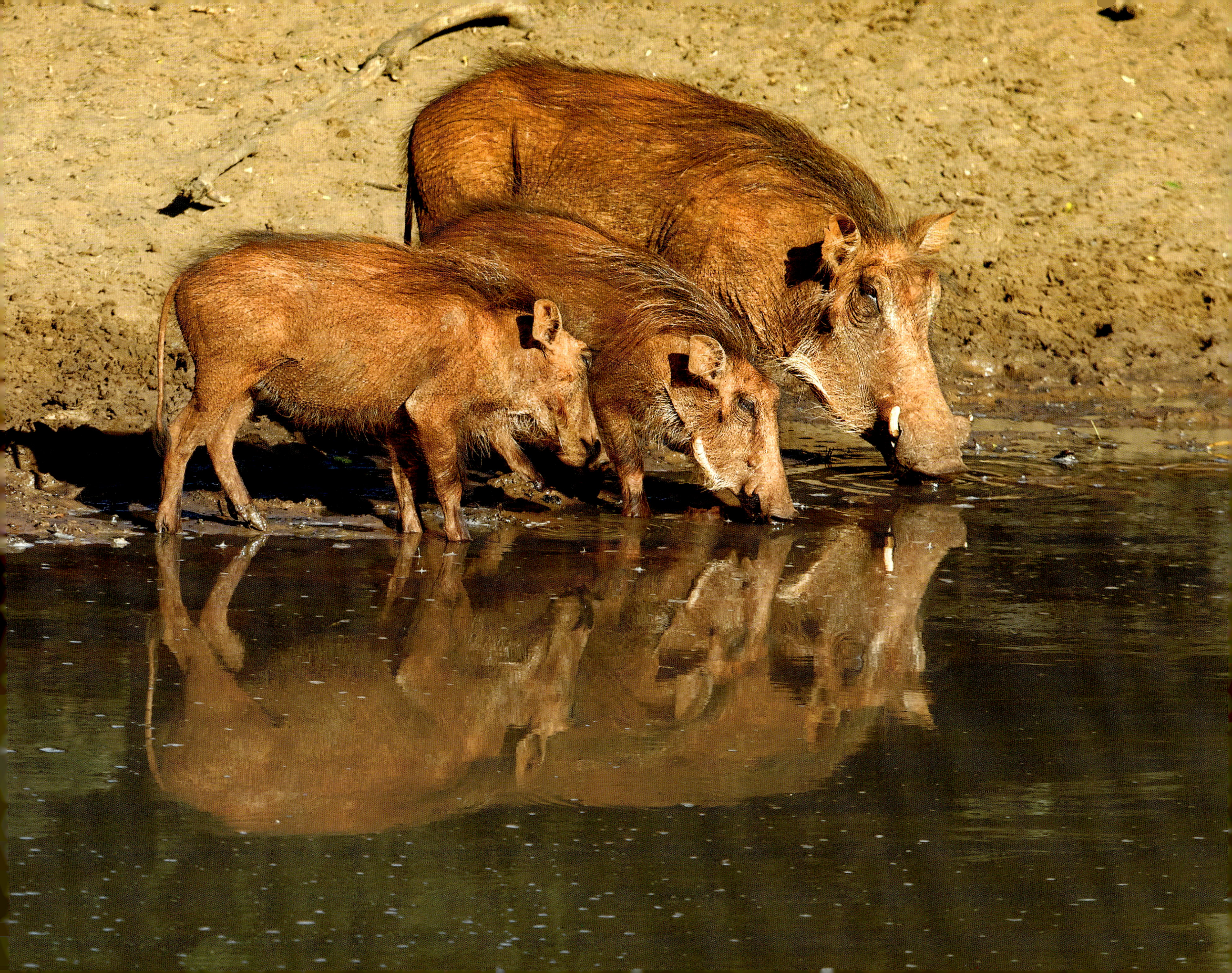

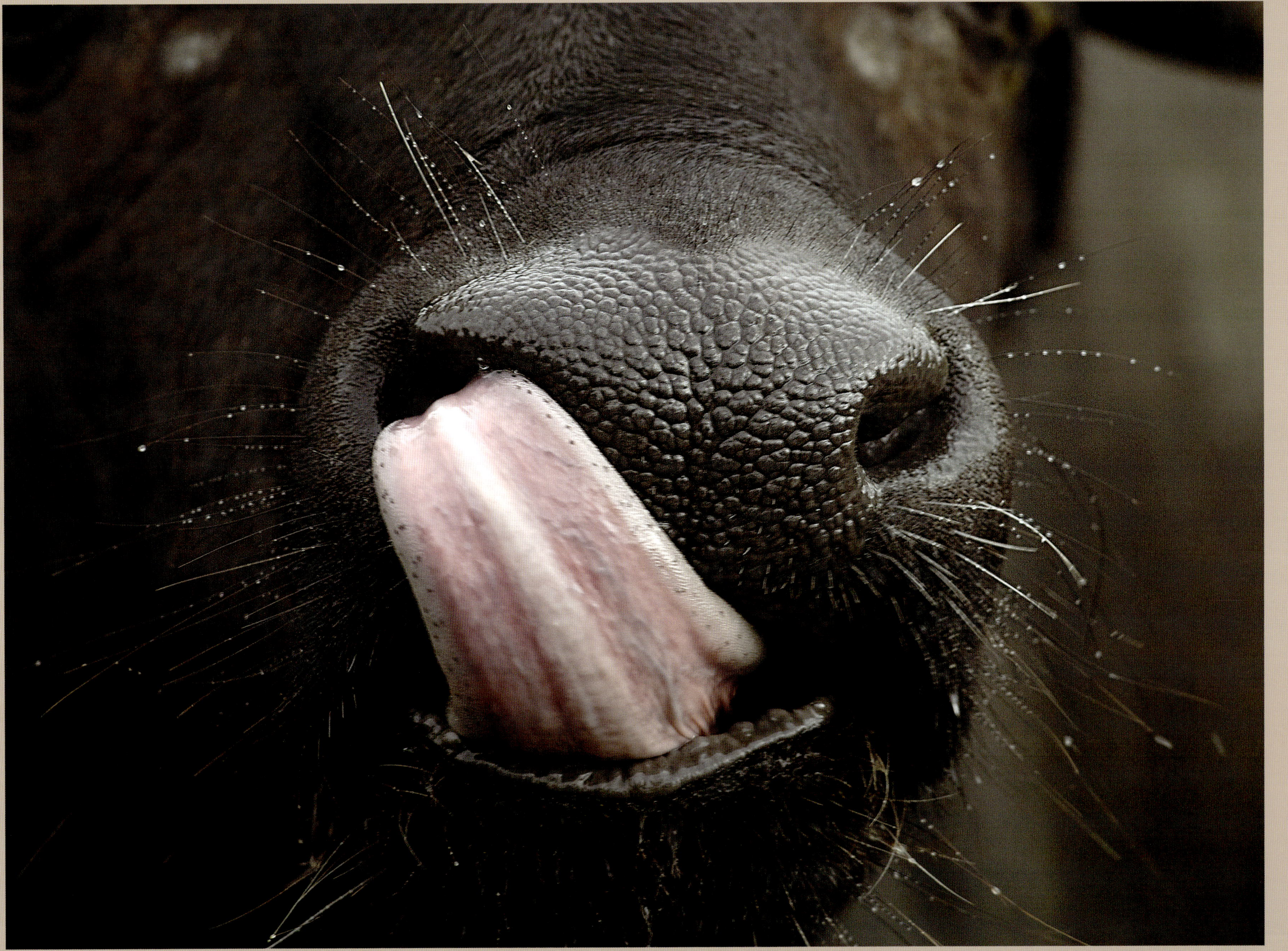

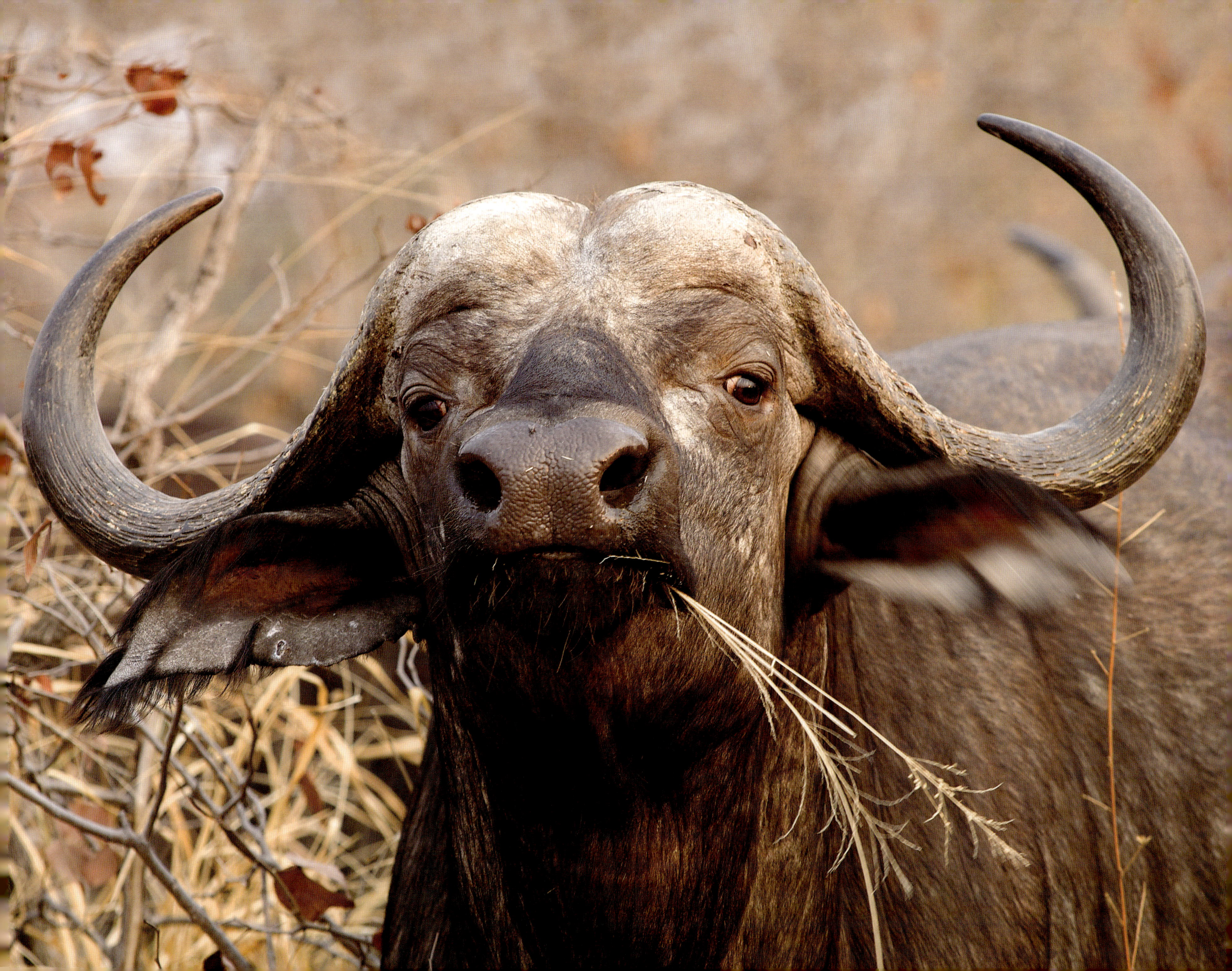

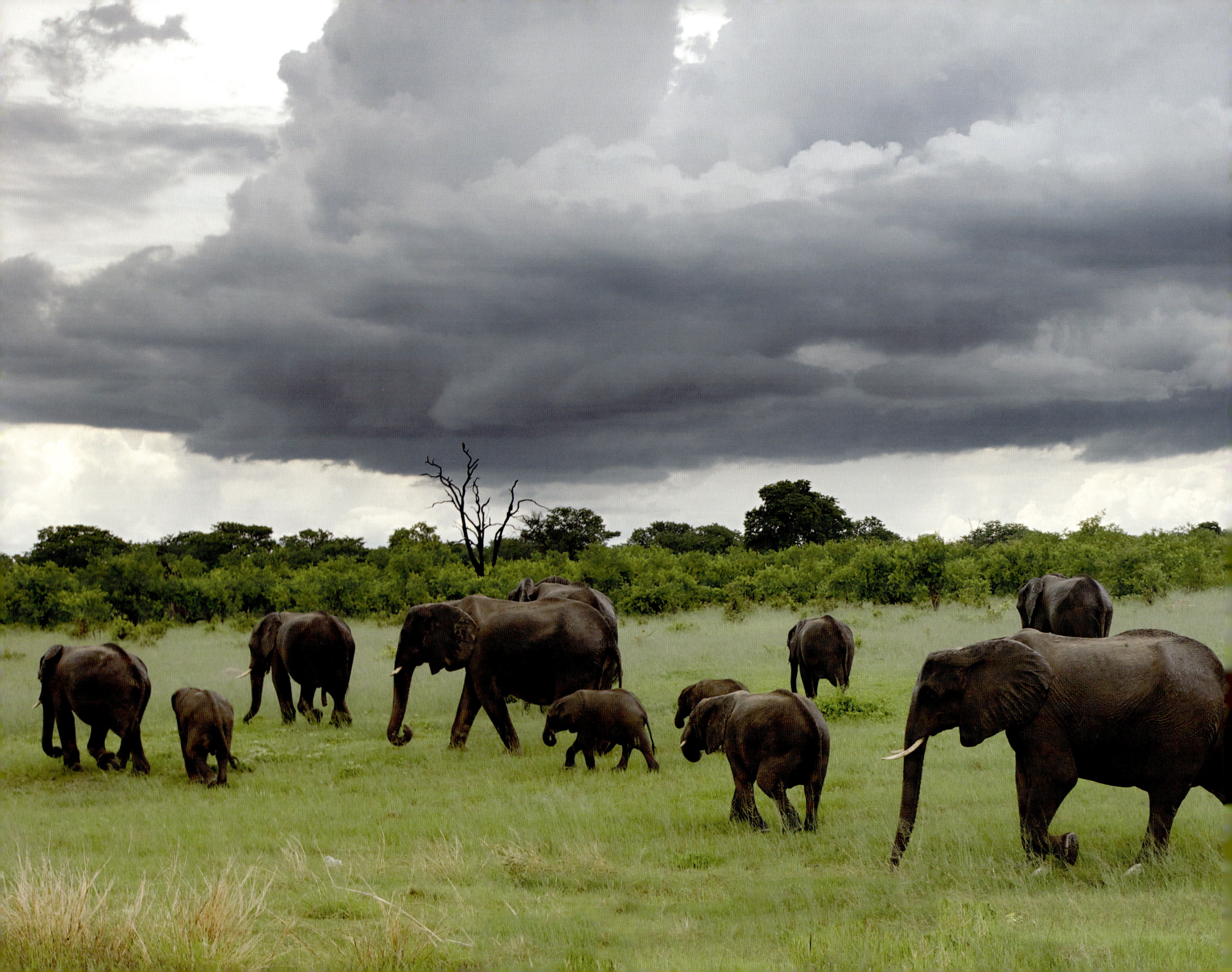

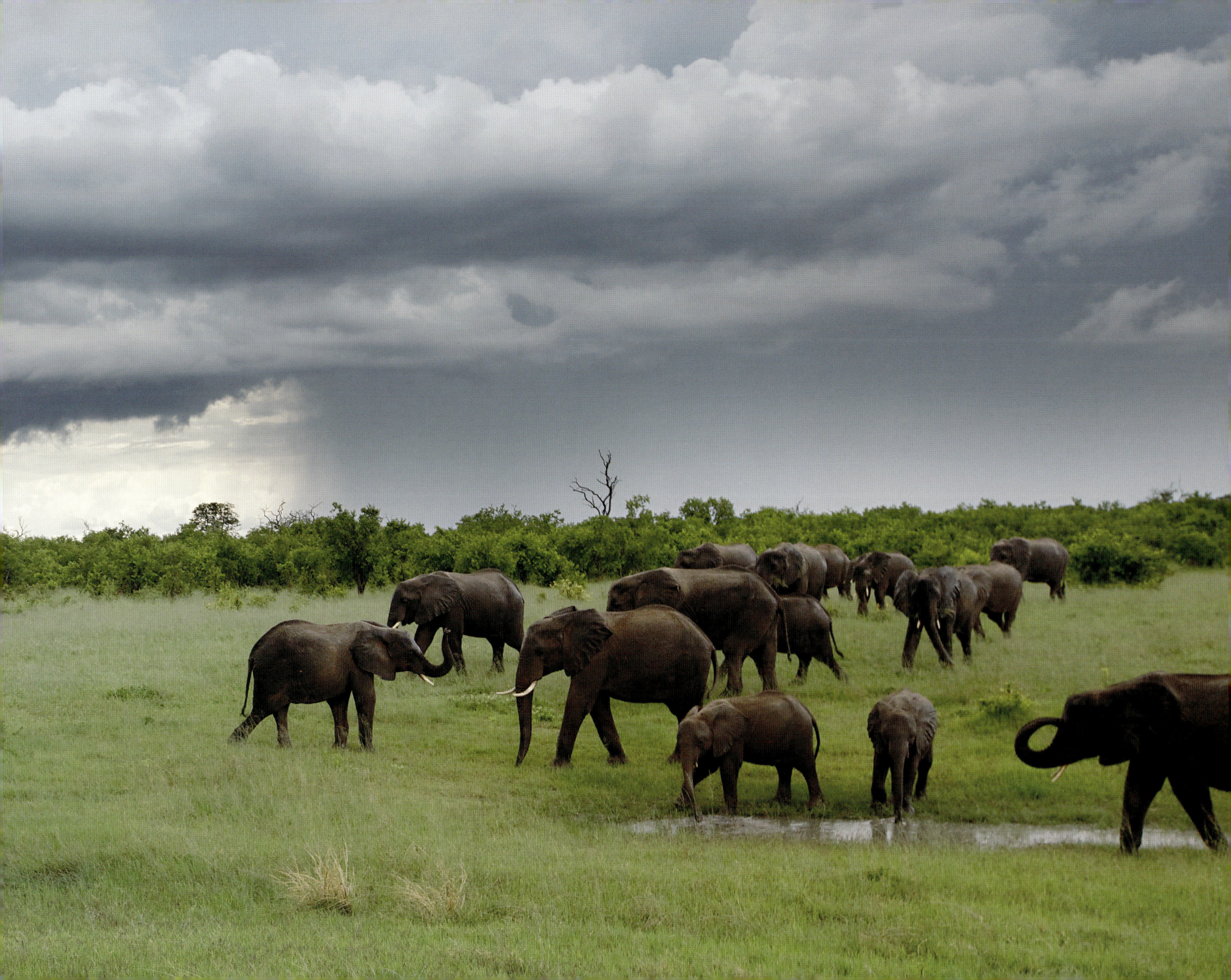

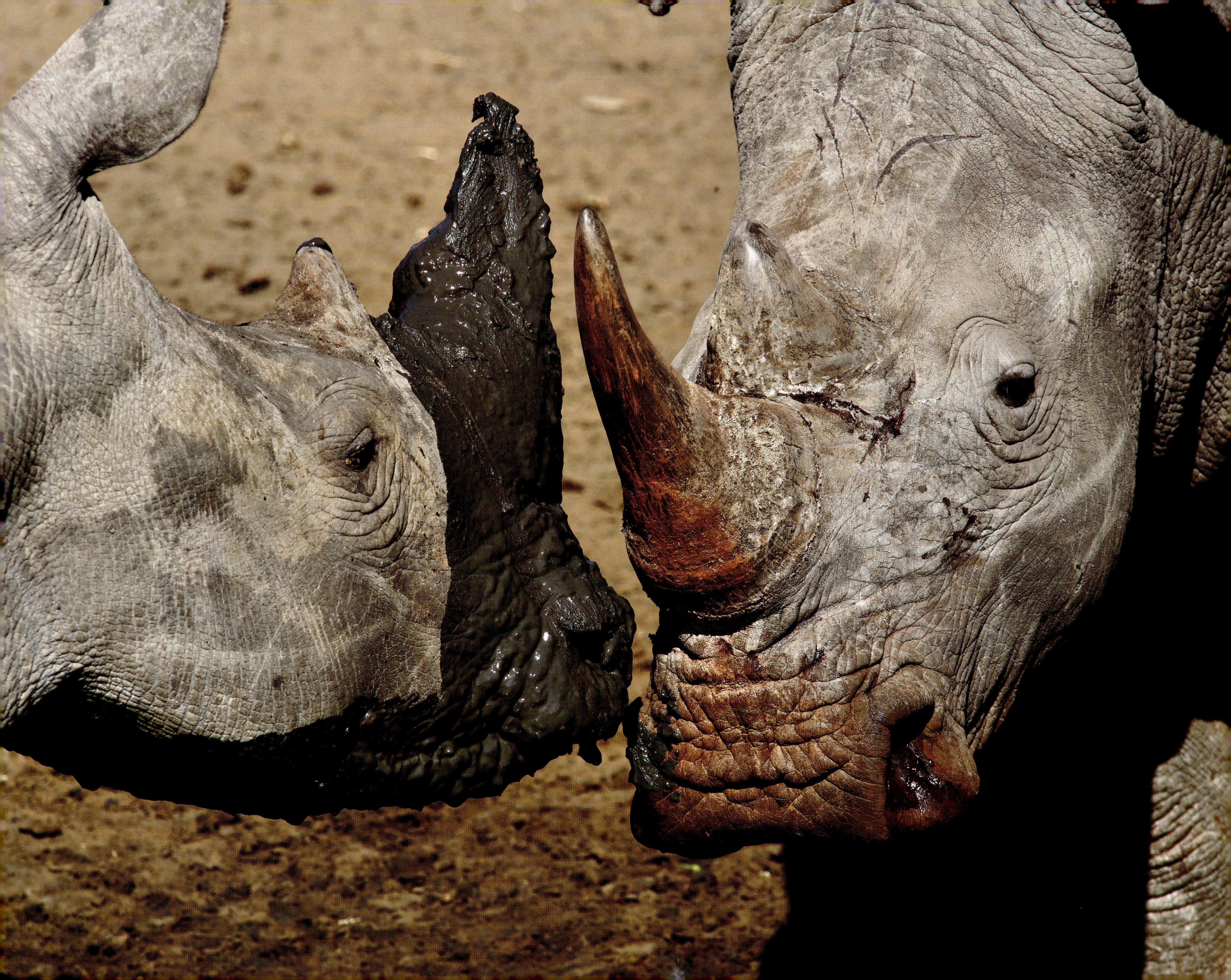

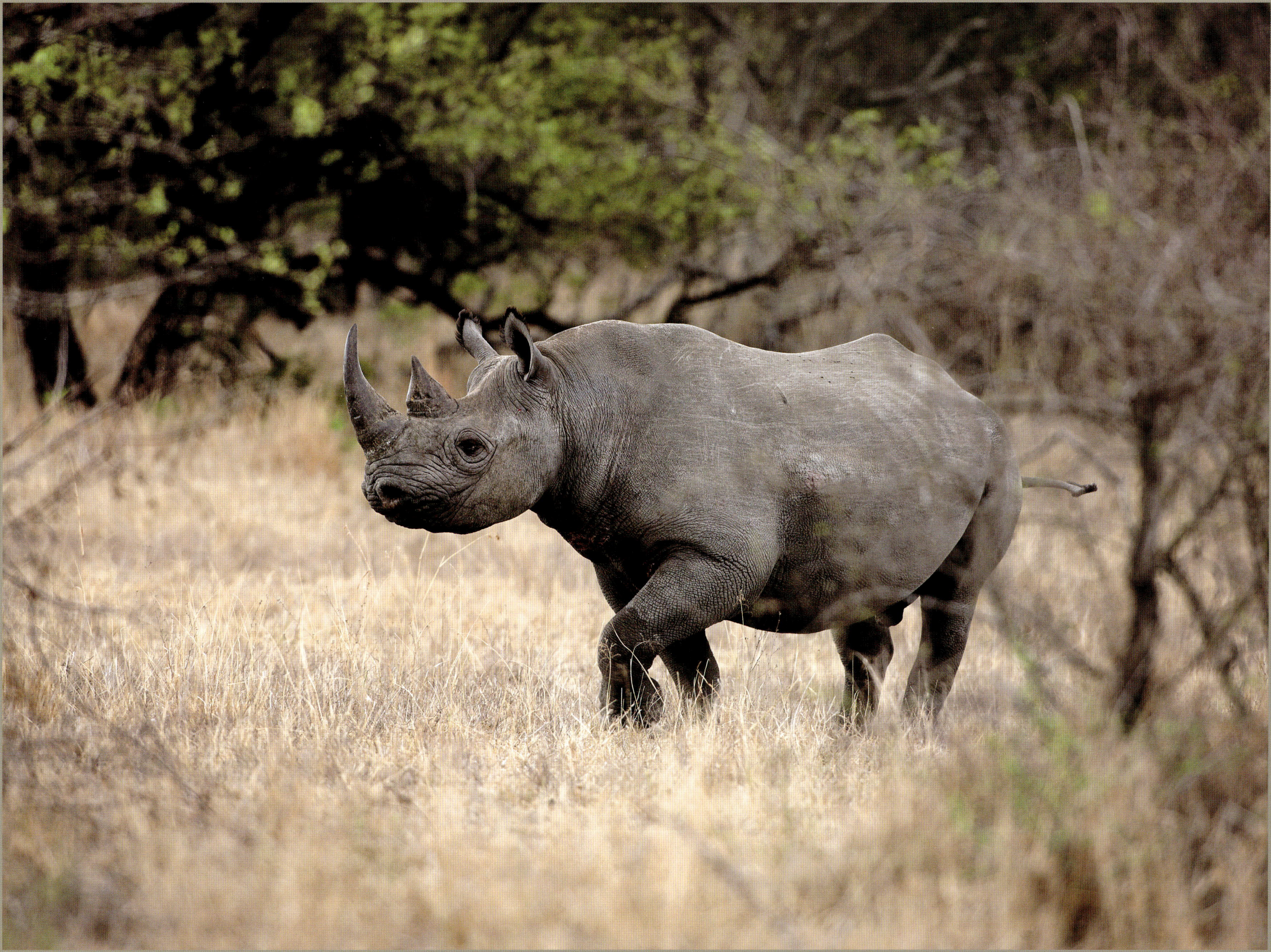

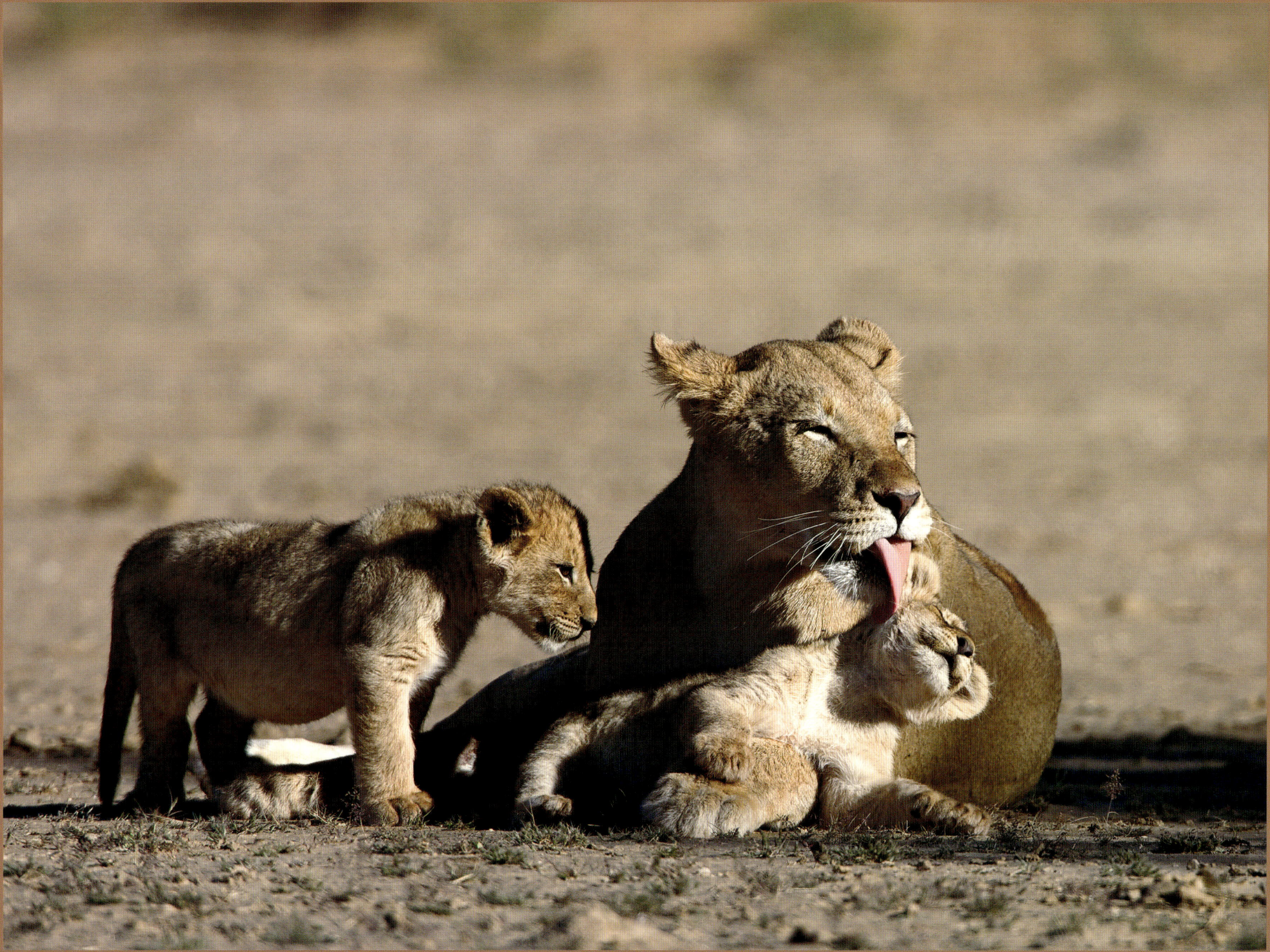

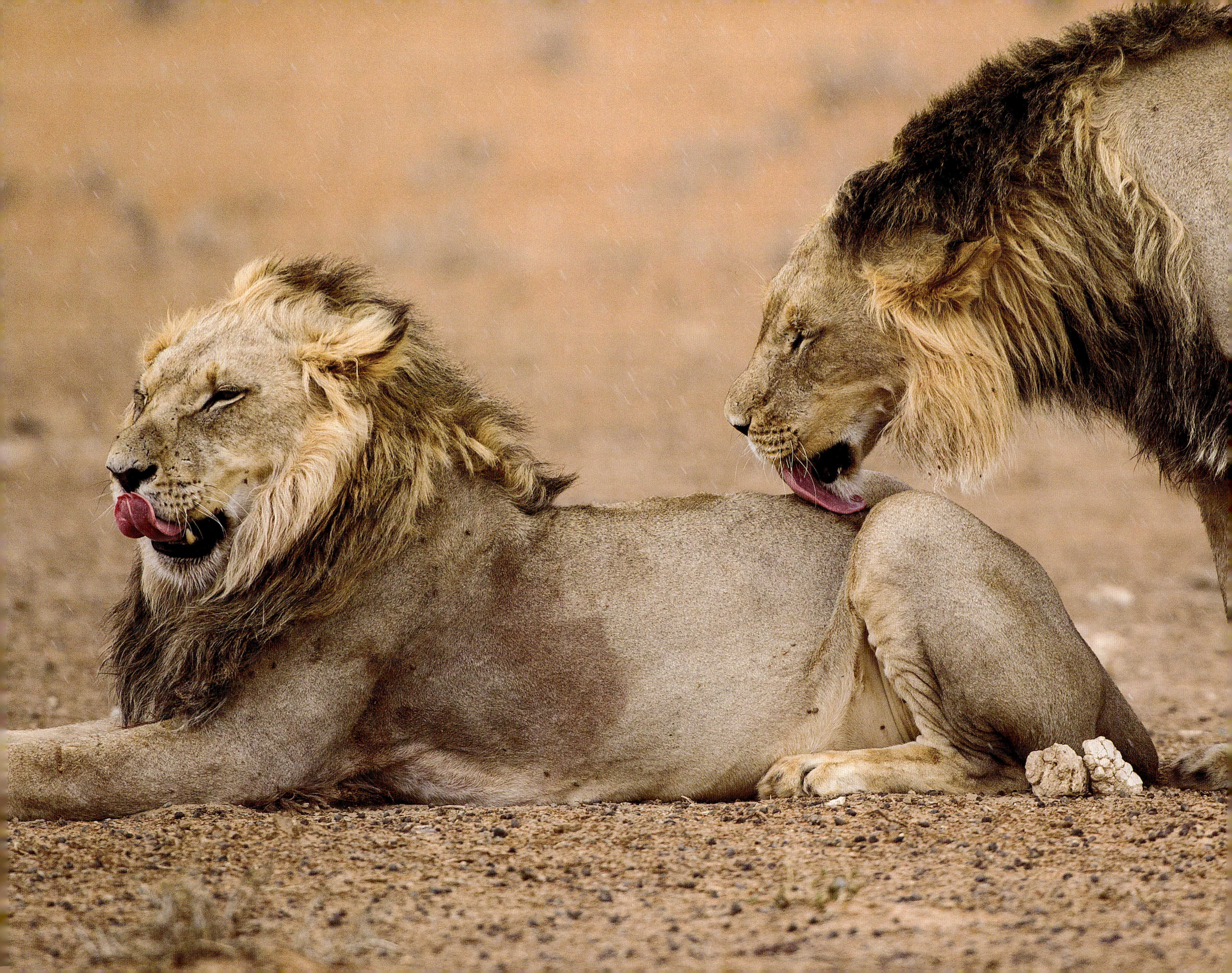

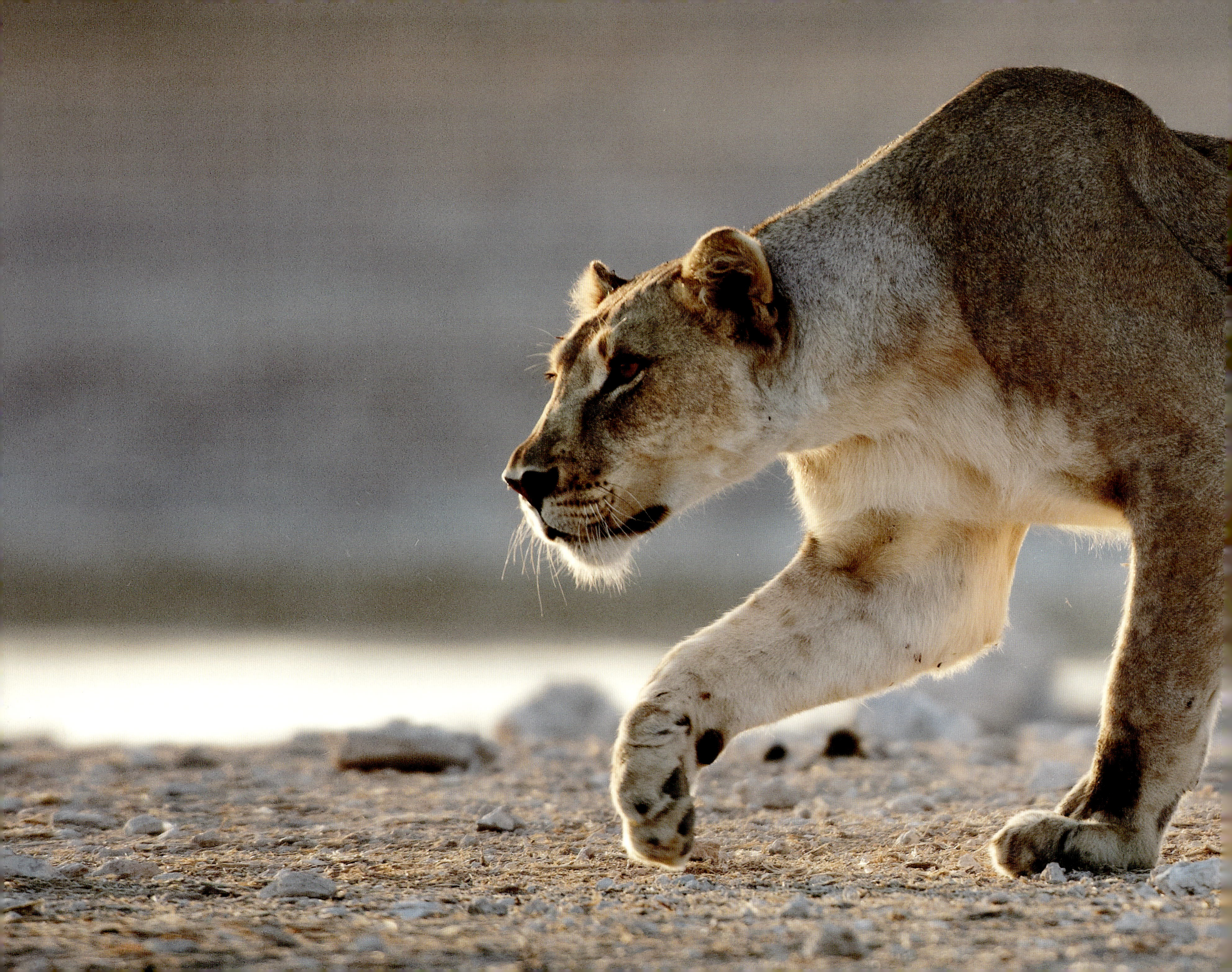

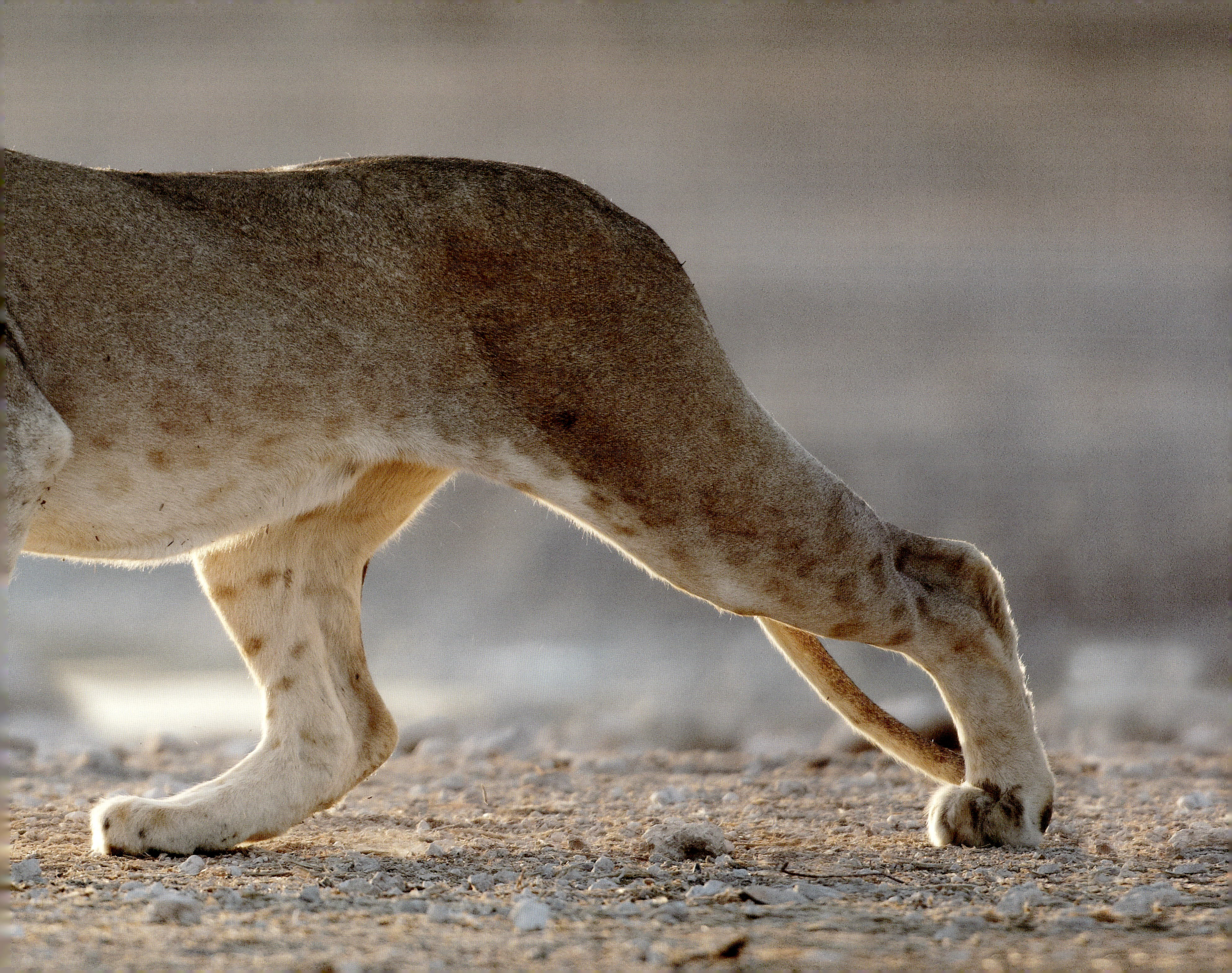

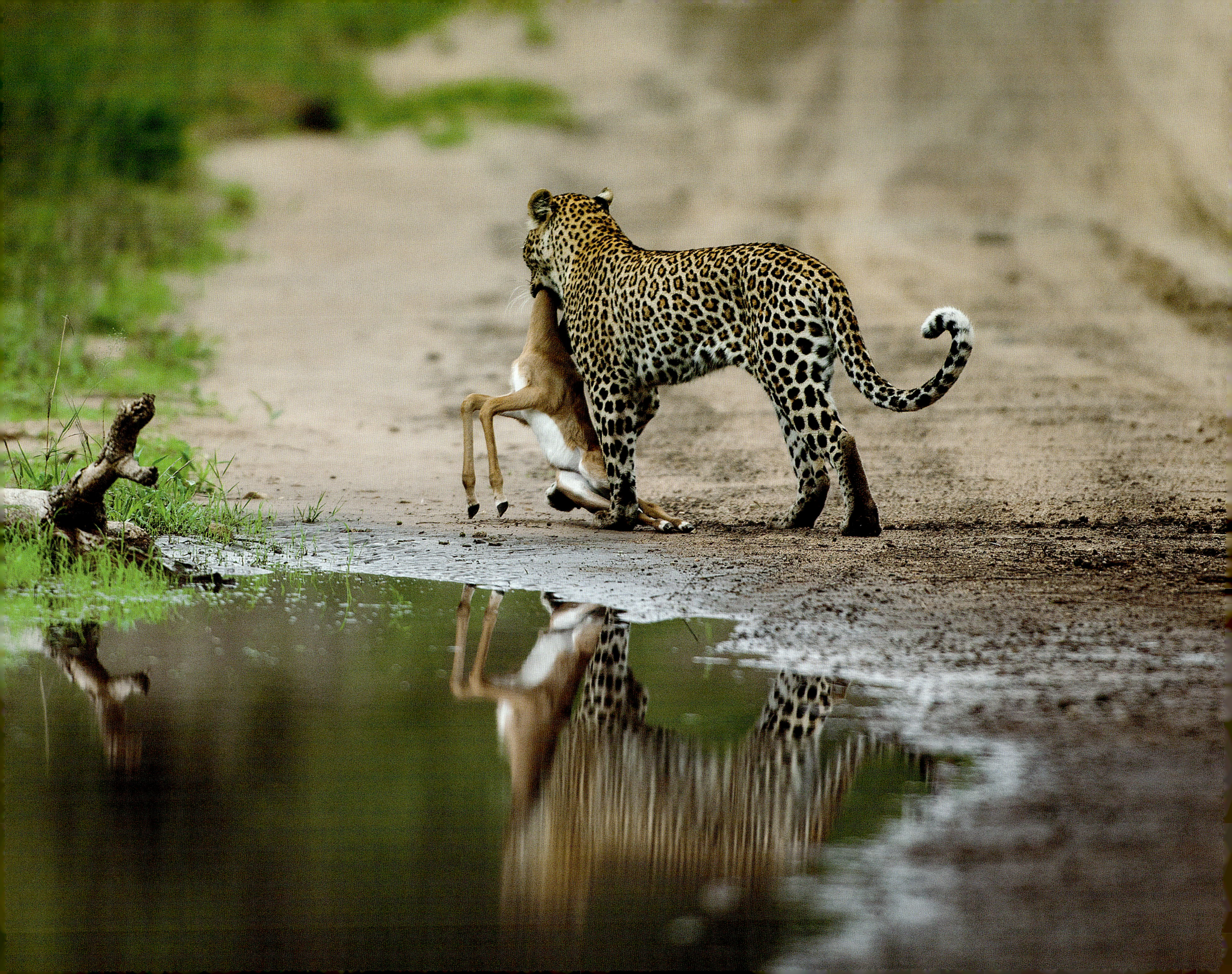

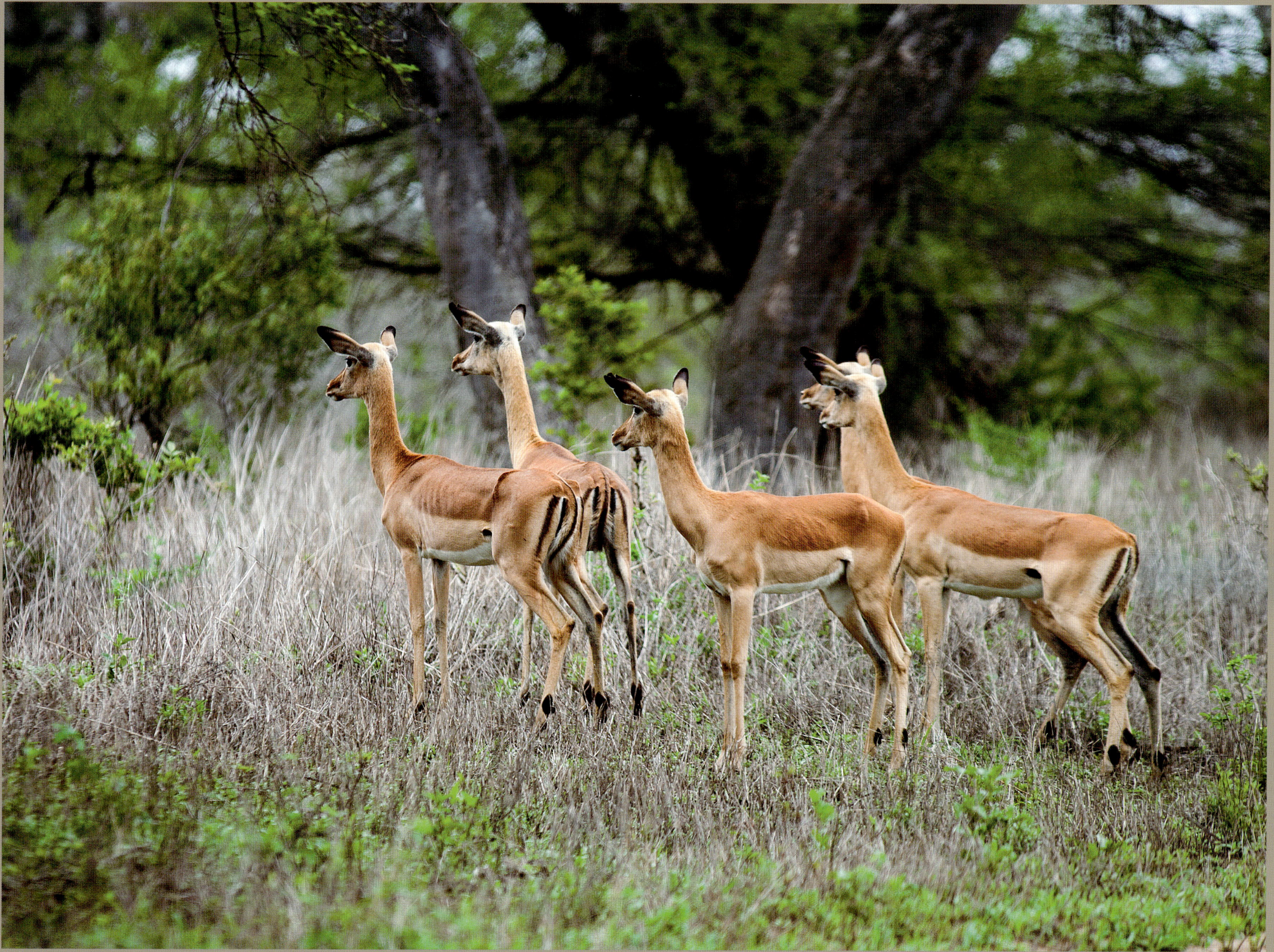

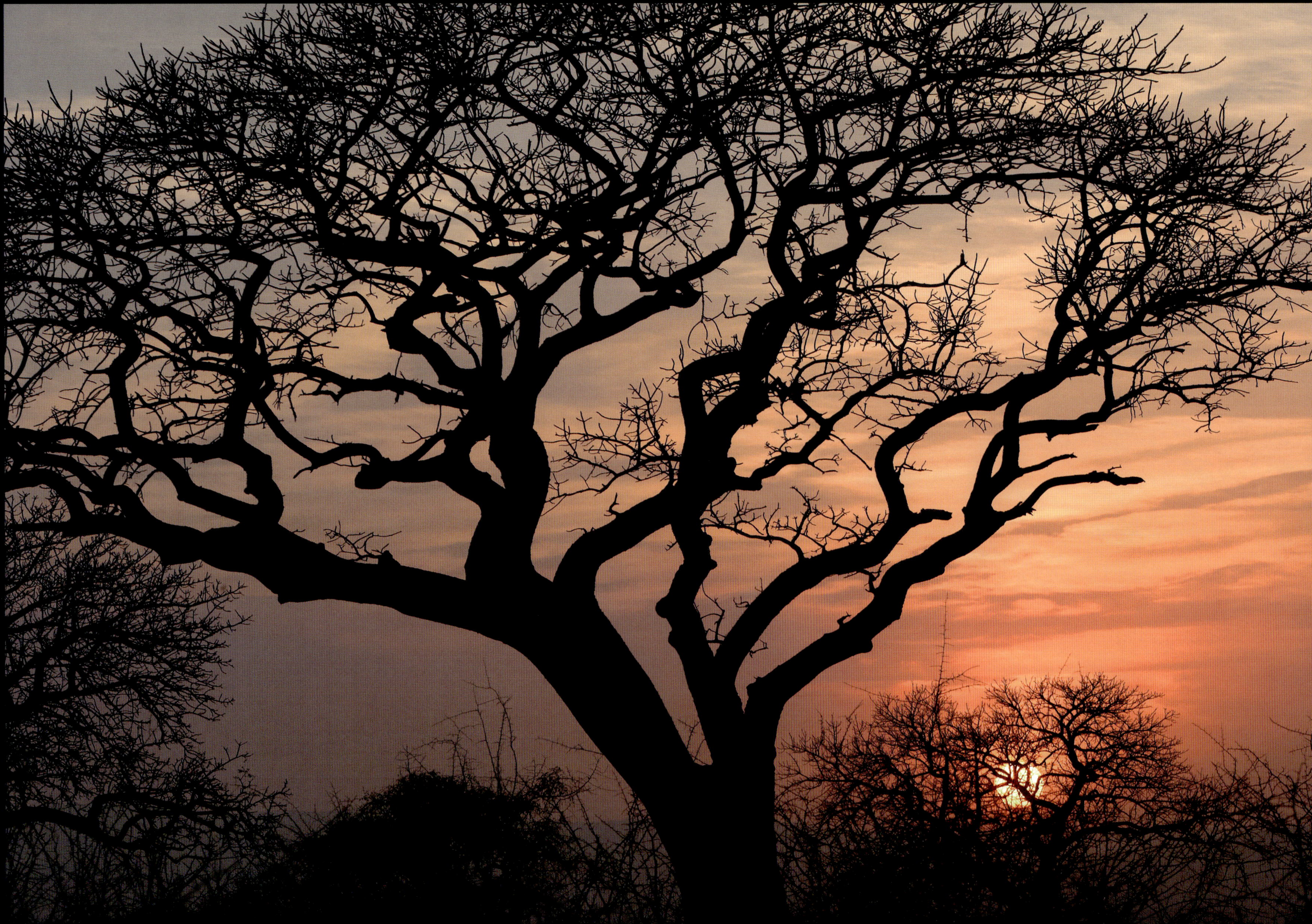

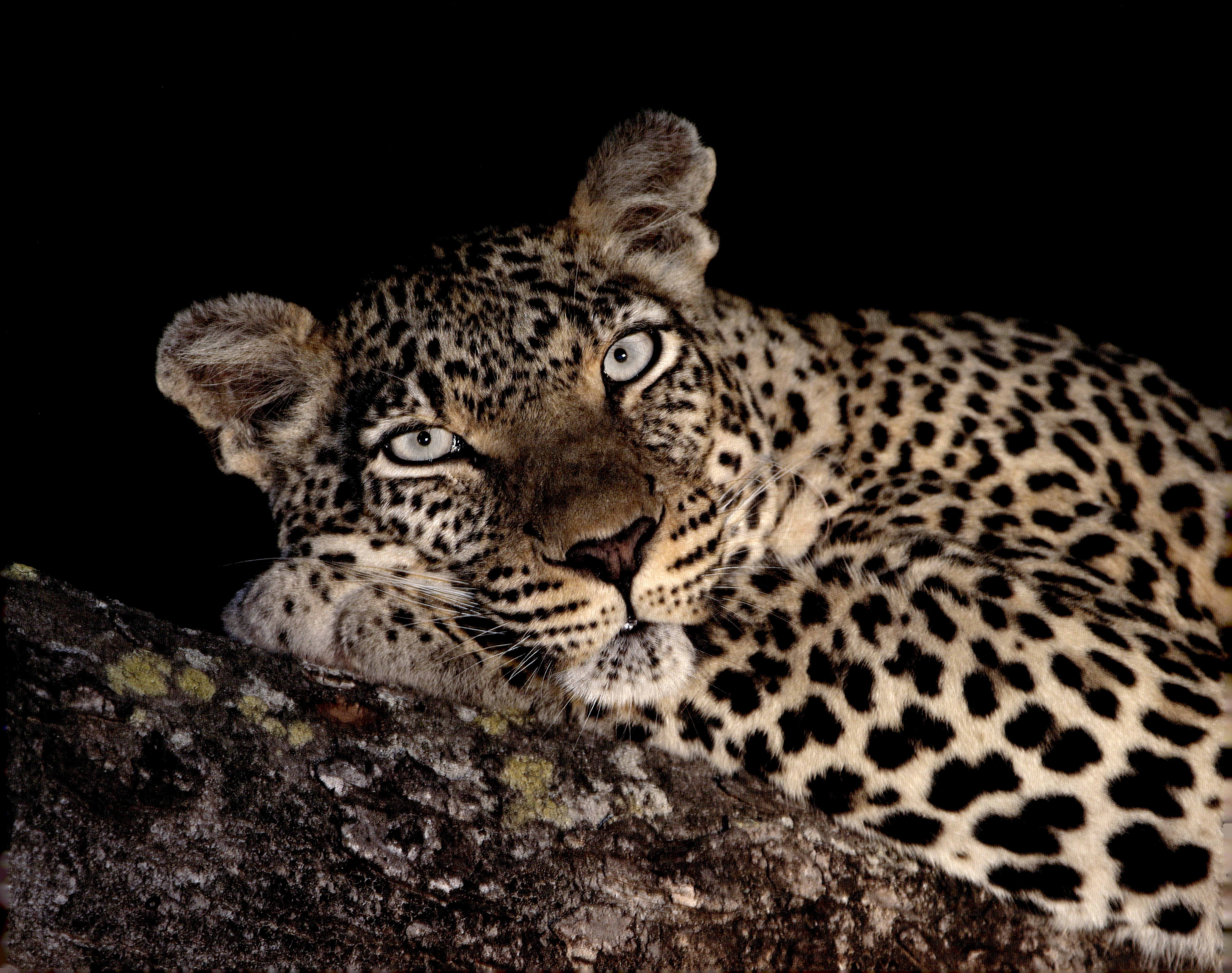

More about the Savannah Biome

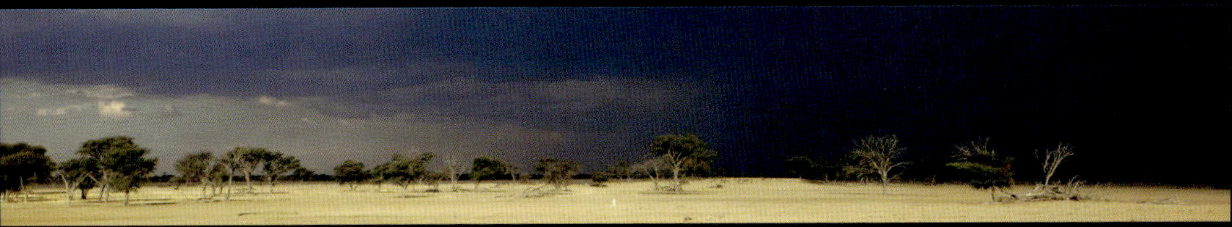

Typical savannah woodland

The term savannah is derived from Spanish and describes an environment that consists of interdependent grasses, shrubs and trees. The ratio between grasses and trees shifts according to the influence of climate, seasons and animals. Of course, soil types and geology then have a marked influence on the plants that grow there. This leads to a 'mosaic' of different landscapes, each with its own general species composition. In the Savannah Biome, as in the Grassland Biome, grass is the foundation of the food chain, with all animal species dependent on its nutrition in one way or another. Interestingly, the plants of the savannah support more large animals than any other biome in the world.

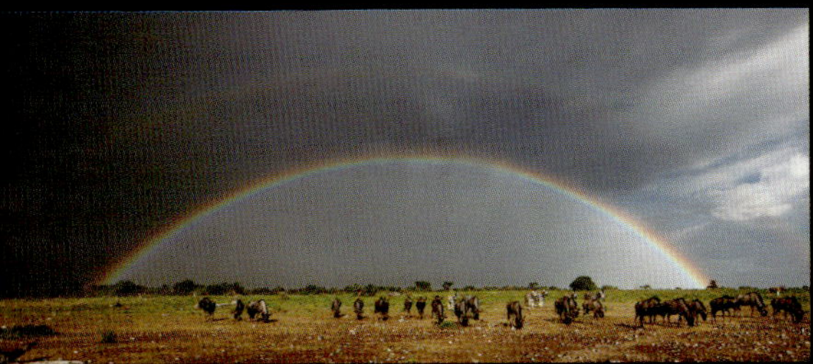

Distant showers and rainbow

At the western extremity, the Etosha National Park is an example of the transition of savannah from rather dense tree stands in the eastern parts of the park to semi-arid savannah interspersed with vast grassy plains that gradually change into a true desert environment towards the coast. Because savannah is summer rainfall area, distant showers often paint the sky with magnificent rainbows.

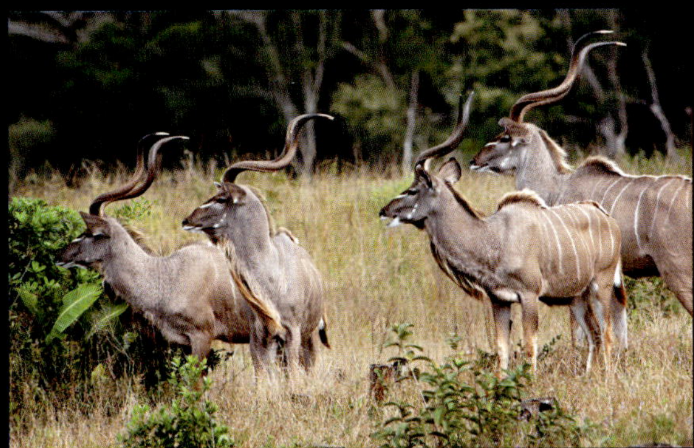

Kudu bulls with their twisted horns

On the coastal plains in the eastern parts of the subcontinent, savannah looks very different. Instead of short sweet grass that remains nutritious and palatable during most of the year, the grass stands are high but mostly unpalatable and poor in nutrients. This is due to a much higher rainfall and sandy soils where minerals are leached away. Kudu, however, are browsers rather than grazers and prefer long grass for cover.

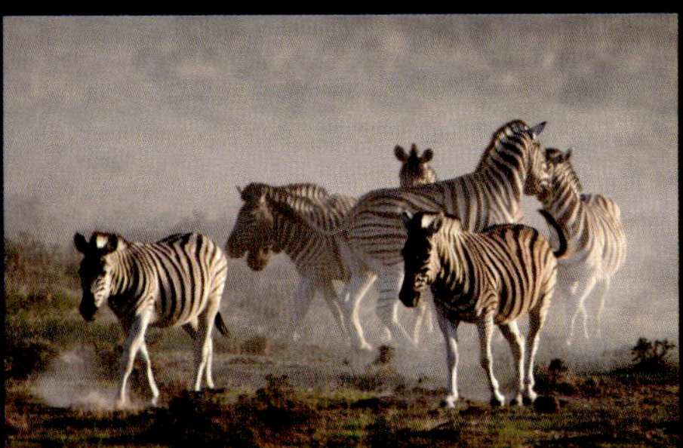

A zebra herd on its way to water

Zebra, on the other hand, are typical grazers. They prefer the scattered grassy plains of the savannah where trees are few and the grass is sweet. They move in groups and often mingle with other species such as the blue wildebeest. They do not compete for food since each species prefers a different part of the grass. They do, however, often stay together because there is 'safety in numbers' and zebra seem to have better eyesight while wildebeest have excellent hearing. In this way they boost their chances of spotting predators in time.

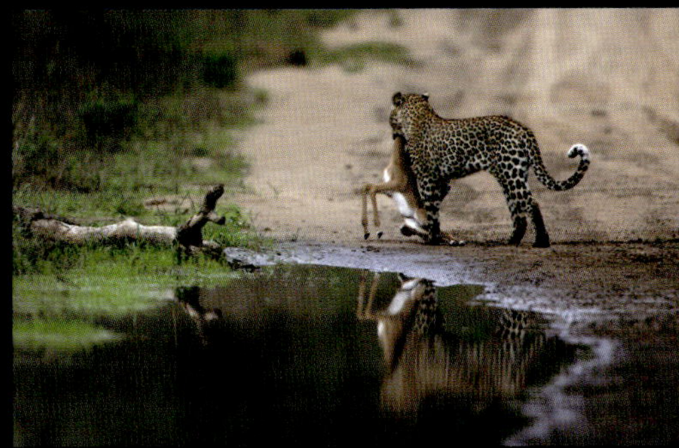

Leopard prey on impala

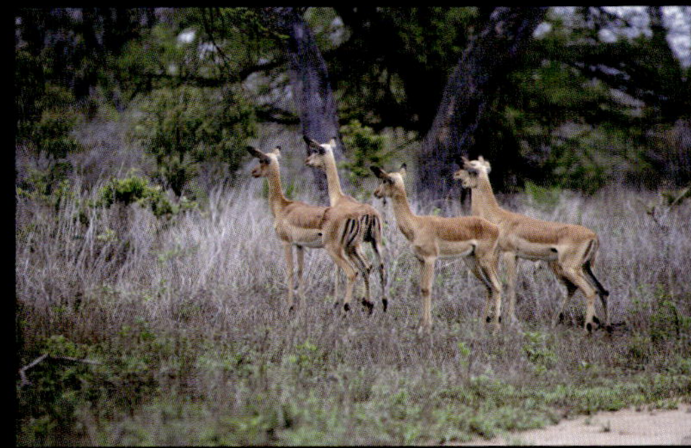

Impala watch as the leopard disappears

Impala are the most numerous antelope in the eastern savannah. Although they are often found in open grassland patches, they are both grazers and browsers. Mixed woodlands are preferred to mopaneveld. They are the favourite prey of many predators such as leopard, in spite of their agility and speed. They can leap over high obstacles and run in bursts of up to 80km/h.

Baboons are mostly vegetarians but they also feed on invertebrates, small mammals and birds. They live in troops of up to 40 individuals with a strict social hierarchy. Mothers tend their babies and regularly give them a ride on their backs.

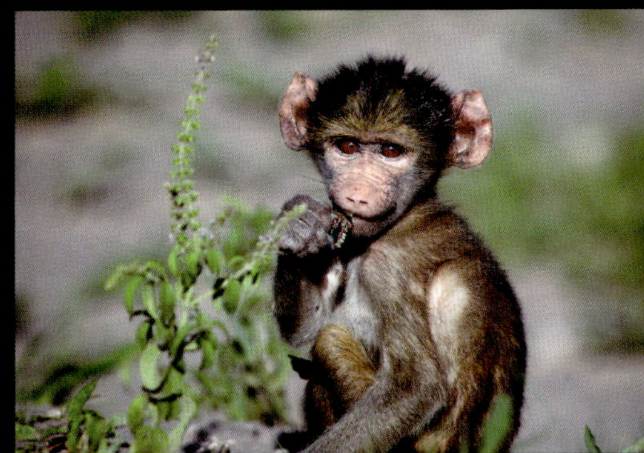

Baboon, typical of the bushveld

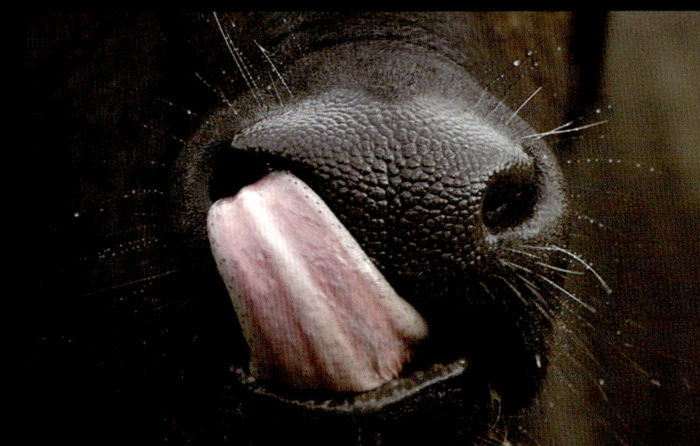

Buffalo

To maintain a healthy balance between grass and bush, savannah needs bulk grazers to remove the coarse fibrous grass to make way for new growth. Buffalo do exactly that. They move in large herds and graze on coarse grass in areas where there is enough grassland, but they also need trees for shade during the heat of the day and access to water. Every adult requires about 21 litres of water daily.

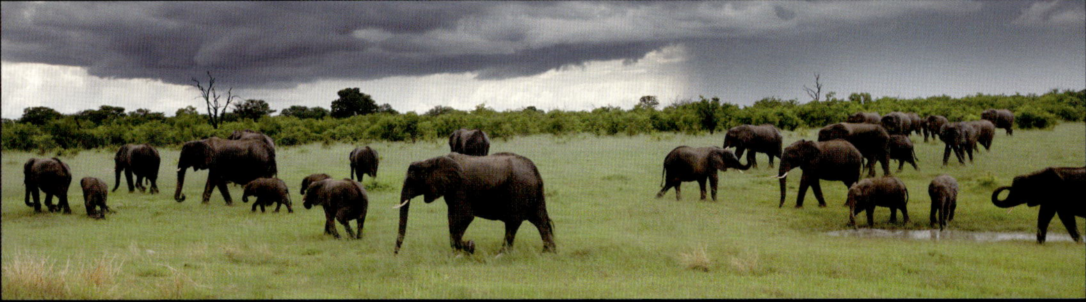

Elephant browse and graze

Since they need large roaming space, they occur mainly in the larger savannah reserves.

The largest mammal in Africa is also both a grazer and a browser and needs up to 200 litres of water a day. Elephant move in large matriarchal herds that include the young and sub-adults, but the males keep to themselves in separate male herds. When an adult bull comes into musth, it will temporarily join a female herd. Over-population of elephant is a real threat since an adult elephant eats about 250kg of plant matter per day and needs extensive roaming space.

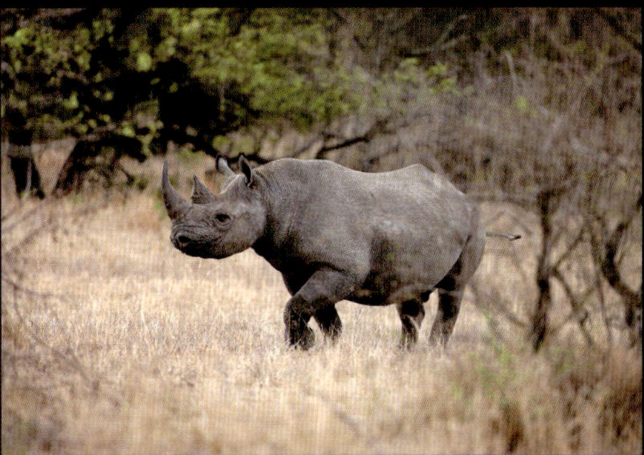

Black rhino

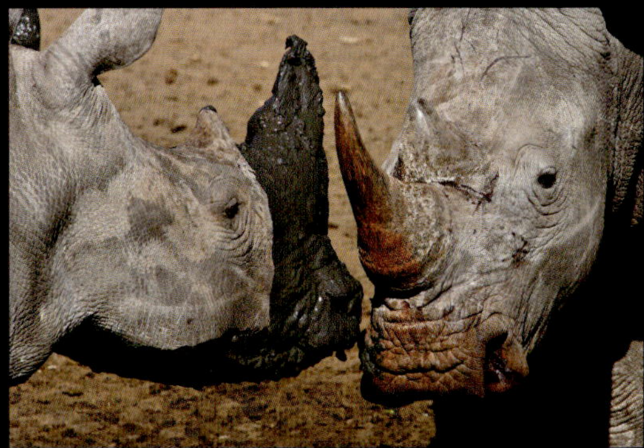

White rhino

Rhino have ensured that they do not compete for the same food resources. The white rhino with its broad, square-lipped mouth is perfectly adapted to 'mowing' grass, while the black rhino with its pointed beak-like lips can pick the tender leaves between sharp spines on branches. Although the white rhino is a grazer, it needs shelter in the form of shade during the hottest part of the day. They therefore do not survive in pure grassland.

Almost all the larger animals of the savannah need to drink water regularly, especially the grazers. Some, like the buffalo, need to drink twice a day, while others can go for a day or more without doing so and yet others, like the gemsbok in the semi-arid regions, hardly ever drink. The tiny steenbok can also survive on the moisture it gets from the fresh leaves it feeds on. Warthog, however, drink regularly and like to wallow in the mud to help keep their bodies cool.

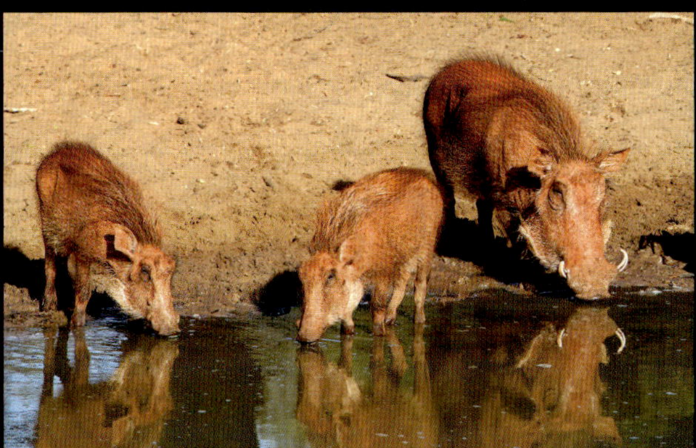

Warthog

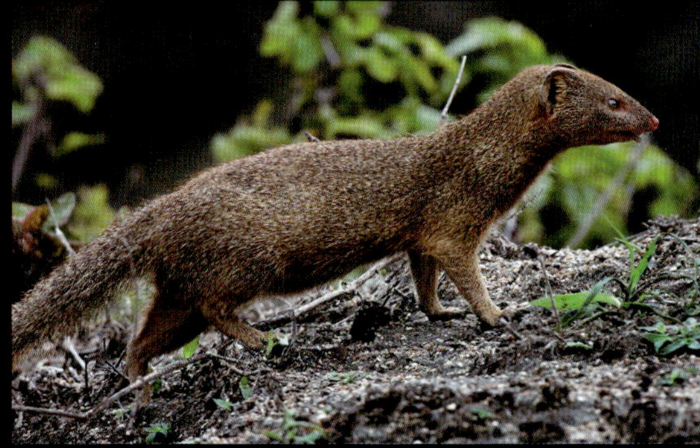

Slender mongoose

Small mammals such as the slender mongoose are easily overlooked, but these little predators can often be seen crossing the tourist roads in search of prey.

Where there is enough prey, predators will follow. Although lion are not specific to the Savannah Biome, the larger savannah parks and reserves are the only places where they still occur naturally. In many savannah parks and others within other biomes they may have been reintroduced. Lion were killed as vermin in the past, but they now also face the serious problem of dwindling numbers. They prey mainly on medium to large hoofed animals such as wildebeest, waterbuck and zebra, and some prides even go for giraffe and elephant.

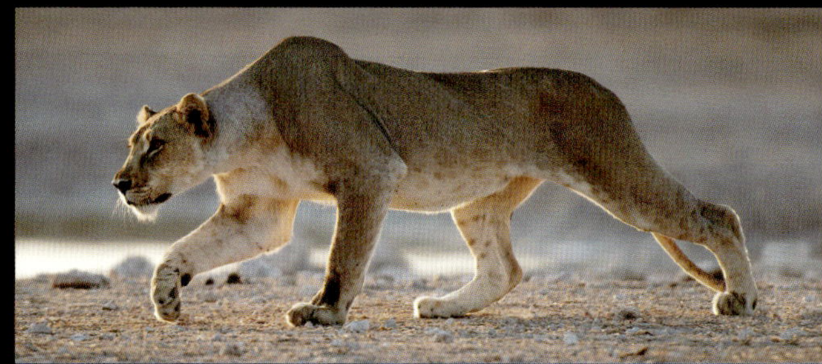

Lioness

In woody bushy areas both predators and prey tend to have camouflage colouration. Leopard have spots that perfectly conceal them in the dappled shade of leaves and branches. They favour the branches of trees such as the marula to rest and watch for possible prey.

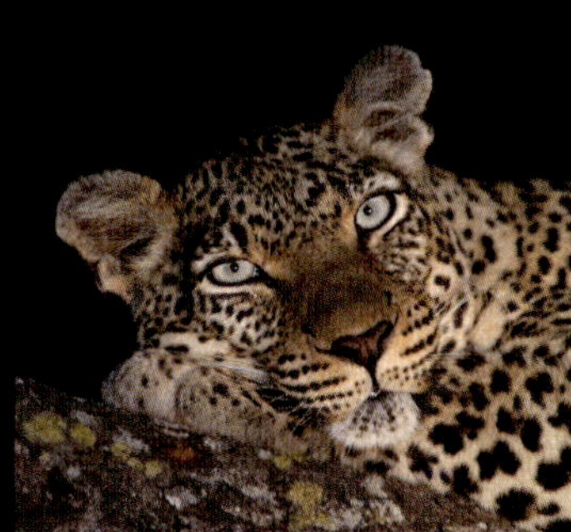

Leopard

The savannah has many thorny trees, especially acacias. There are many different acacia species, but all of them have compound leaves, which are favoured by browsers and heavily thorned as a protection against browsing. They shed their leaves in winter. Broad-leafed species include bushwillows, apple leaf, silver cluster-leaf, torchwood, mopane and several others. The well-known marula is a large upright tree that makes a beautiful silhouette against the sunset when devoid of its leaves. At other times of the year it is heavily browsed, with its fruit a favourite of elephant.

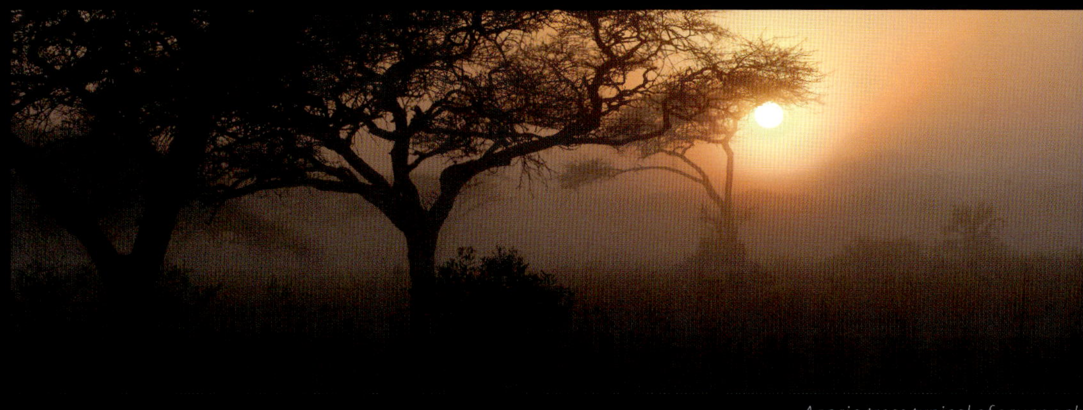

Acacia trees typical of savannah

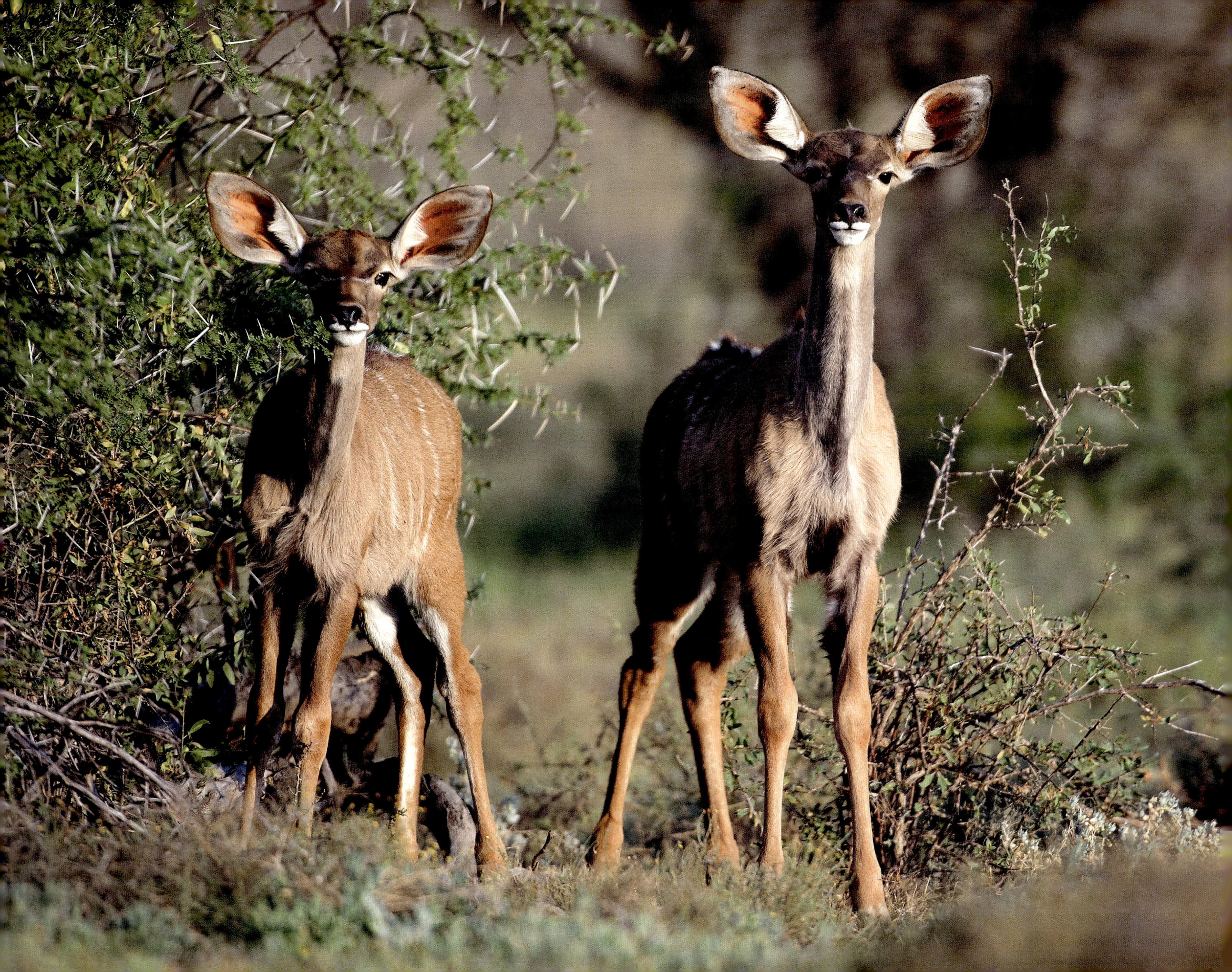

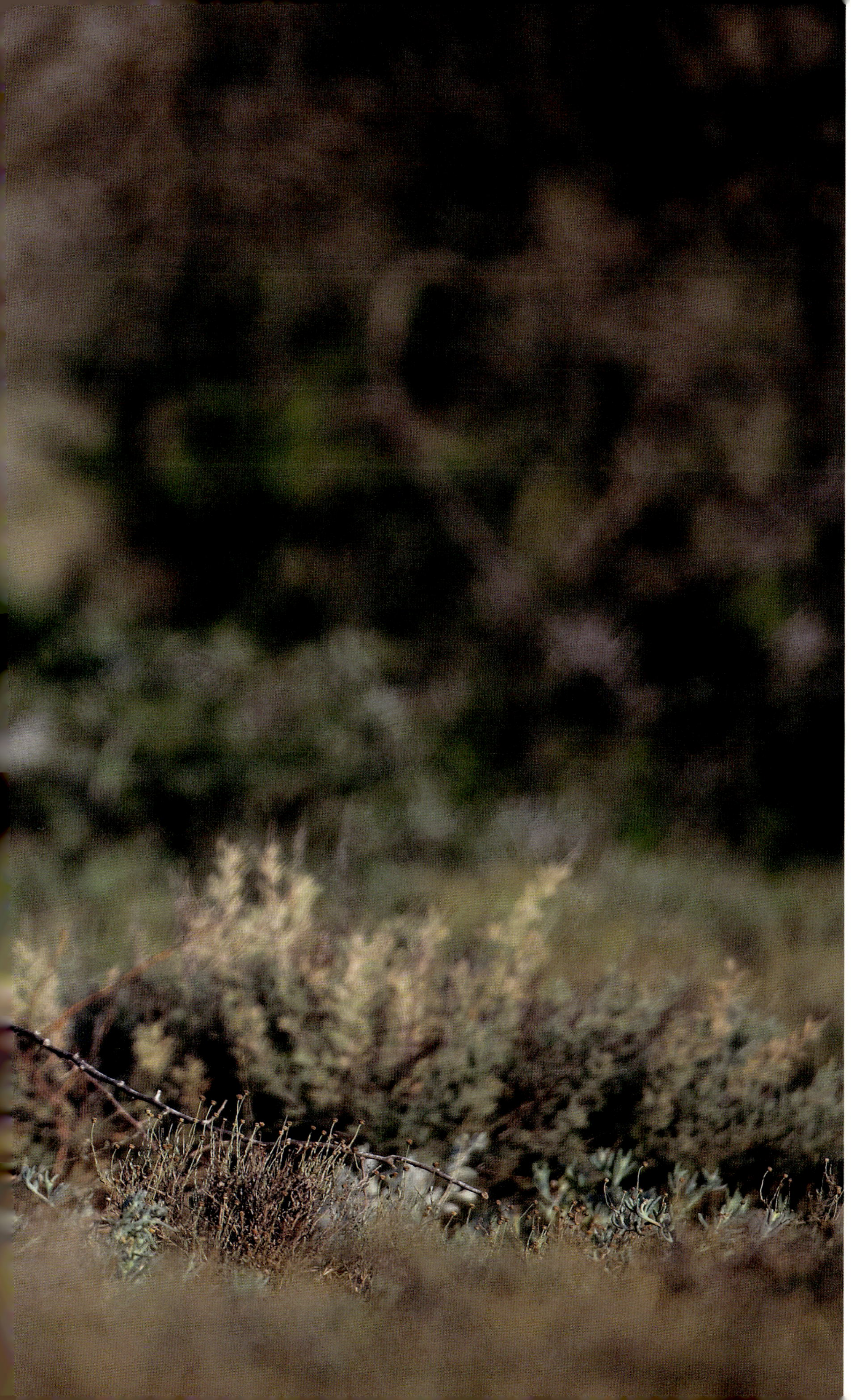

Albany Thicket

The in-between biome

In parts of the Eastern Cape of South Africa, there is a vegetation belt that is neither savannah nor forest. It is more of a shrubland since the 'trees' are succulents, shrubs or vines. Many have vicious thorns that make this type of vegetation almost impenetrable. The rainfall is too low for a true forest, with growth not reaching the proper height of forest trees. There is also no evidence of various strata below the canopy. As there is no real grassy ground layer, it is not savannah either. It borders on no fewer than seven other biomes, combines elements from all and is indeed rather a transition between these various major biomes. This is called Albany Thicket or valley bushveld, but it is not recognised unanimously as a separate biome.

There are a few endemic plant species, mainly succulents, but thicket shares most other species with various surrounding biomes. Among over 600 different kinds of plants, the 'spekboom' is the most extraordinary and conspicuous with its succulent leaves and masses of tiny pink blossoms. These appear in spring after good rains as do other brightly coloured succulent flowers of the *mesembryanthemum* species. It is also the home of pink pelargoniums and blue plumbago, which contribute to the creation of a vibrant wilderness spring garden.

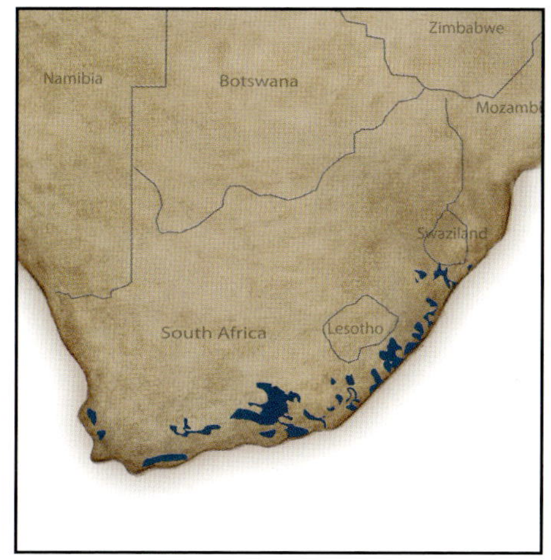

Thicket has an extraordinarily high carrying capacity for browsers, is always green and supports vast numbers of animal species. This kind of vegetation is impressive in its ability to provide concealment for game – even huge animals like elephant can simply disappear from sight in a moment – making them invisible at a distance of a mere metre or two.

The flightless dung beetle, associated with the dung of elephant and other mammals that occur in the thicket, is an endemic insect that has been threatened with extinction and is now classified as 'vulnerable' in the red data book of threatened species.

The main mammal species associated with thicket are browsers such as the Addo elephant, black rhino, bushbuck, kudu and to a lesser extent the warthog that feeds on roots.

Good places to visit in thicket regions

The Addo Elephant National Park is a prime destination for seeing Albany Thicket vegetation and was proclaimed especially to preserve not only this kind of vegetation but also the few Addo elephant that remained after they were ruthlessly hunted in the early 20th century. Other well-known private game reserves nearby are Shamwari Private Game Reserve, Lalibela Game Reserve, Amakhala Game Reserve, Kwandwe Game Reserve near Grahamstown, Pumba and Inkwenkwezi Private Game Reserve. Much of the thicket vegetation has been eradicated to make way for agriculture and other farming practices. However, pockets of thick bush have remained on many private farms, especially on the mountain slopes and in deep ravines, and can be seen from the national roads in the area.

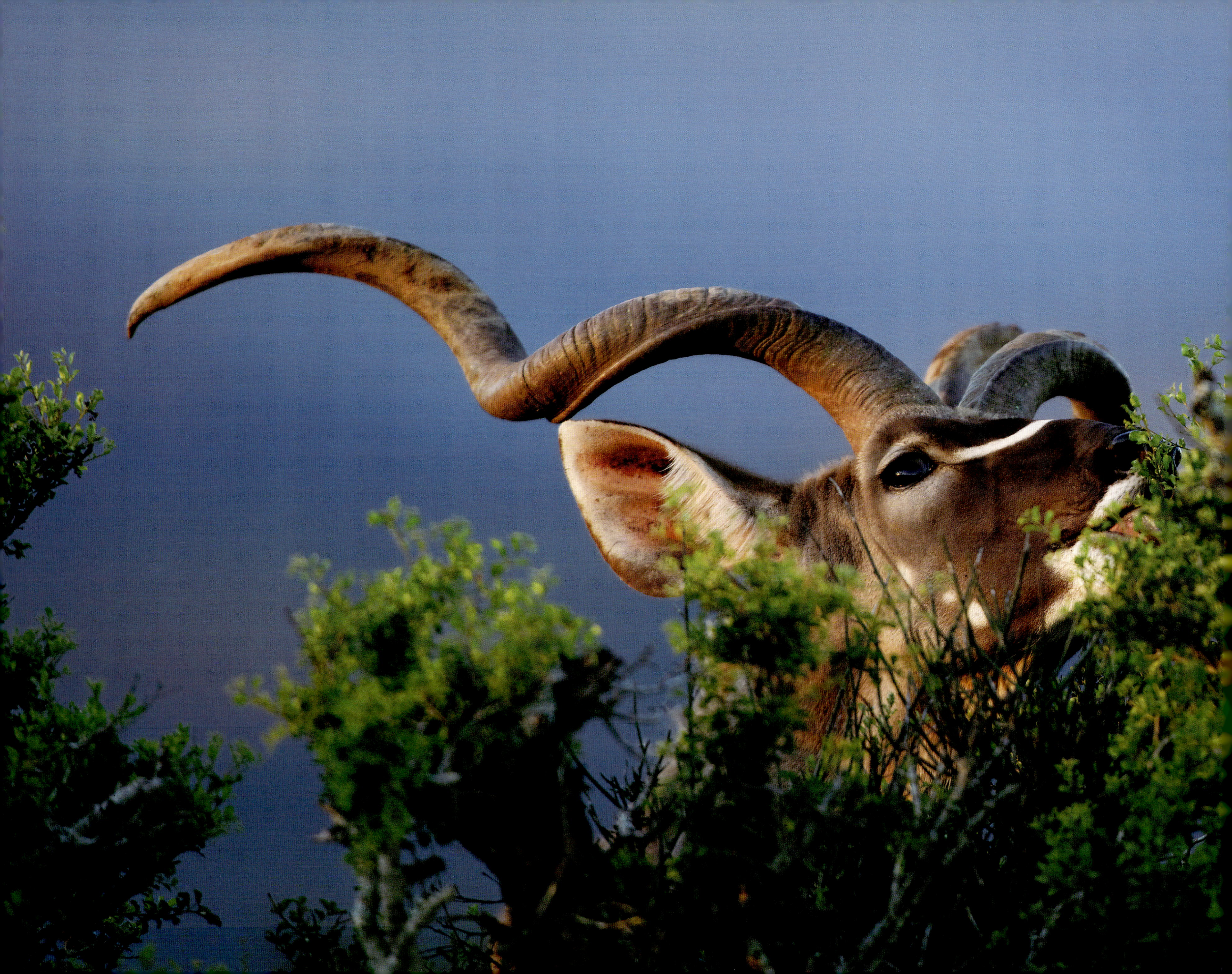

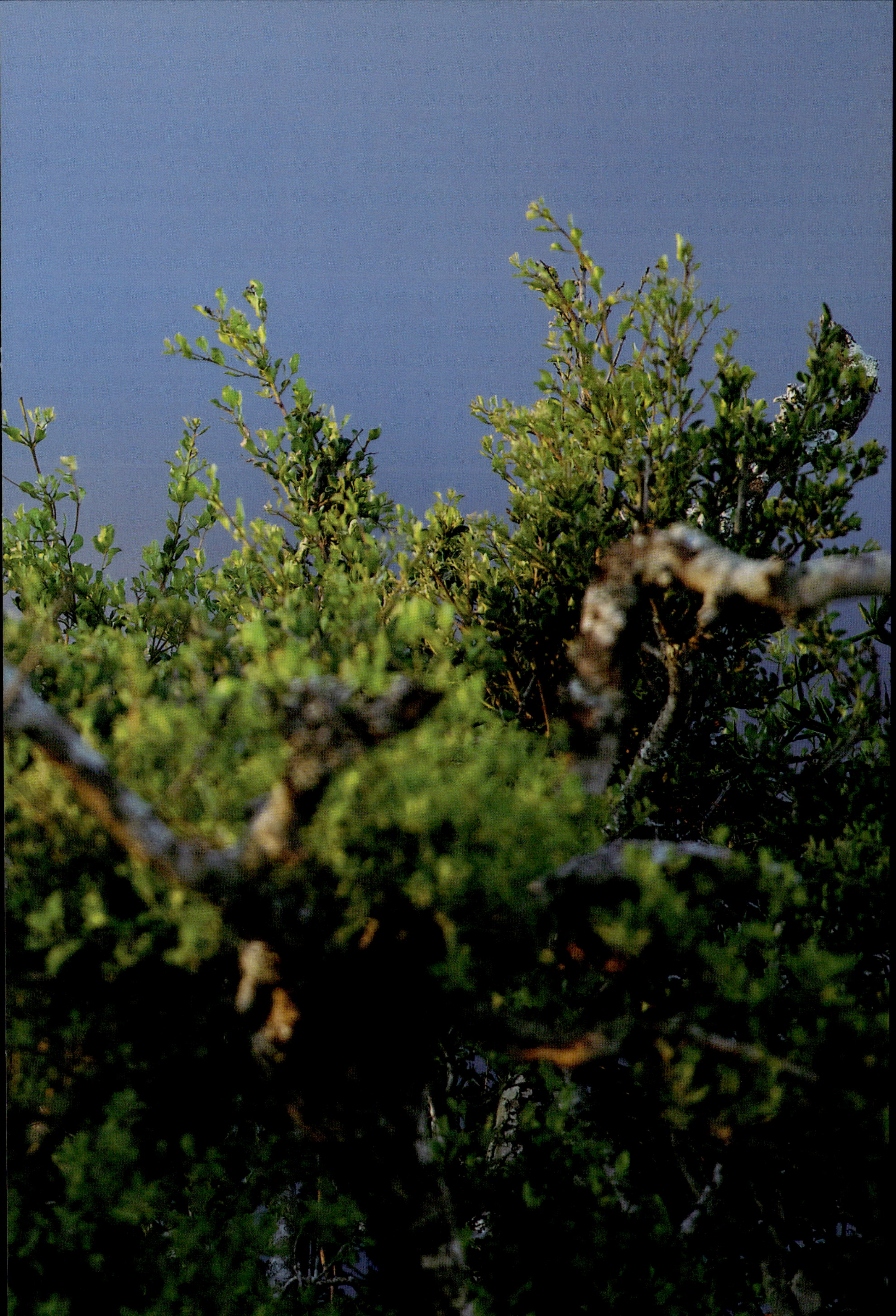
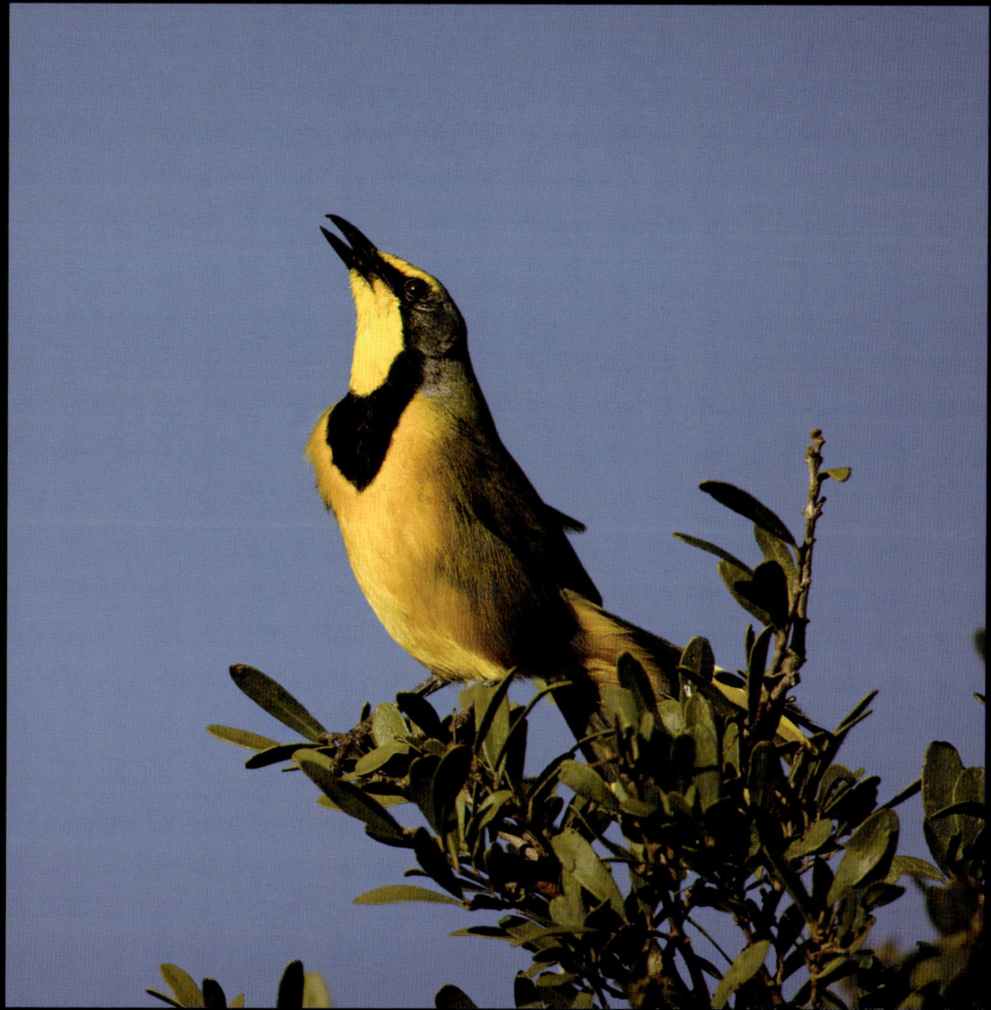
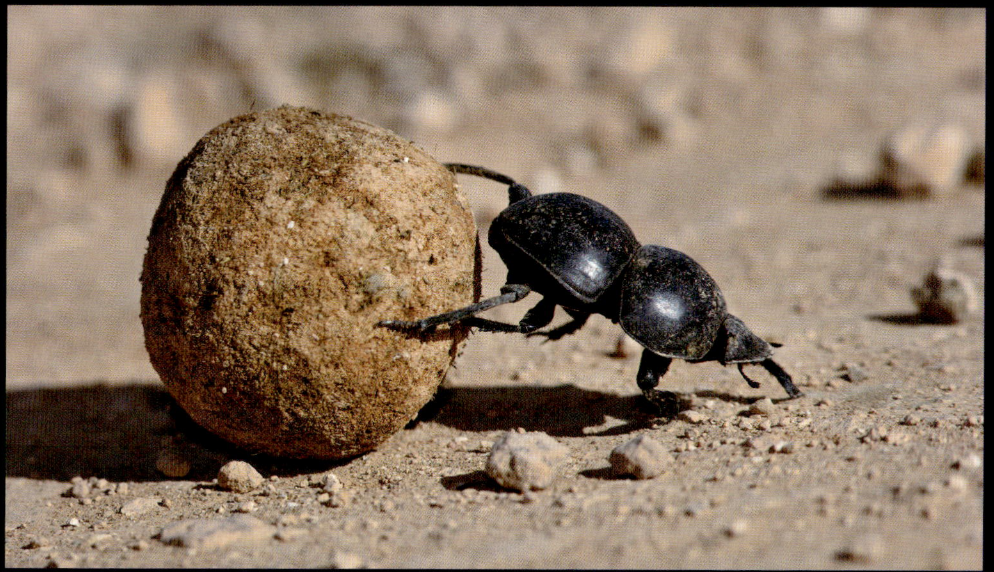

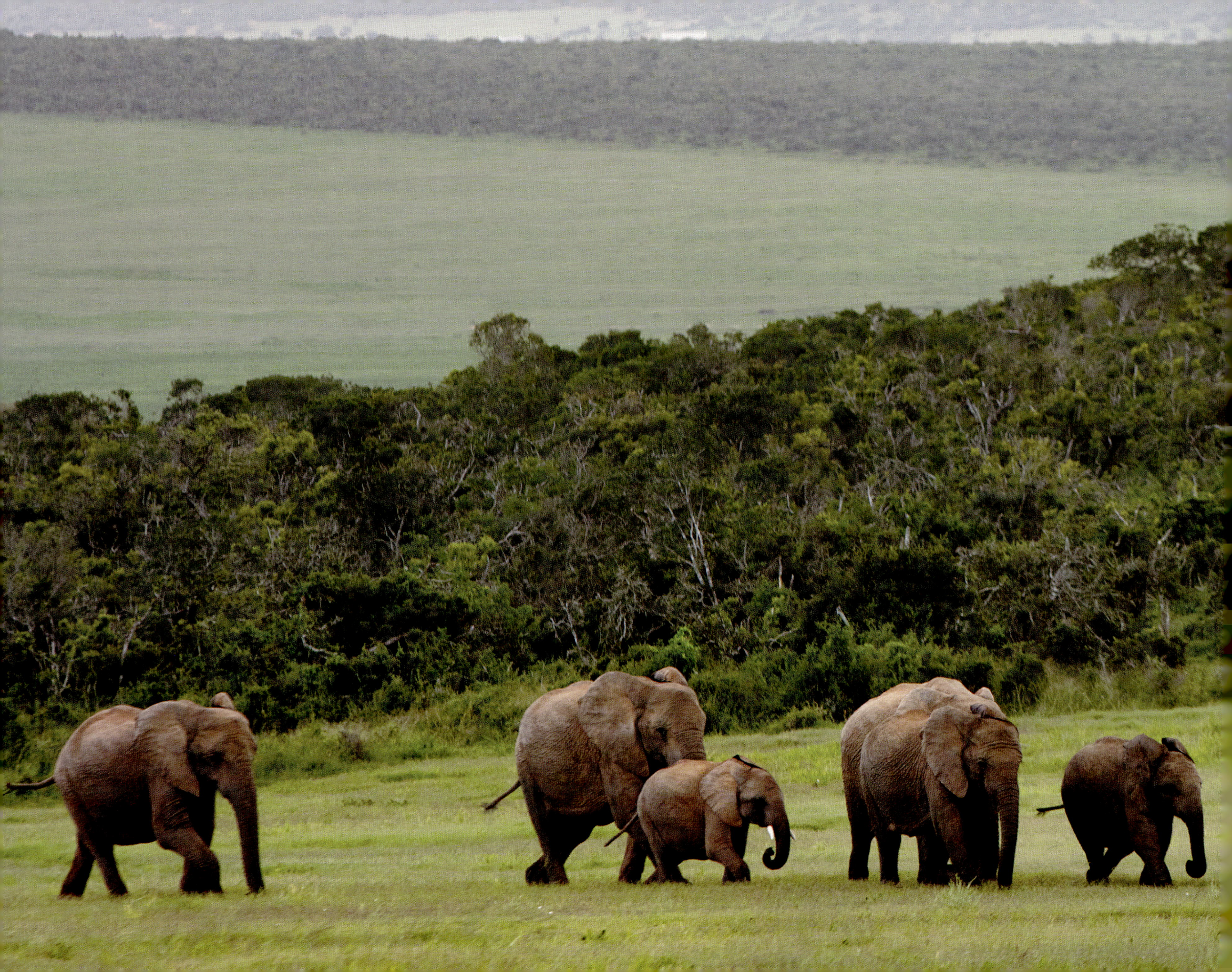

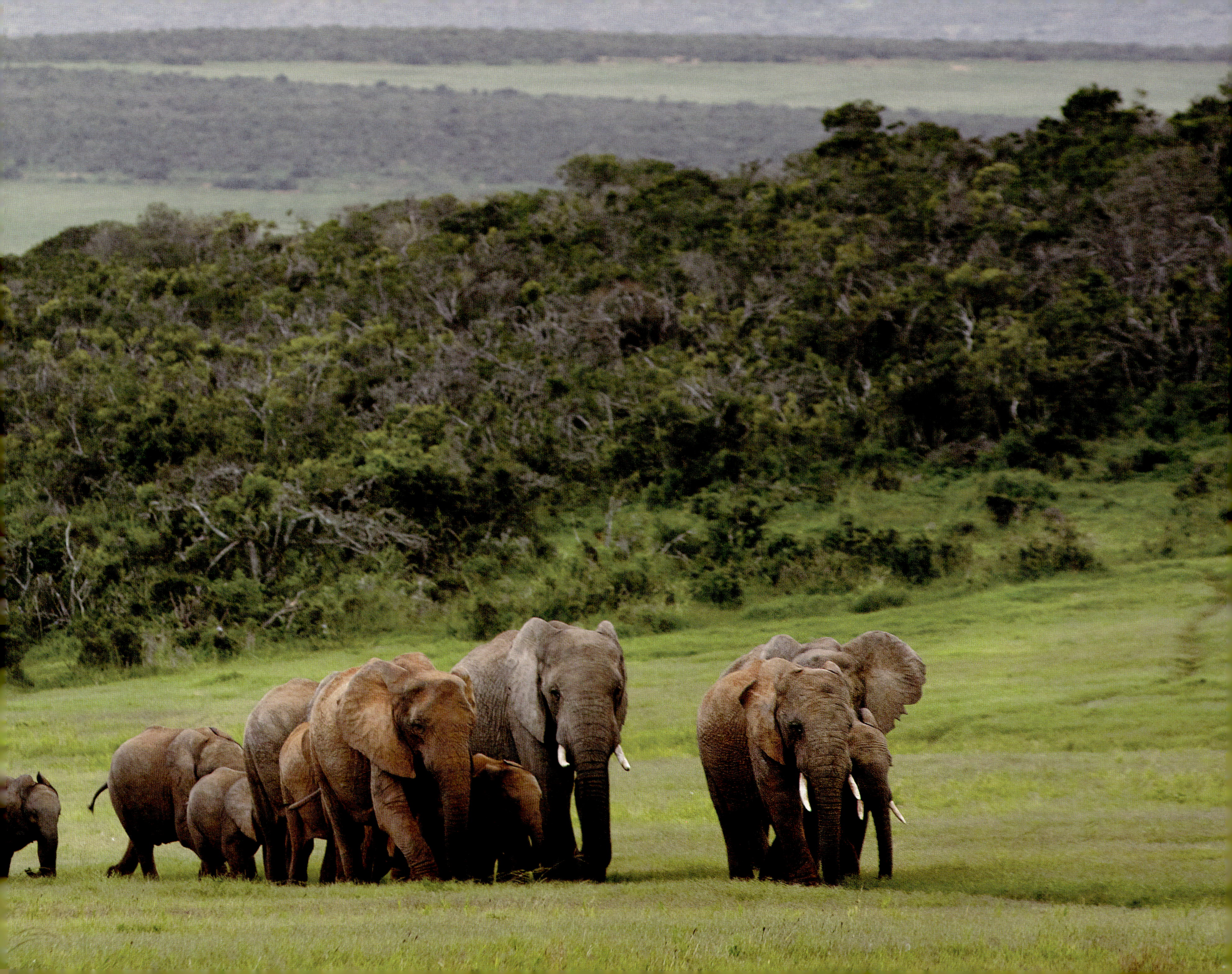

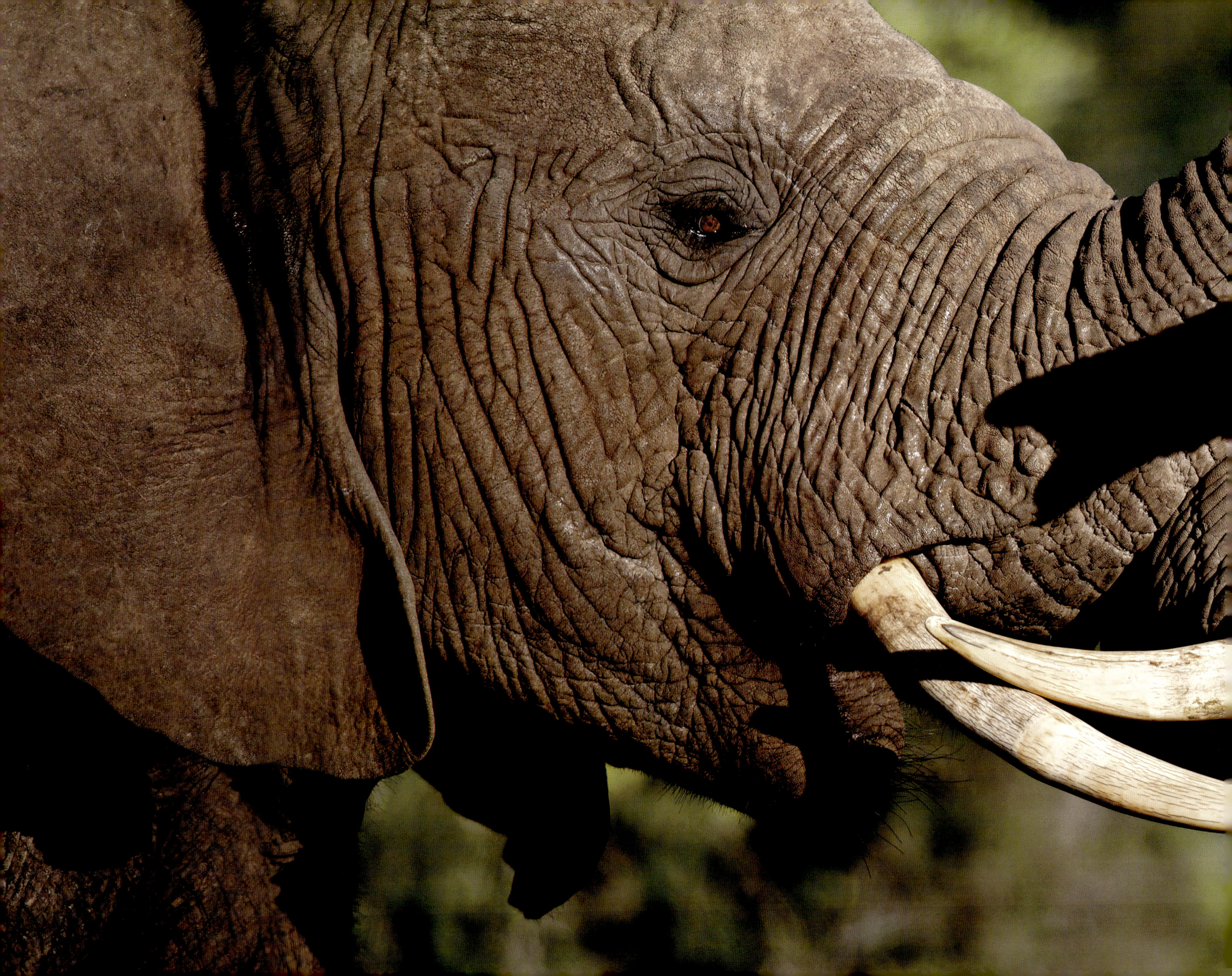

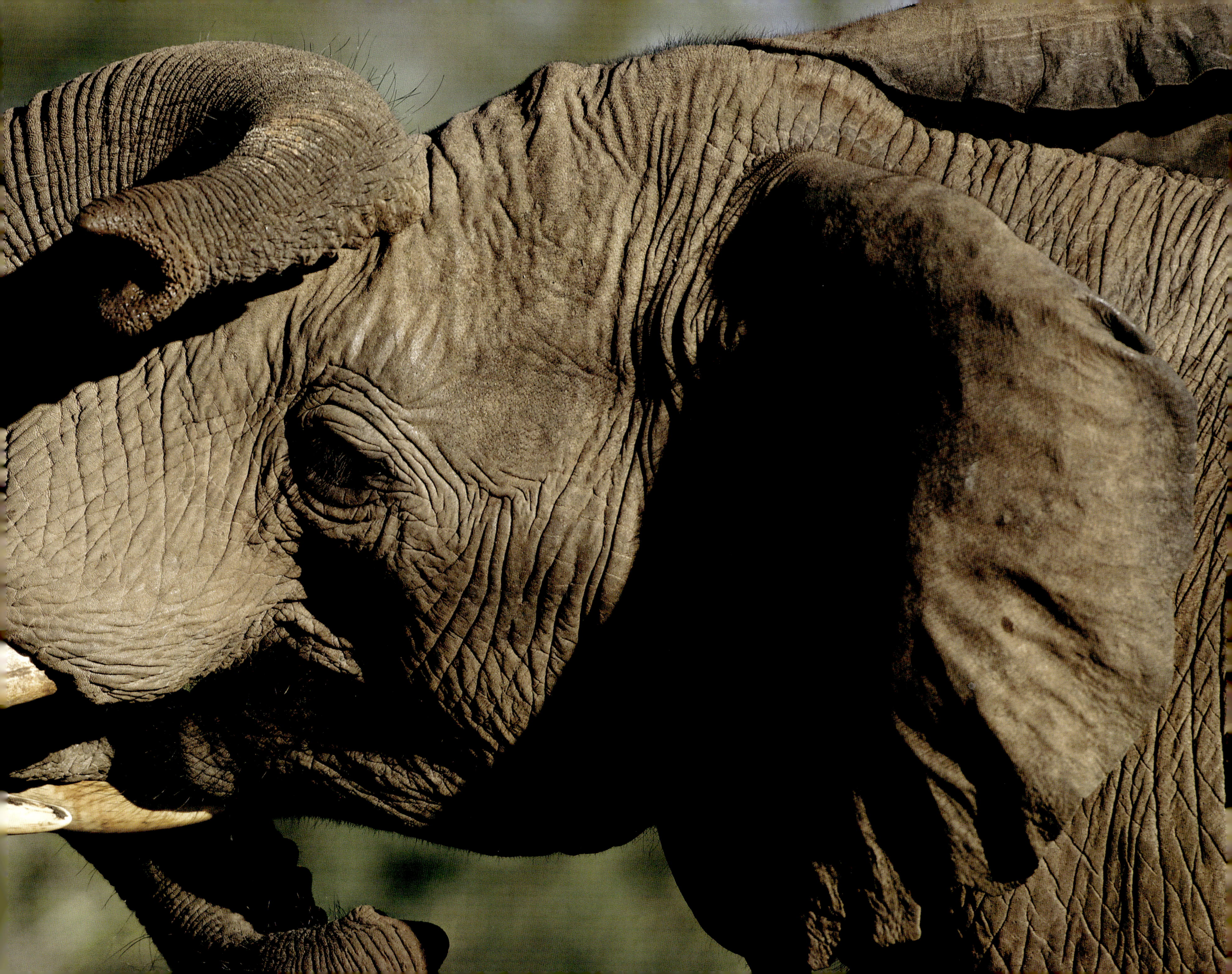

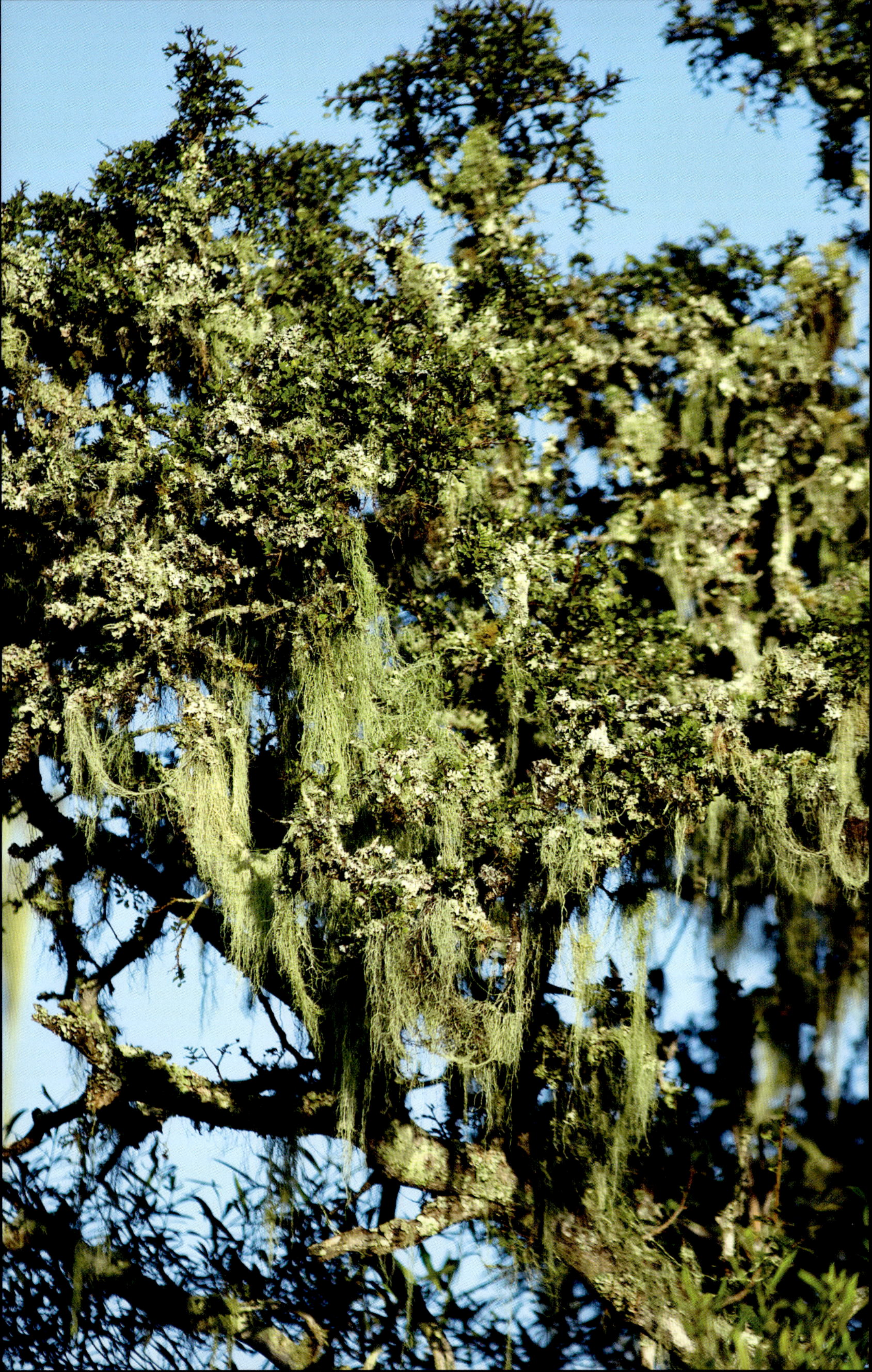

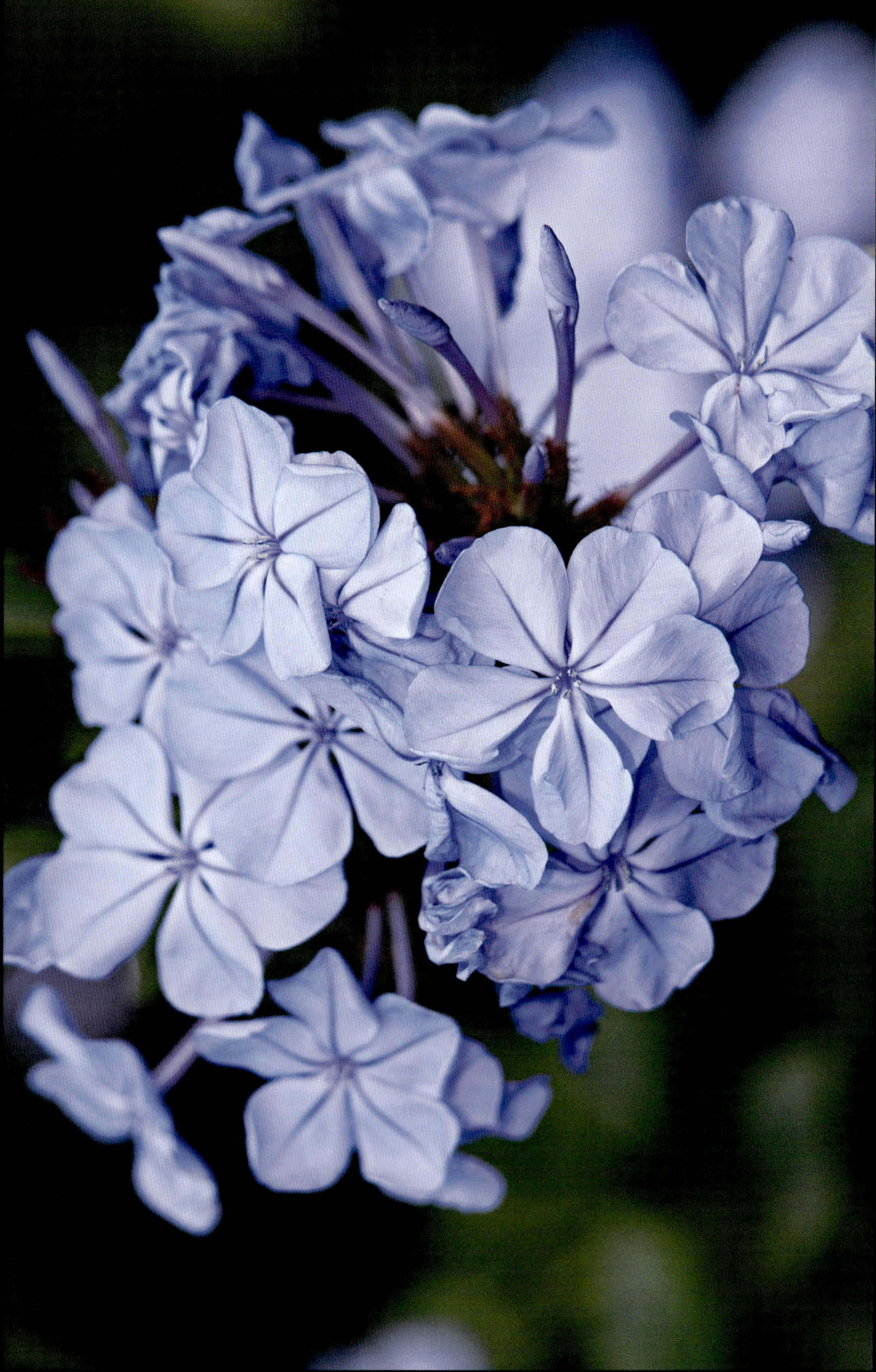

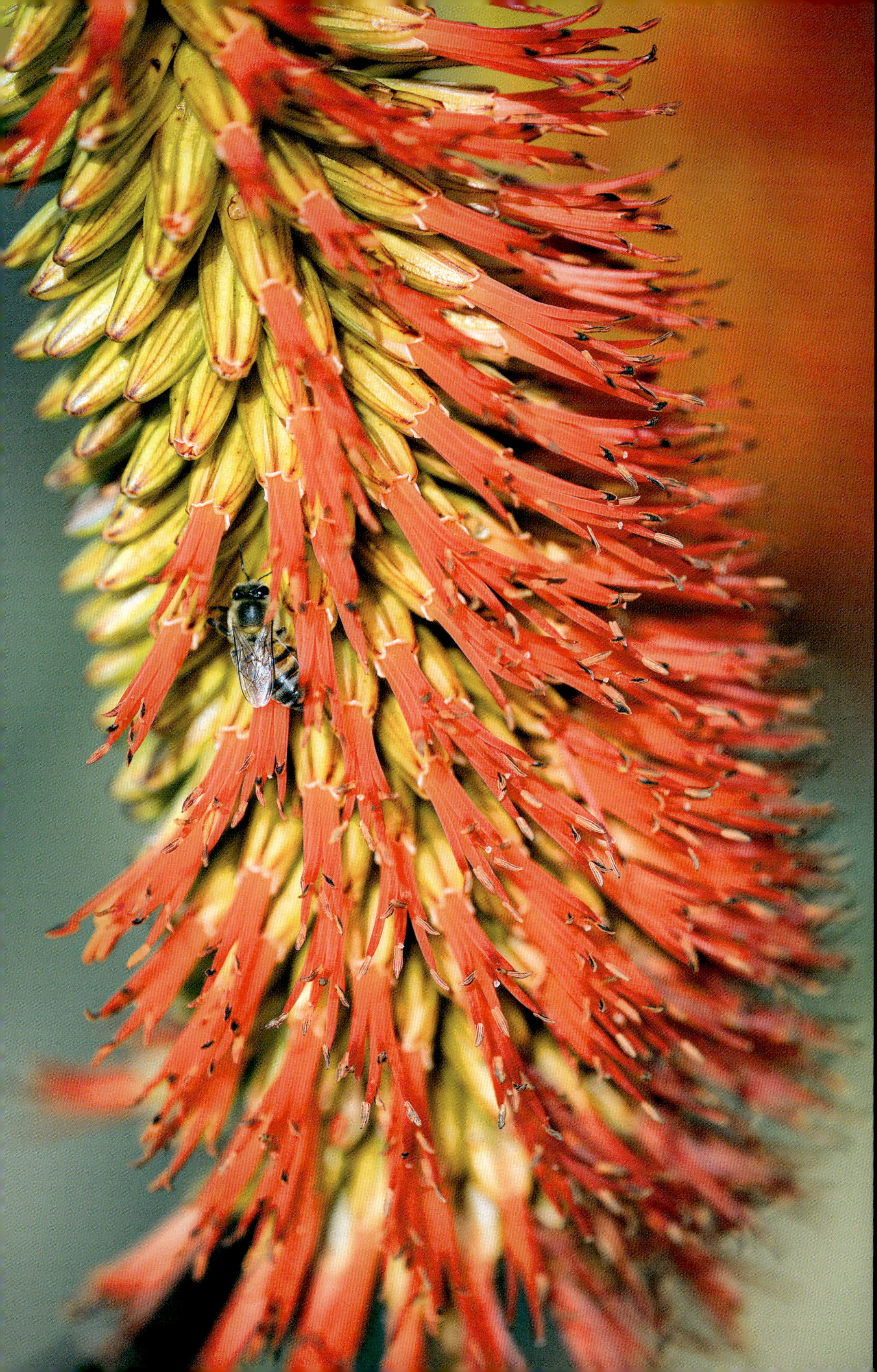

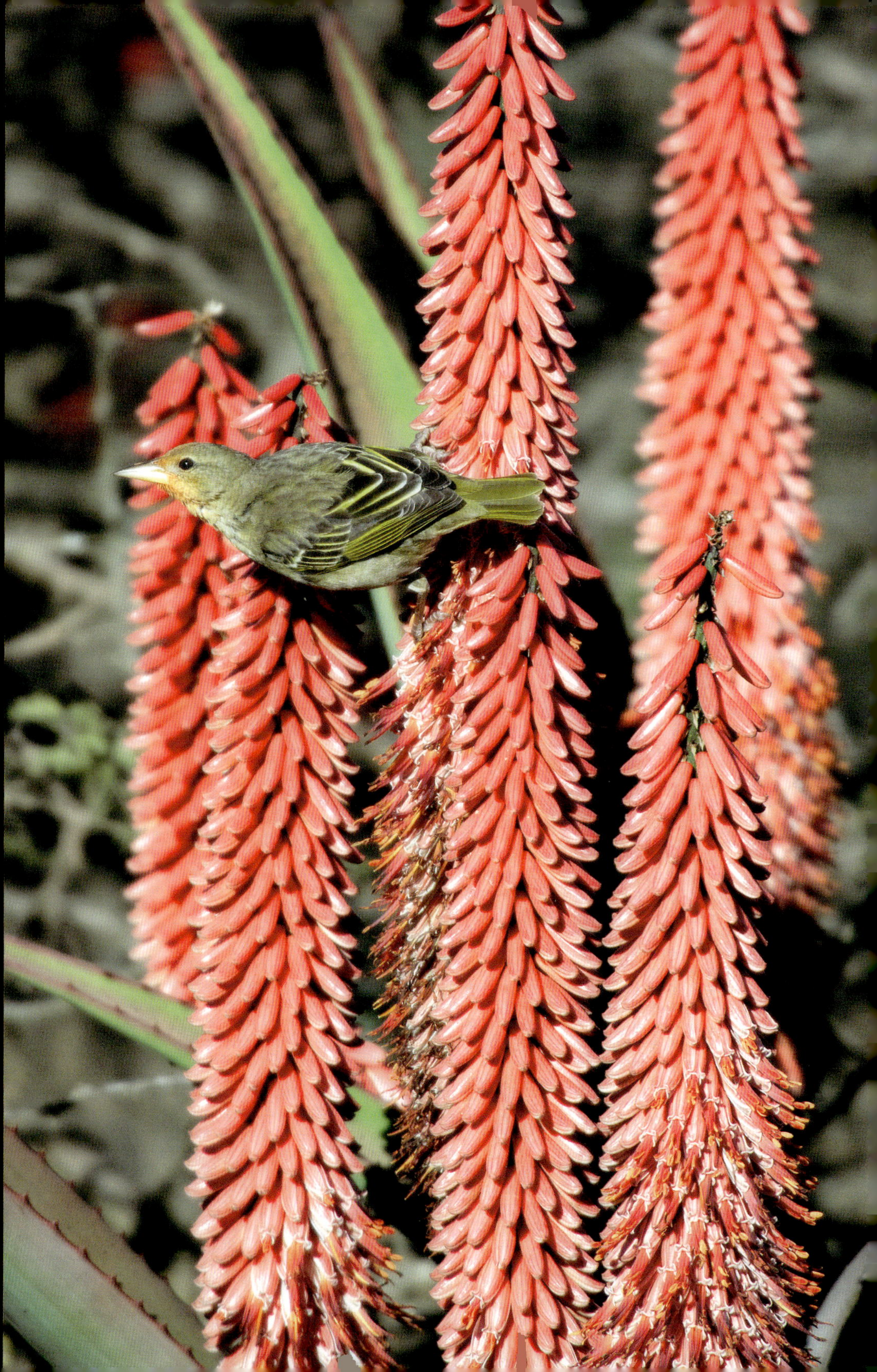

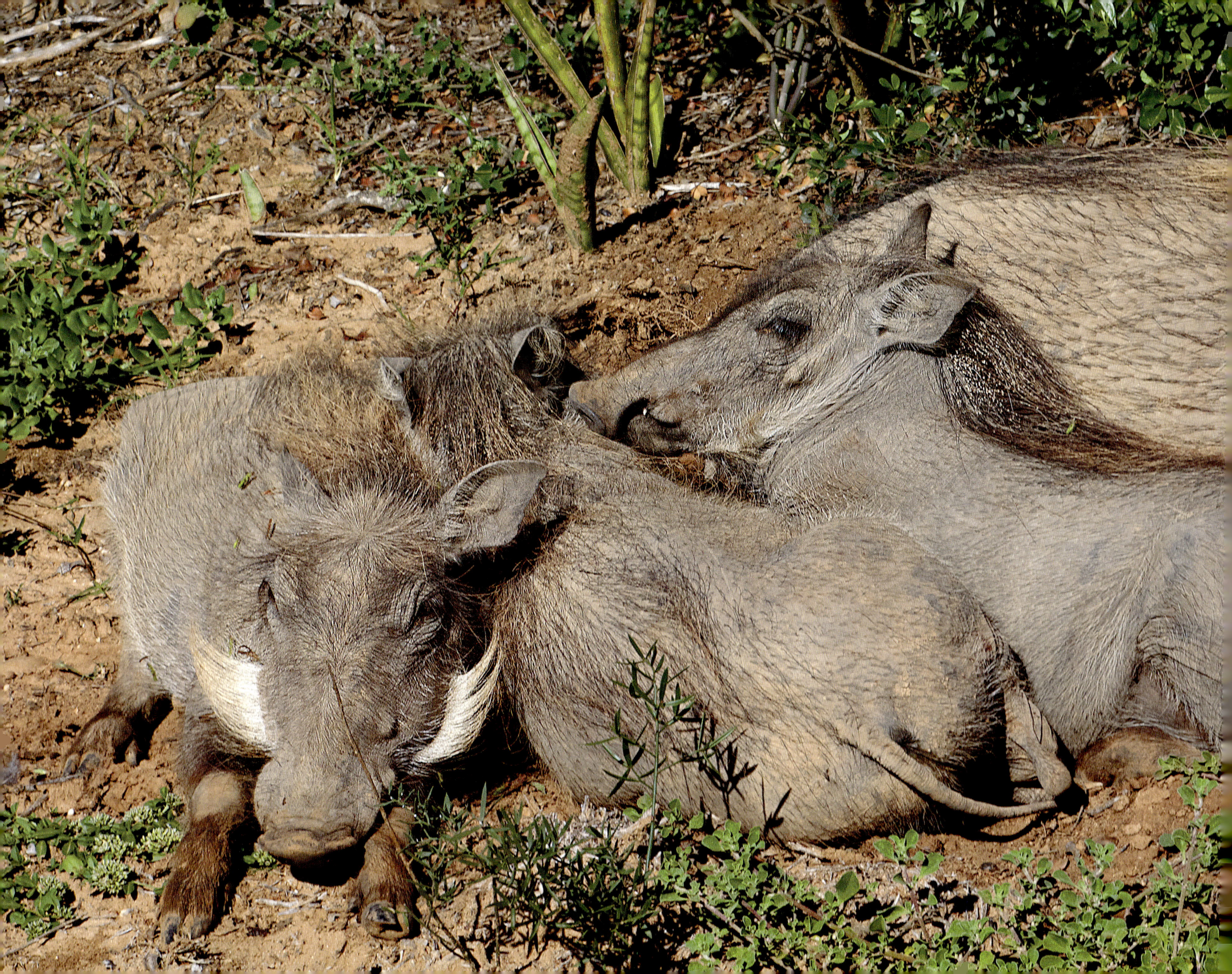

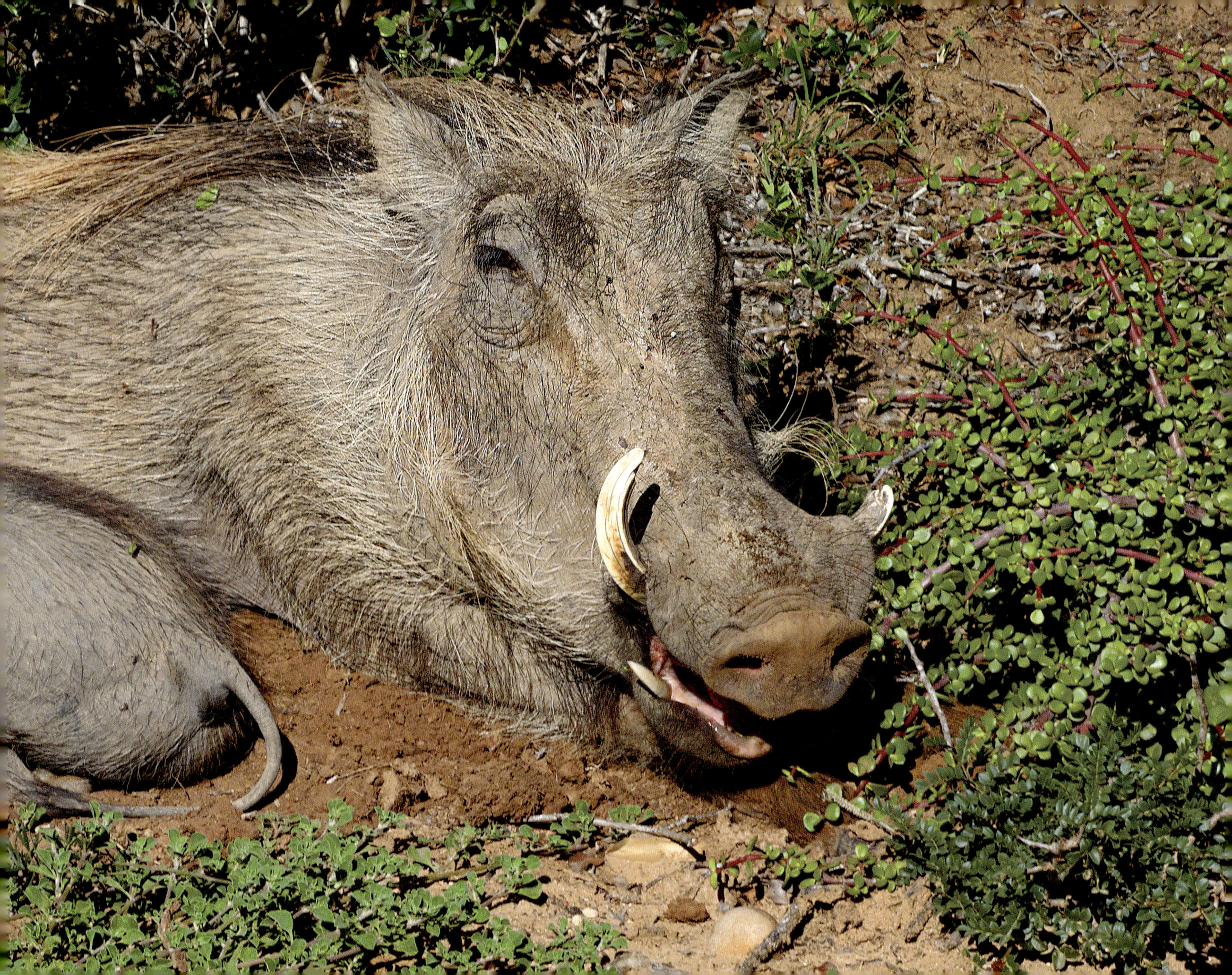

More about the Thicket Biome

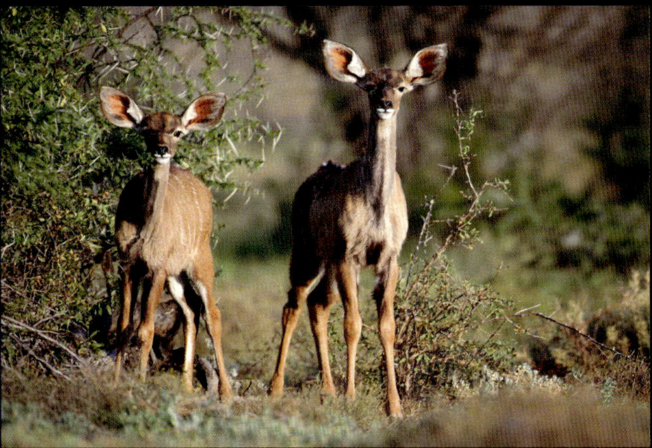

Kudu cows

Addo Elephant National Park is set deep within the dense valley bushveld of the Sundays River region in the Eastern Cape, South Africa. Here the Albany Thicket averages 2m to 3m in height and in its pristine state is almost impenetrable. Yet antelope like kudu thrive in this environment. They have the cover of the thicket, are not affected by the thorns or the dense undergrowth and derive enough moisture from the plants they eat, which means they are not dependent on drinking water during times of drought. The adult males lift their heads when running through the dense thicket and so lay their long spiral horns flat on their shoulders to avoid their horns getting stuck in branches.

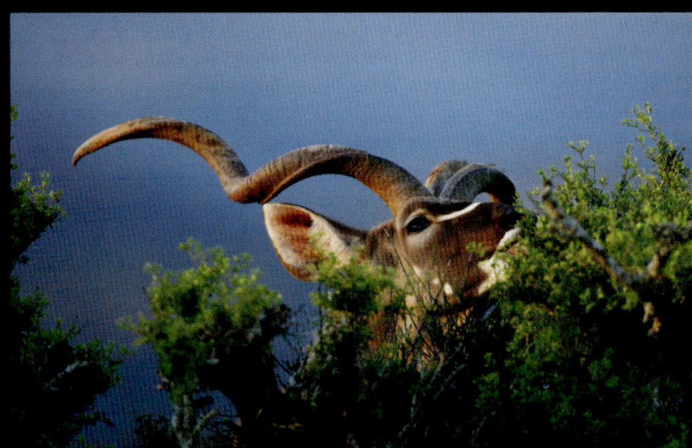

Kudu bull browsing

While elephant were widespread in southern Africa prior to European settlement, their numbers had dropped drastically in many parts by 1900. In the Eastern Cape, a population of around 130 animals remained. This was further reduced during the period 1919 to 1920 by a professional hunter who had been contracted by local farmers to kill the rest, since they were a nuisance and interfered with farming activities. By 1931 only 11 elephant remained in the thicket and the Addo park was created in an attempt to preserve that small population. However, they were still a problem, breaking out and causing havoc, until an elephant-proof fence was built in 1954 to keep them in the reserve. By December 2000 the population had grown to 324 and in recent years males have been translocated from the Kruger National Park to add to the gene pool. The population grew rapidly and today over 600 elephant roam the enlarged Greater Addo National Park. As they are browsers, the thicket vegetation suits them perfectly.

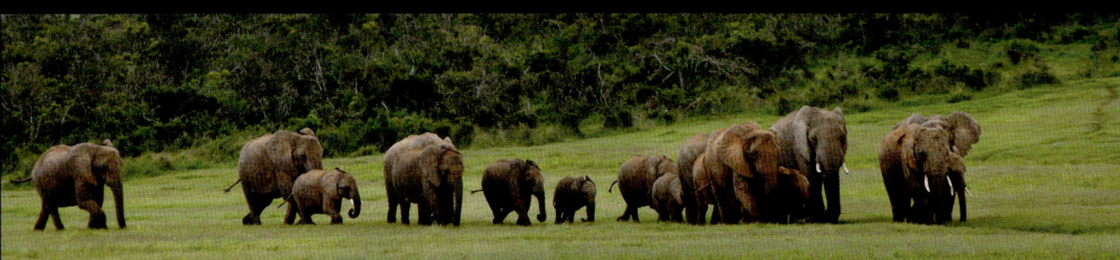

An elephant breeding herd in Addo Elephant National Park

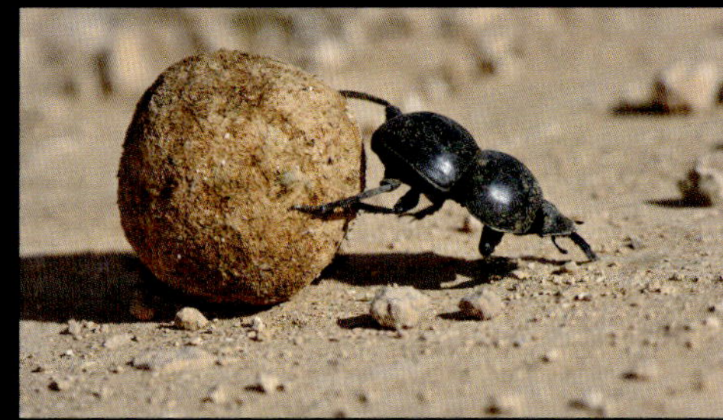

The flightless dung beetle rolling dung

Originally widespread, the flightless dung beetle, *Circellium bacchus*, is endemic to the Addo region and today in danger of extinction. Its vulnerability is exacerbated by a number of other factors, including the fact that its habitat is under threat by agriculture and other human activity. Another contributing factor is that it has a low breeding capacity and in addition, being flightless, it is slow to change location and redistribute itself. Its survival is strictly dependent on the dung of a number of vertebrates, including elephant and buffalo.

A wide range of growth forms, including all kinds of succulents, deciduous and semi-deciduous shrubs, dwarf shrubs, geophytes, annuals and grasses contribute to the plant diversity of this vegetation type. Among the more spectacular shrubs is the bitter aloe, *Aloe ferox*, with flowering candelabras rising above an array of ferociously prickly leaves. The orange to red flowers are visited by insects and birds for their nectar and have discernible protruding stamens.

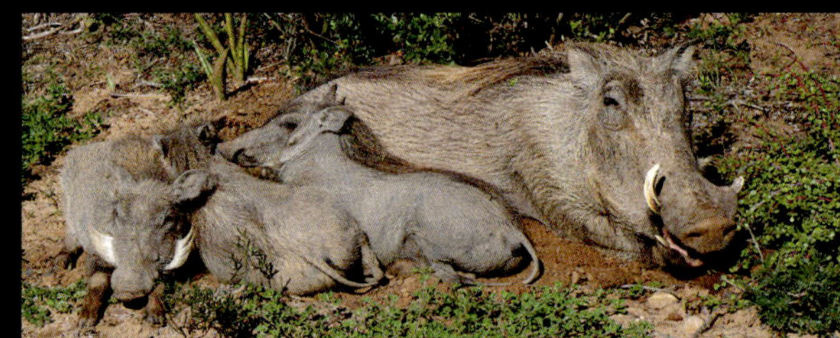

Warthog

Scramblers (climbers with thorns) and shrubs are common and amongst these the plumbago, *Plumbago auriculata*, and the Cape honeysuckle, *Tecoma capensis*, are particularly attractive and also used as garden plants. Like many thicket plants, they can survive in drought conditions. Sunbirds are attracted to the honeysuckle flowers for their nectar.

Elephant food, *Porticulata afra*, or colloquially also known as 'spekboom', is the most extraordinary and conspicuous plant in thicket vegetation. Evergreen and widespread, it is excellent fodder for browsers. The succulent leaves are bright green, fleshy and egg-shaped but as the tree ages the colour turns yellowish. In spring it is covered with masses of tiny pink flowers.

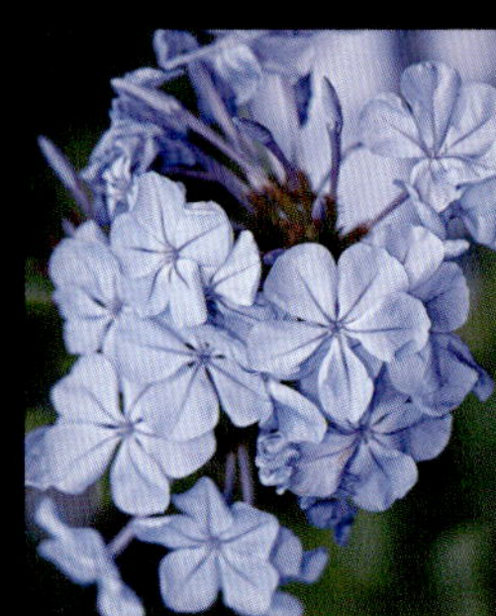

Plumbago auriculata

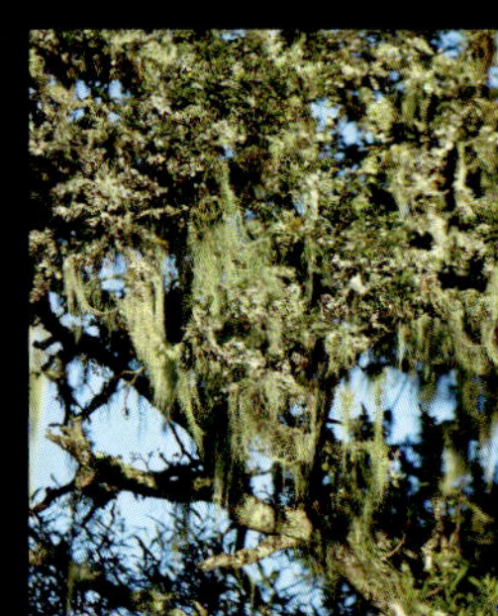

Old man's beard lichens

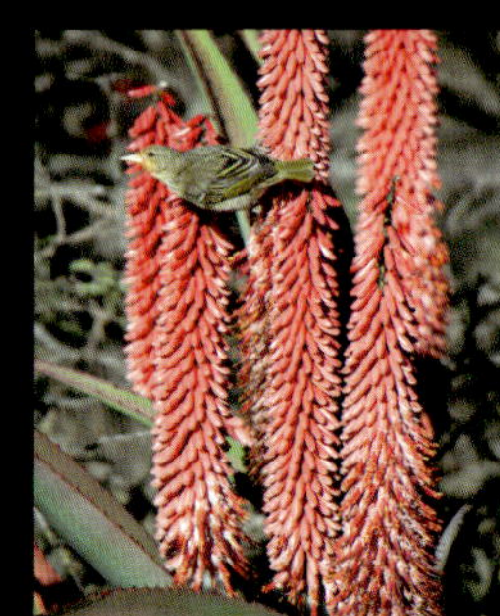

Aloe ferox

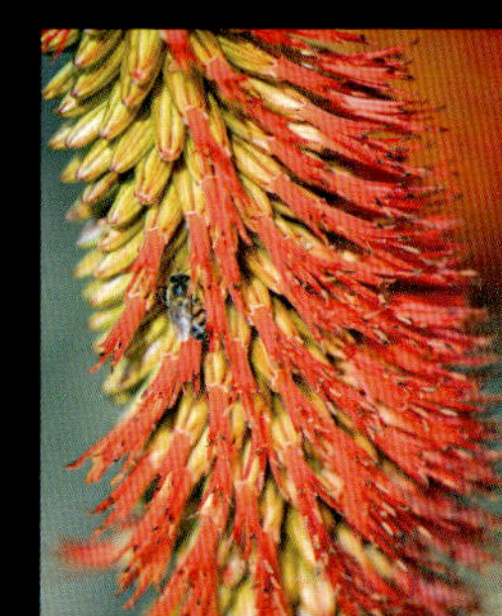

Aloe ferox close-up

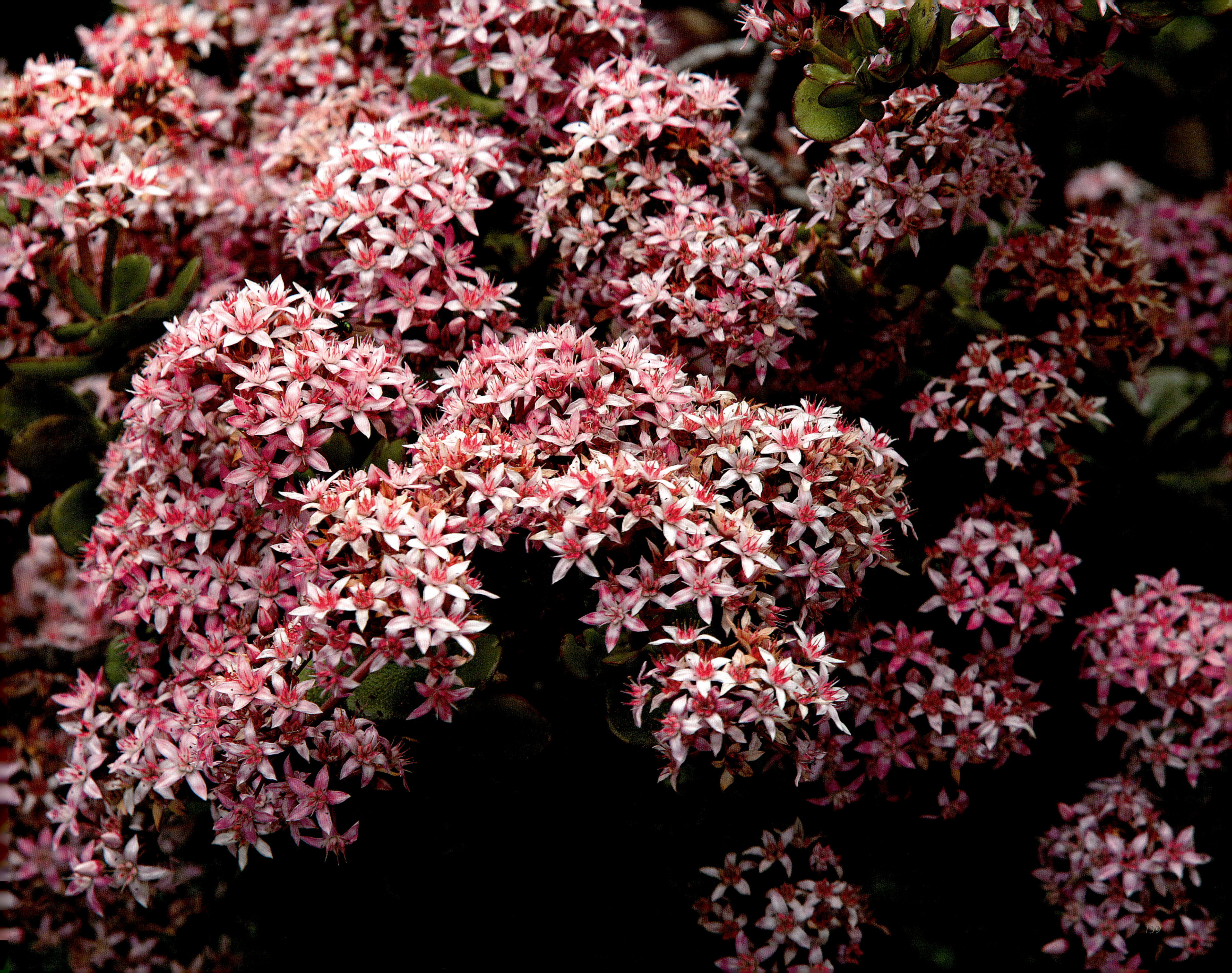

Forest

Places of silence and peace

Compared to other parts of the world, southern Africa is not richly endowed with indigenous forests. This biome is the smallest on the subcontinent and includes the indigenous evergreen and semi-deciduous closed forests of the coastal lowlands and escarpment slopes. In South Africa alone, less than 0.5% of the land surface is covered by closed canopy forest. In Mozambique, forest covers much more, but only 2.3% of it is under conservation.

Most indigenous forests in southern Africa are small, except for those of the Knysna area and other lesser-known inaccessible ones in Mozambique. These forests occur in patches such as those along the Drakensberg and Zimbabwe-Mozambique escarpment, and on the KwaZulu-Natal and Mozambique coastal plains. It is estimated that natural forests occur in at least 16 000 fragments within southern Africa and hundreds of these are tucked into the many valleys and gorges of the mountain ranges.

Forests need lots of rain, more than 525mm in the winter rainfall region and more than 725mm in the summer rainfall areas. Rainfall is a determining factor for the forest's potential to exist, but forests may occur from sea level up to an altitude above 2 000m, however, they cannot tolerate frost and are sensitive to fire. Indigenous forests usually do not readily burn, but fire around the fringes can prevent them from expanding.

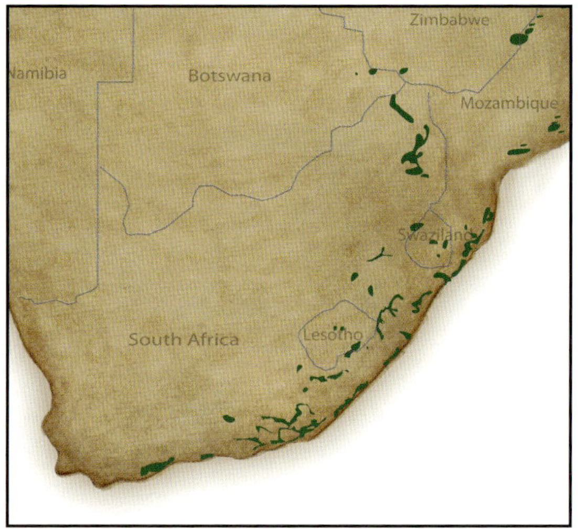

The forest canopy is continuous and mostly evergreen, which prevents light from penetrating down to the forest floor. In the strata below the canopy, ferns, lianas and epiphytes are common, but a distinct ground layer of herbs is usually absent. Grass is altogether absent, except for a very few shade-loving species.

The wet forests of East Africa extend down through parts of Mozambique along the Indian Ocean coastal belt as far south as the Kei River in South Africa. They are clearly sub-tropical with a high annual summer rainfall of between 820mm and 1 200mm. Among the different forest types are mangrove, swamp, sand and riverine forests.

Indigenous forests attract attention, probably because of their special ambience or since they are often so inaccessible, rare and beautiful.

Good places to experience forests

Easily accessible and larger forests of South Africa are confined to the coastal provinces of KwaZulu-Natal, the Western Cape and the Eastern Cape. Around Storms River Village in the Tsitsikamma Forest and along the Garden Route there are various opportunities to walk along forest trails and see the century-old yellowwood trees.

One of South Africa's fastest-growing eco-adventures is the canopy tour that presents an opportunity to enjoy a bird's eye view of the forest scenery in one of the indigenous forests located in Magaliesberg, Tsitsikamma, Karkloof, Magoebaskloof, Drakensberg or Malolotja (Swaziland).

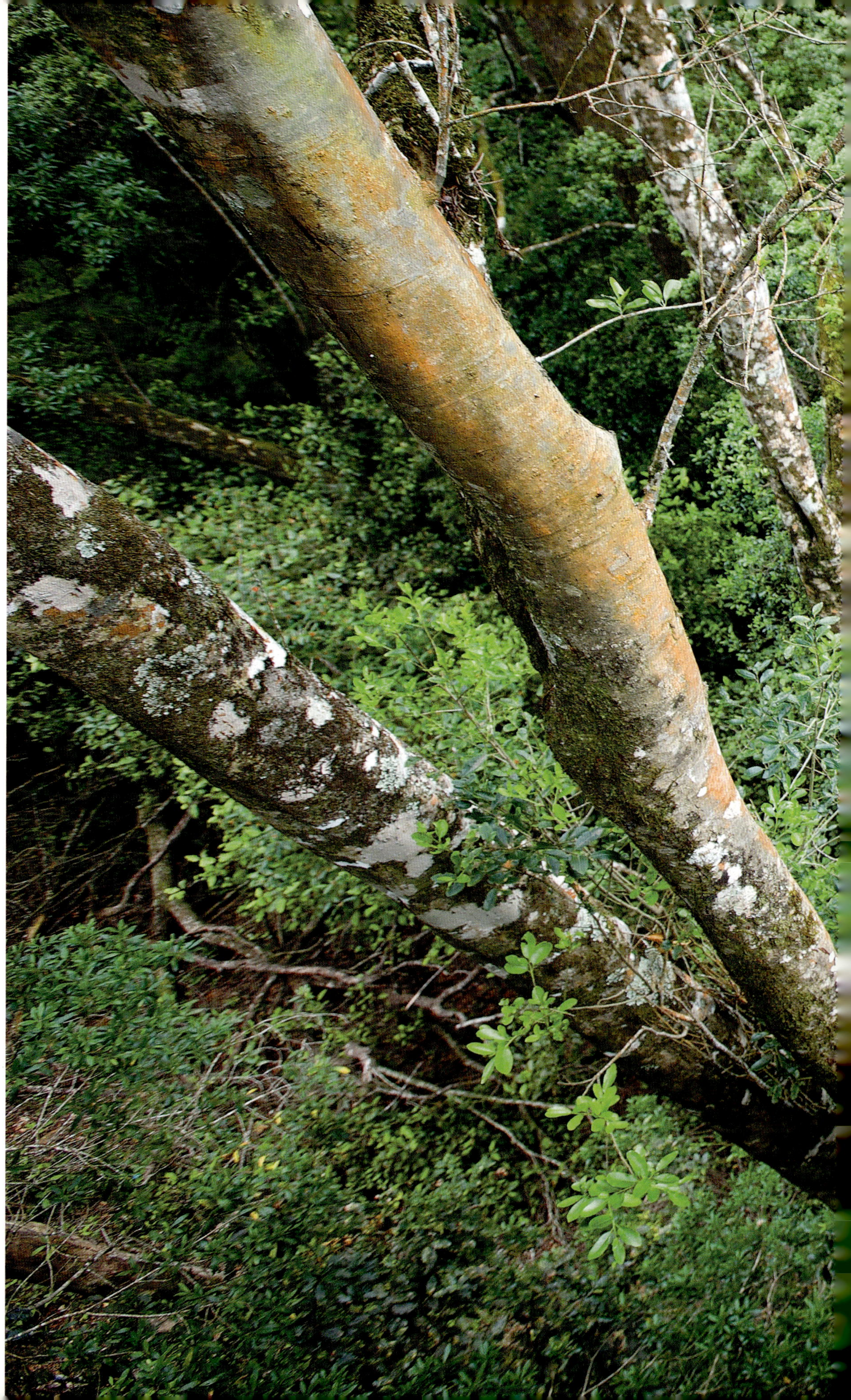

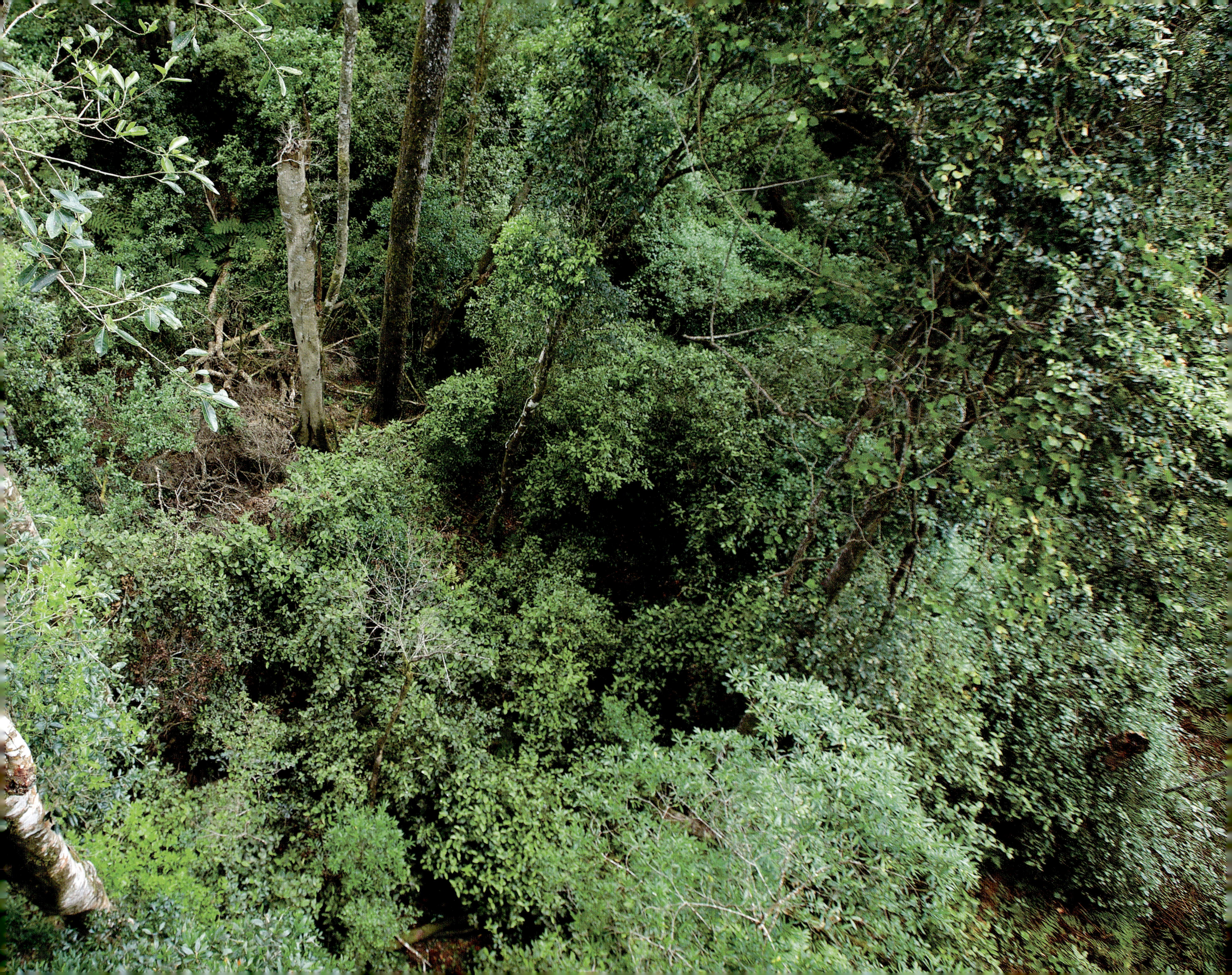

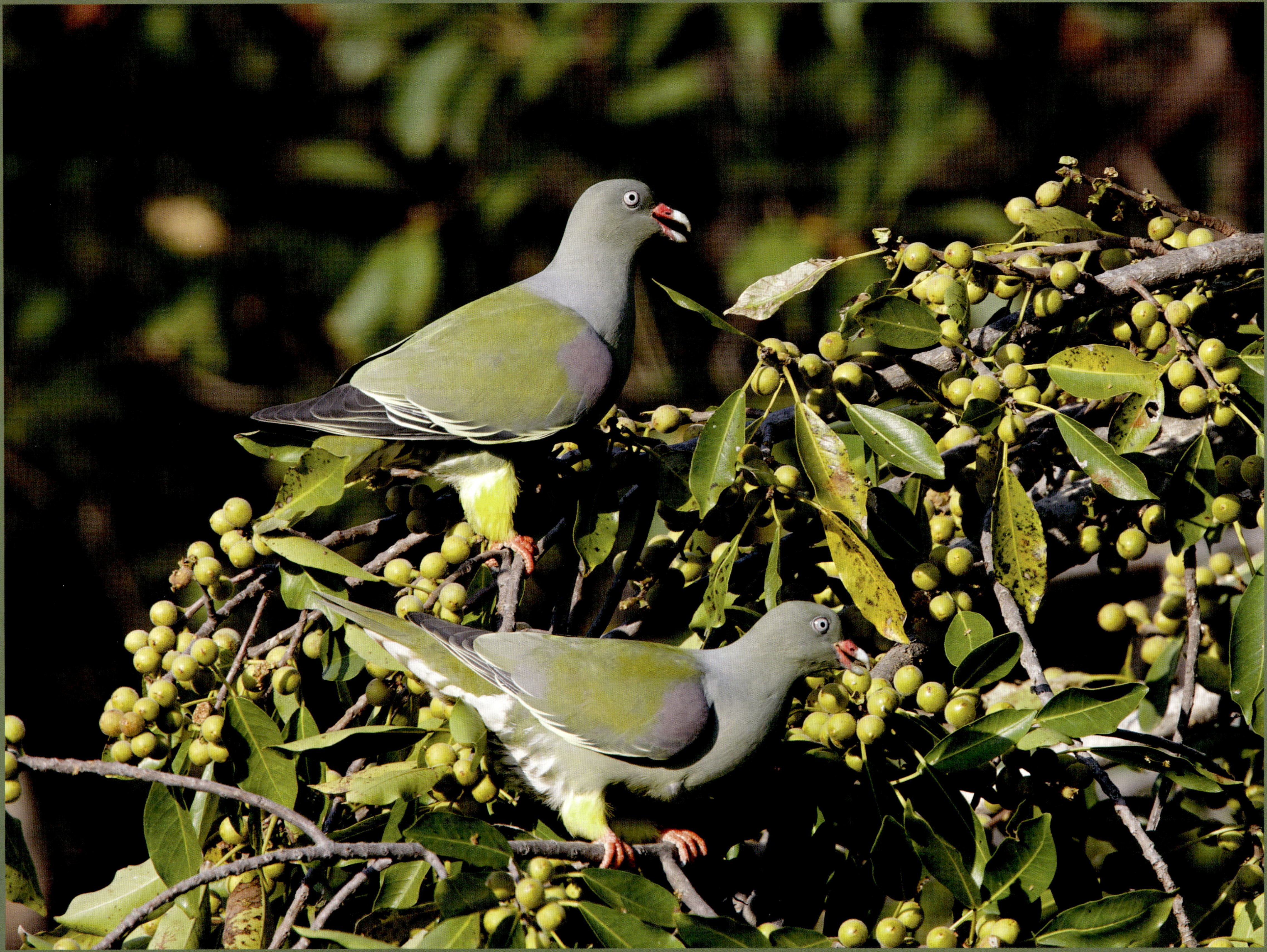

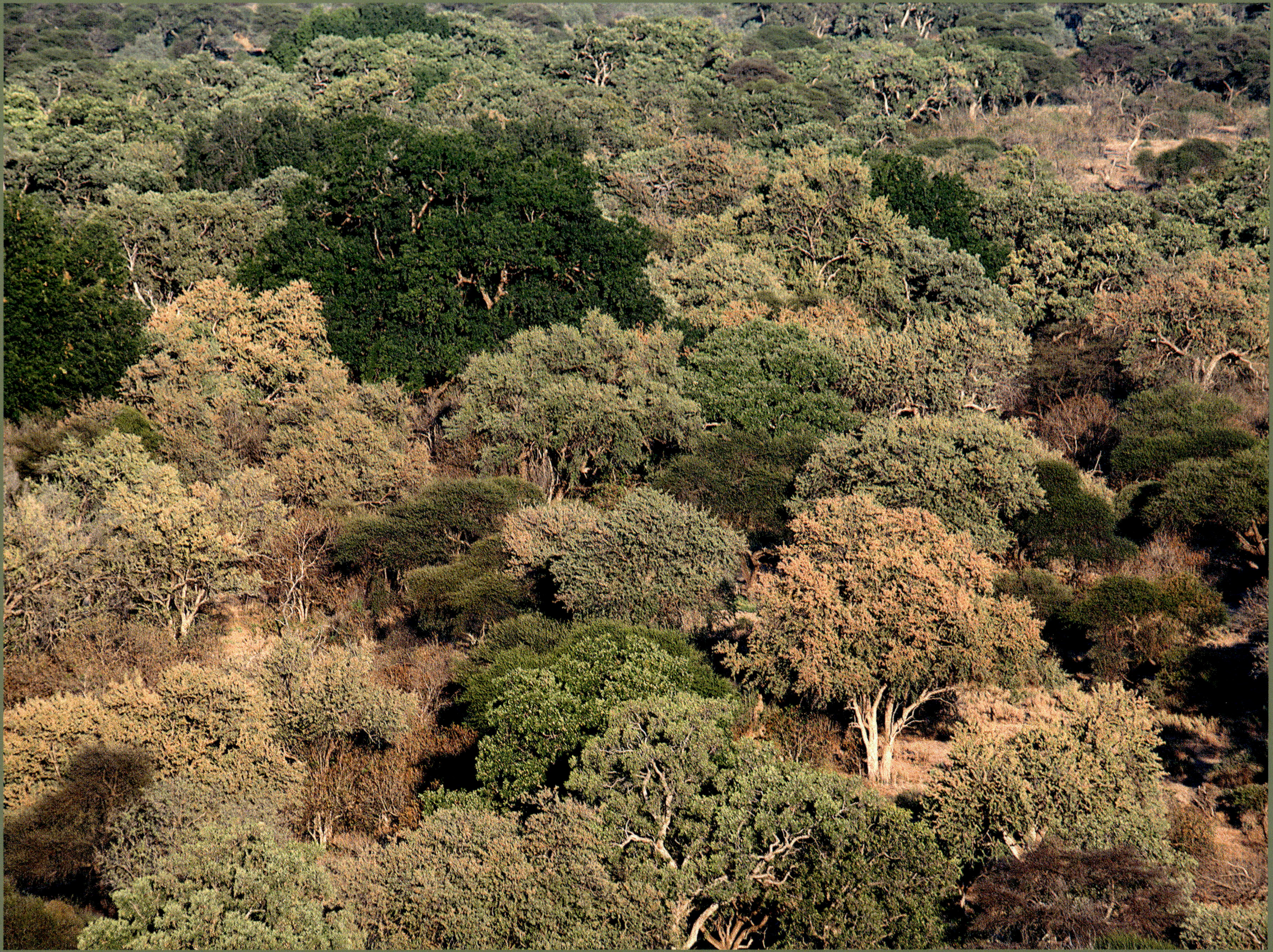

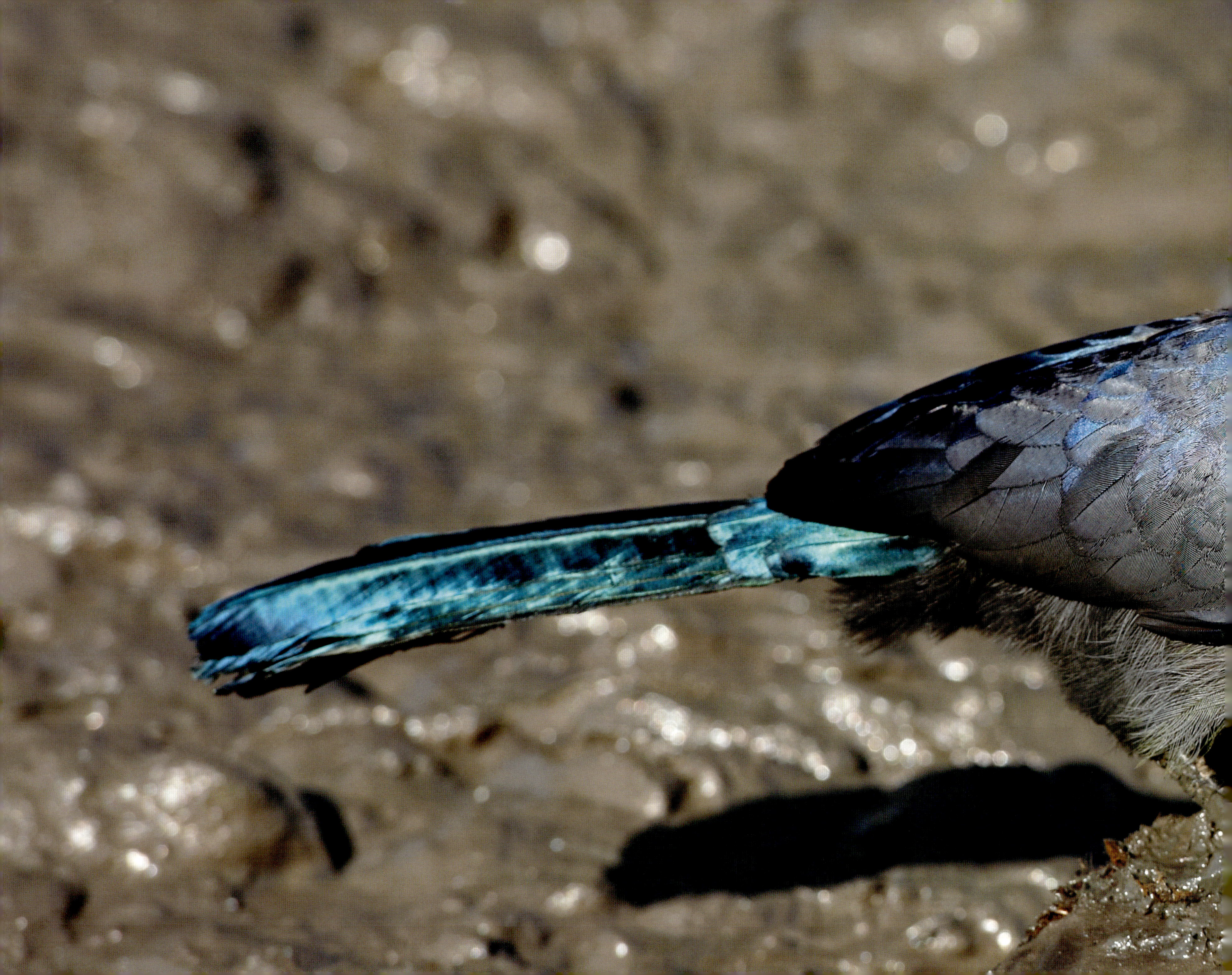

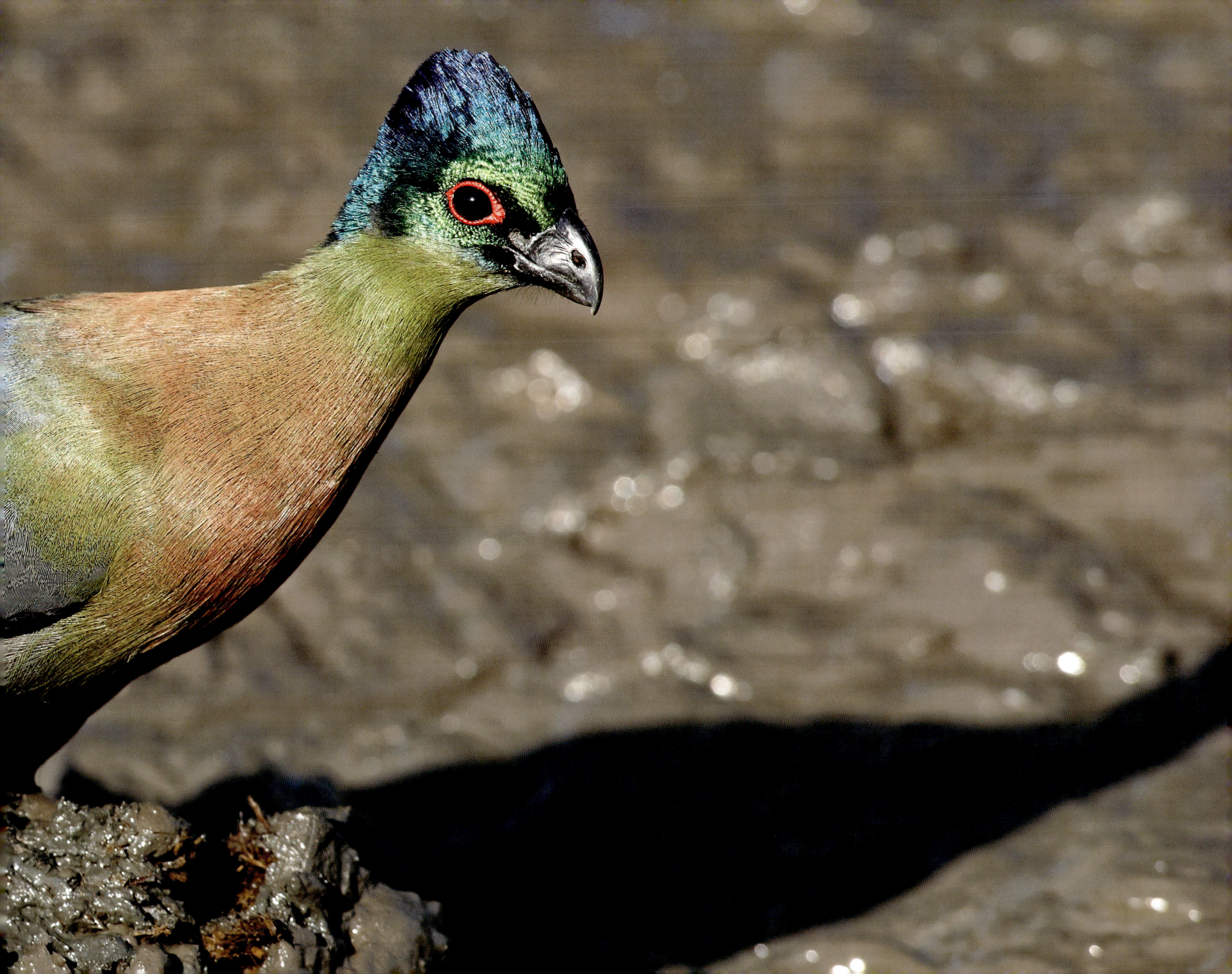

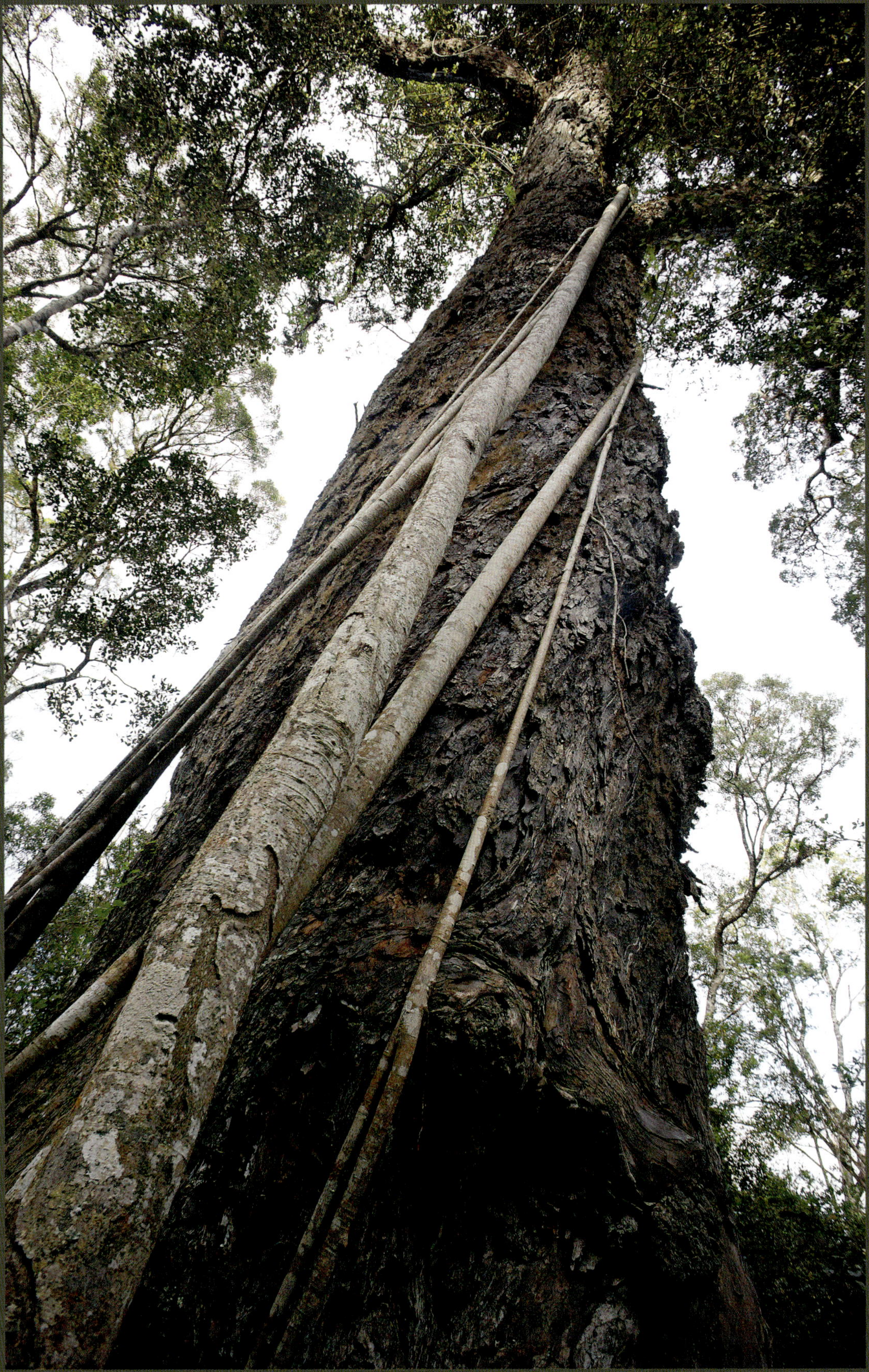

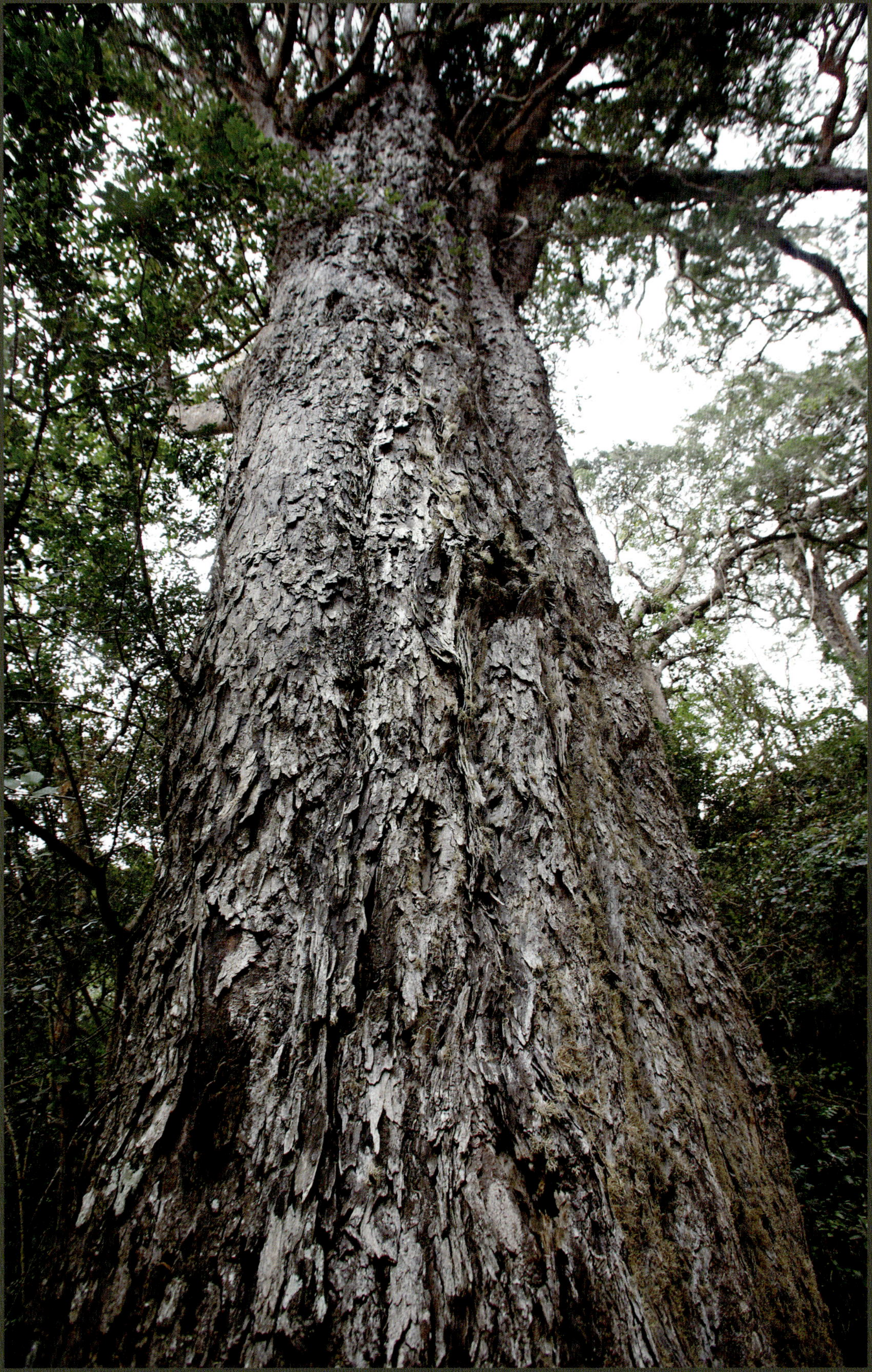

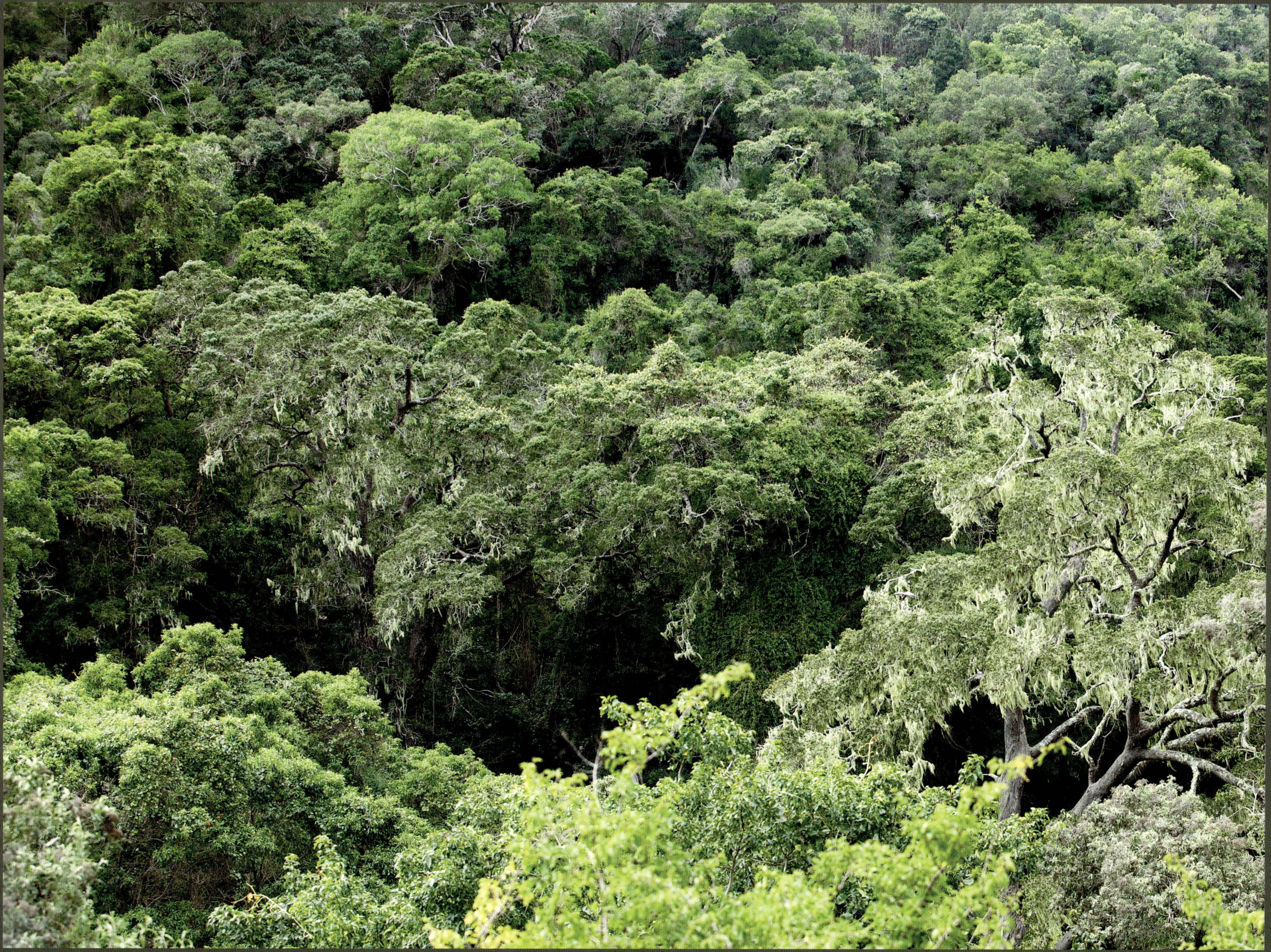

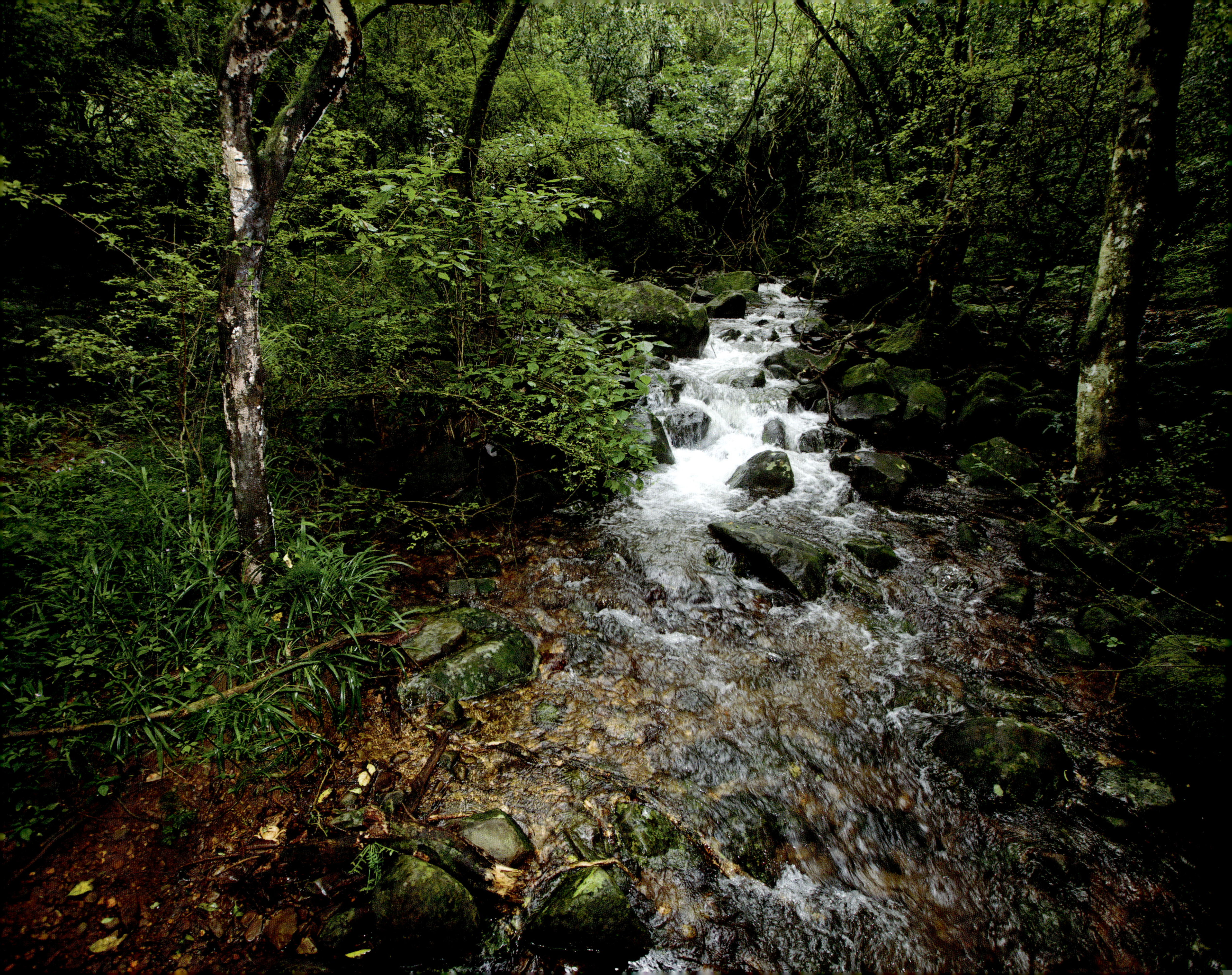

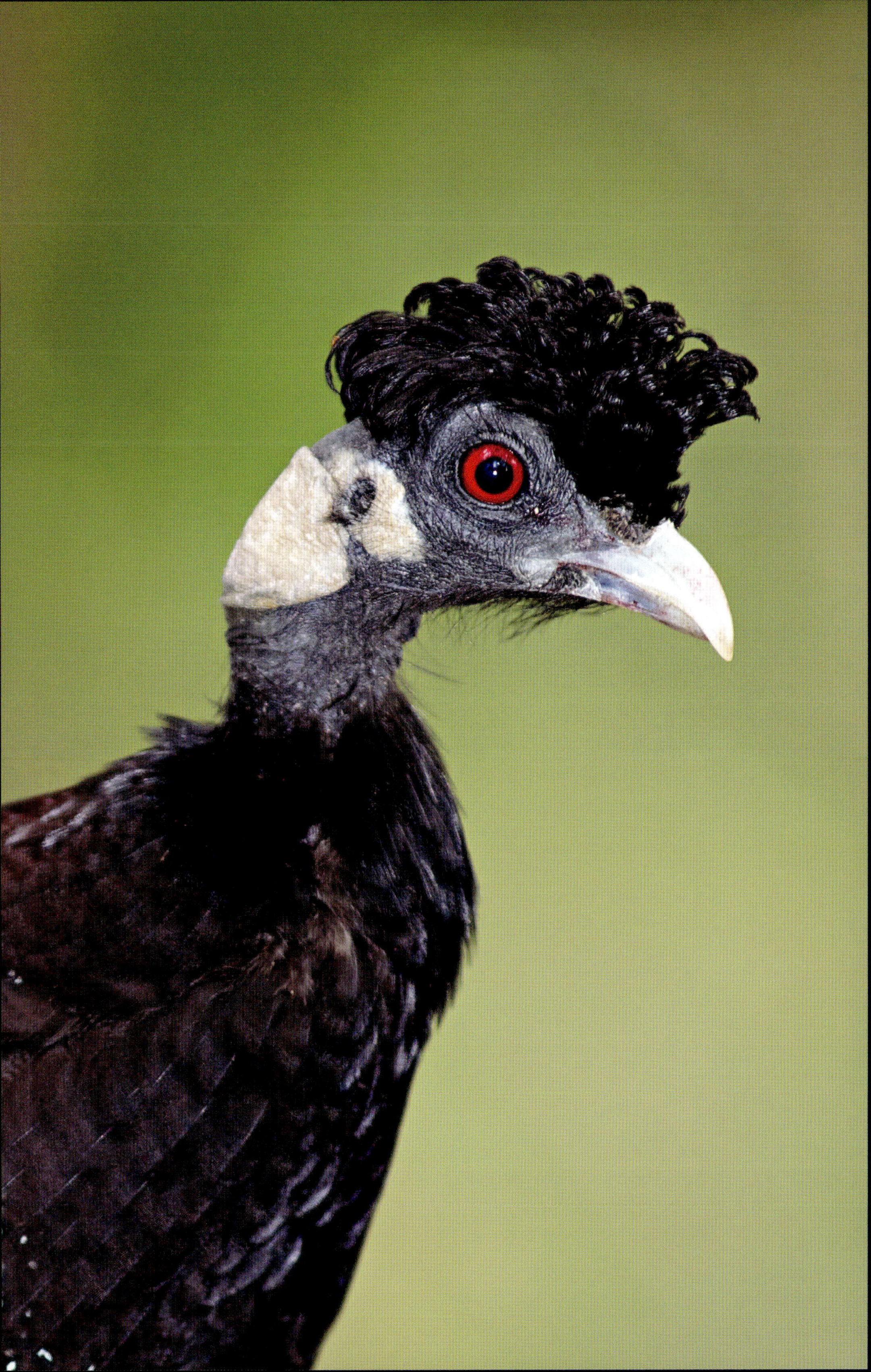

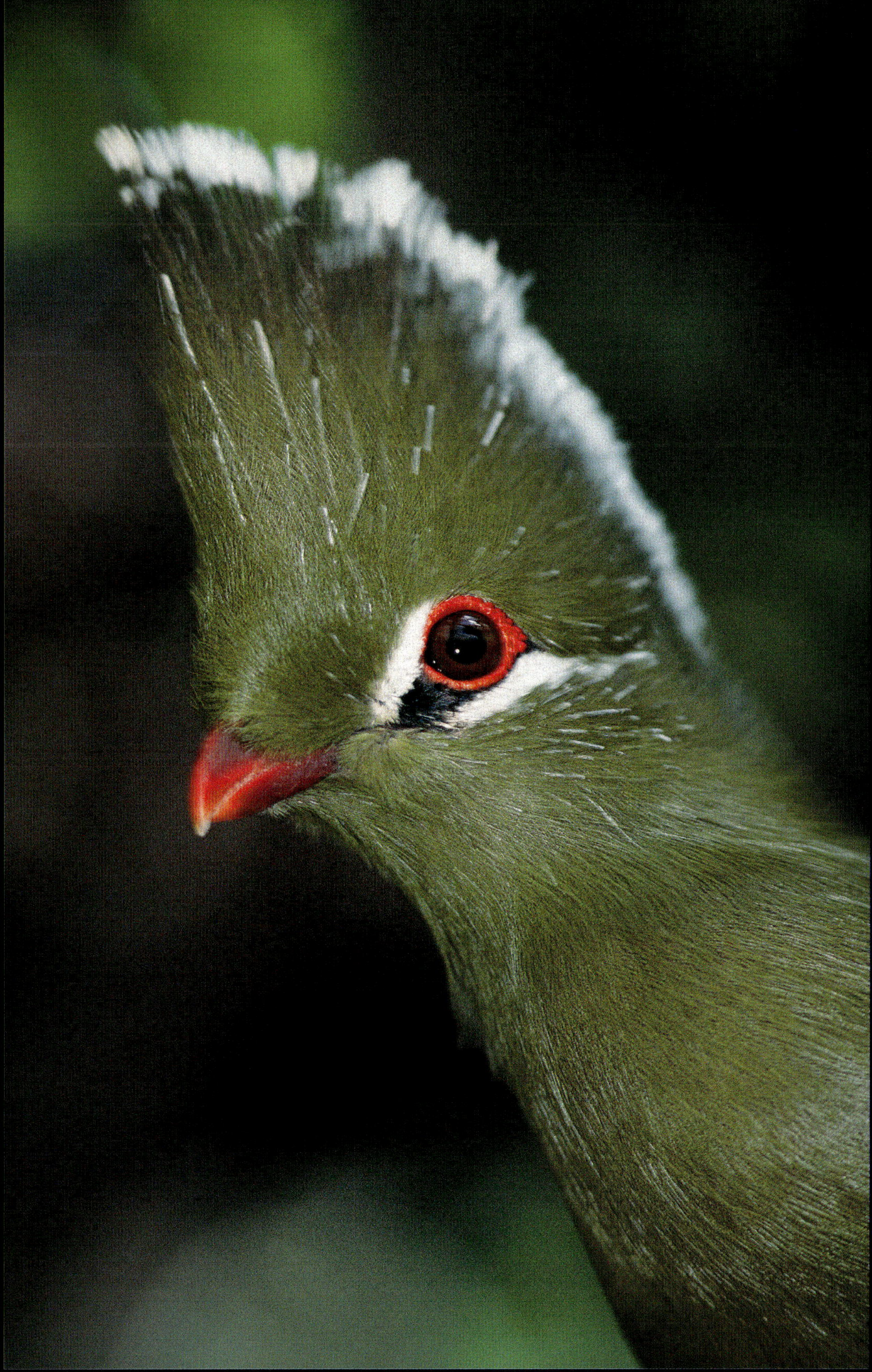

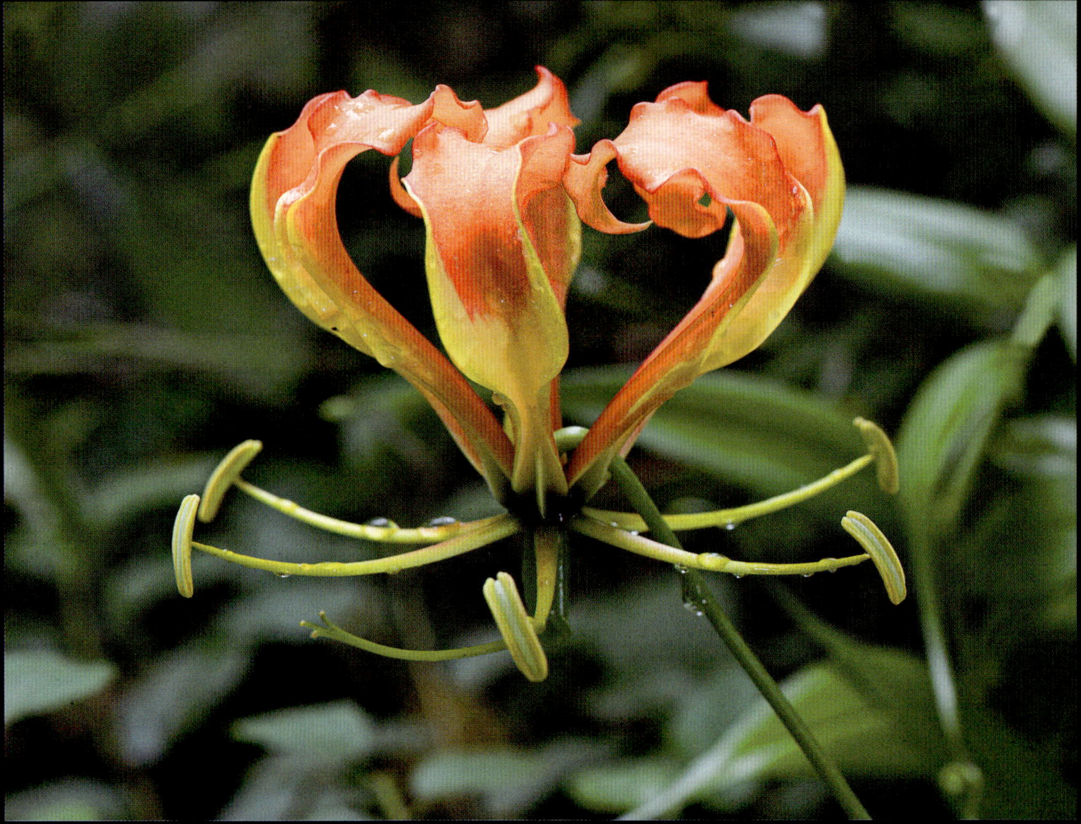
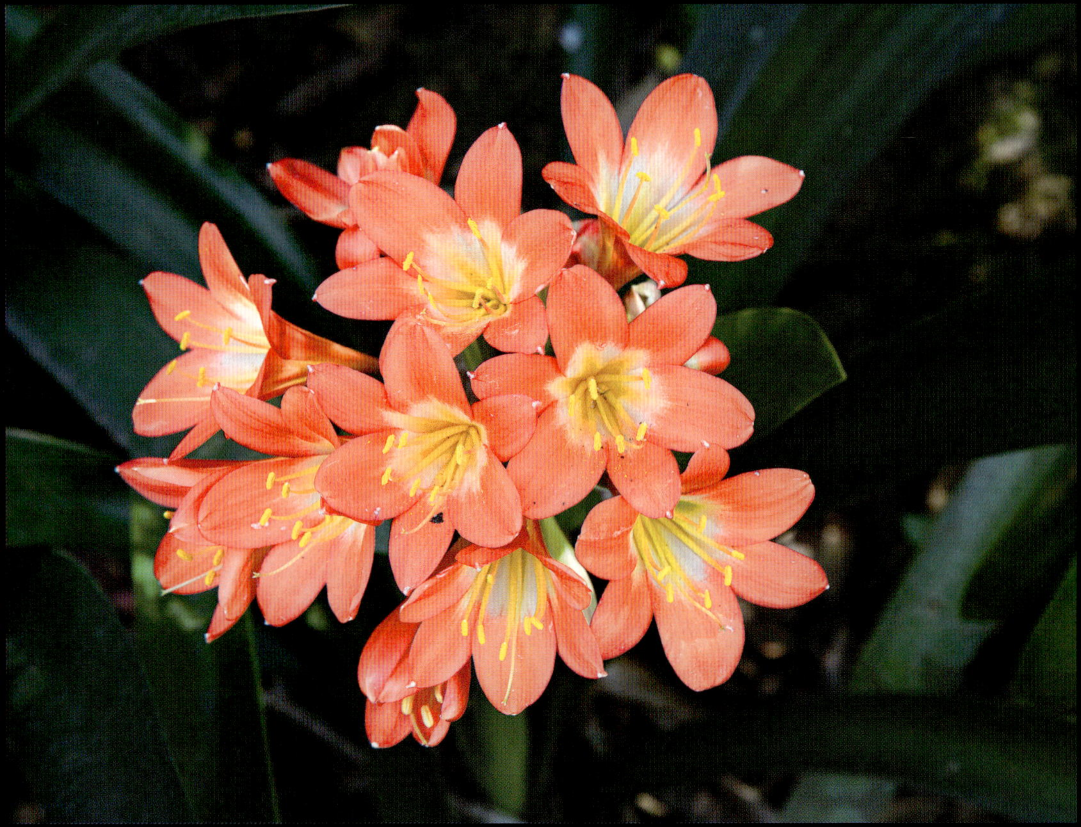
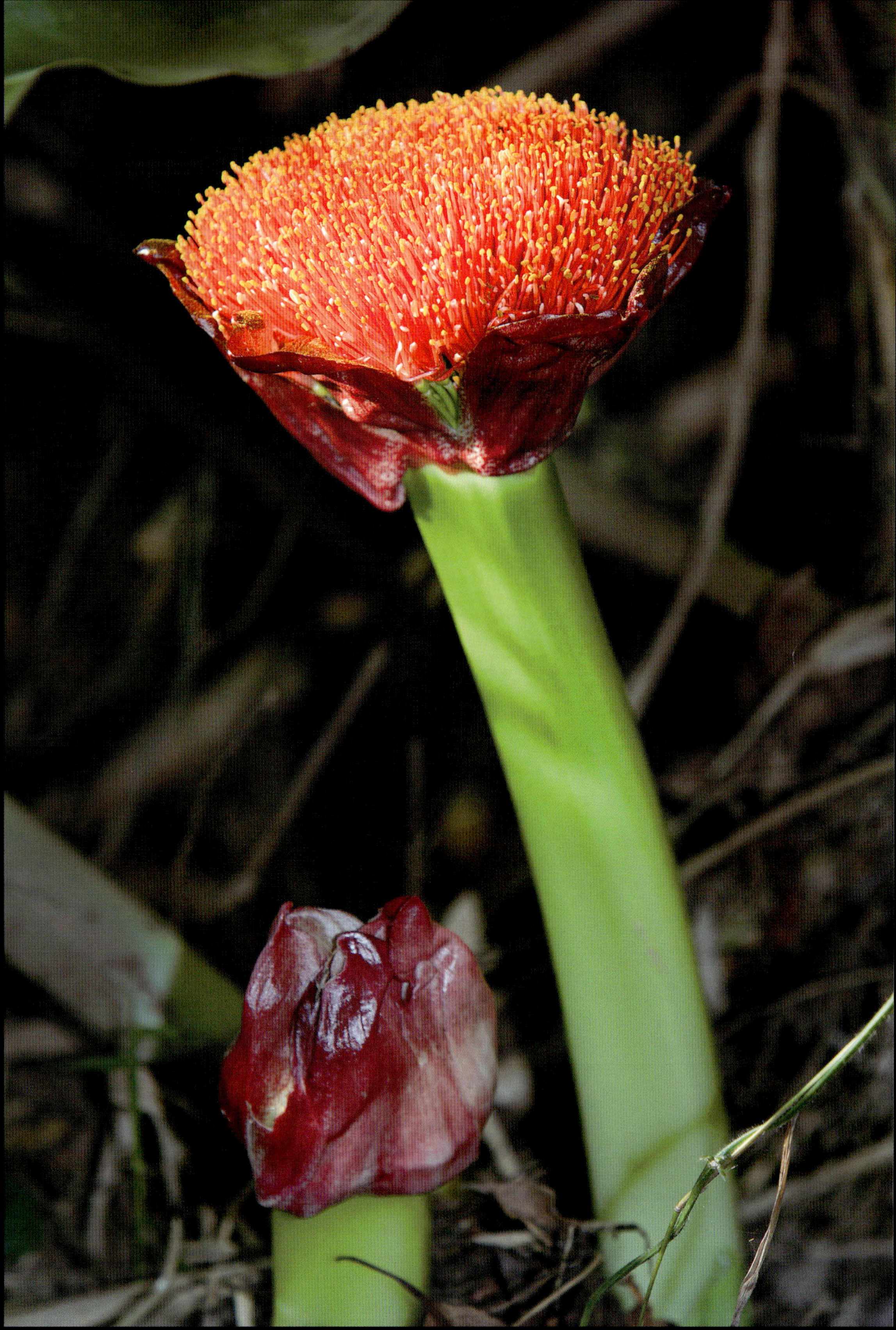

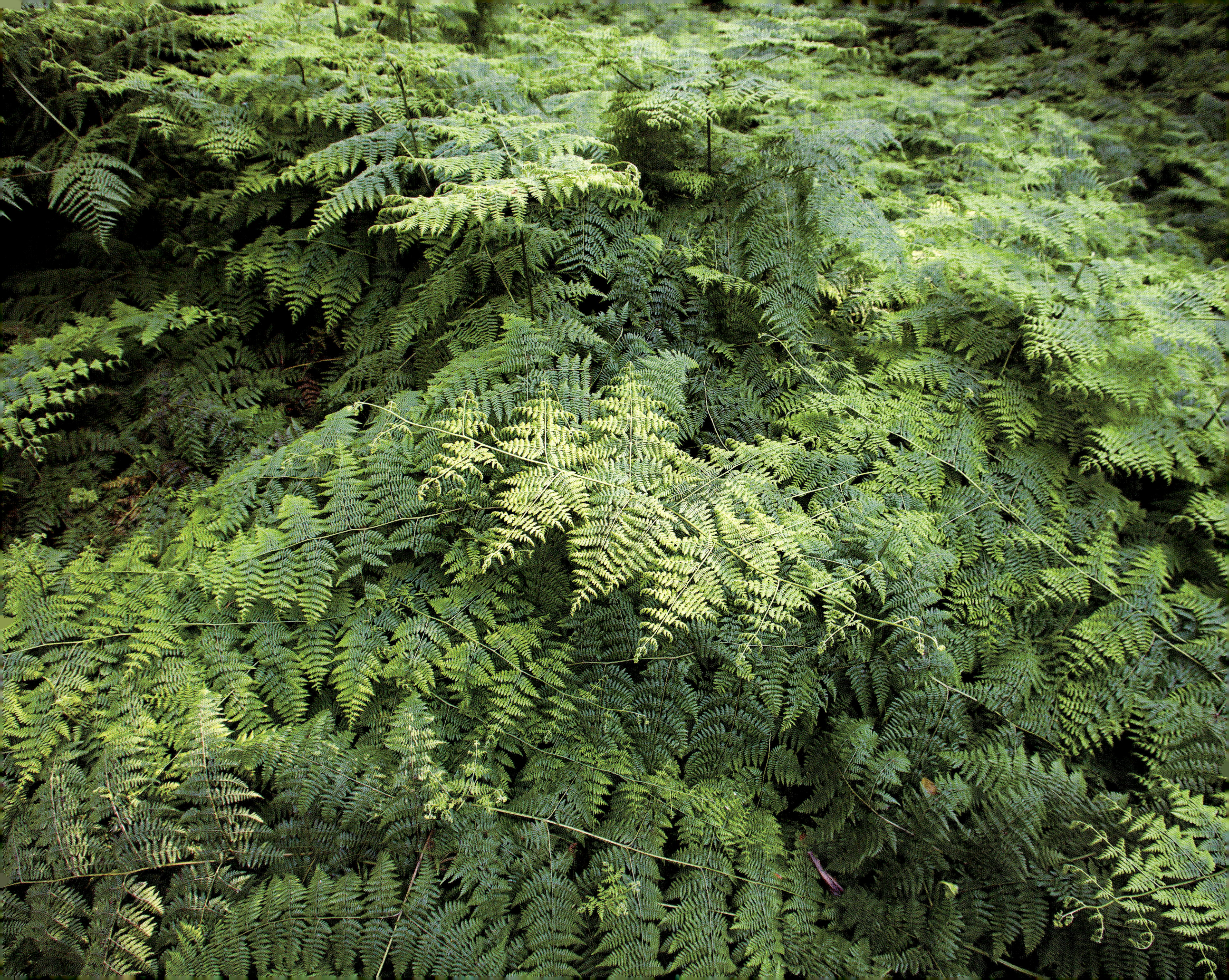

More about the Forest Biome

An indigenous forest is a complex ecosystem comprising the forest soil with all its biological components, including microbes and microfauna, animals of all groups and, of course, plants. The larger mammal species associated with forests include antelope such as bushbuck, red and blue duiker, the tiny suni of the sand forest, bushpig, porcupine, leopard, badger and several smaller ones.

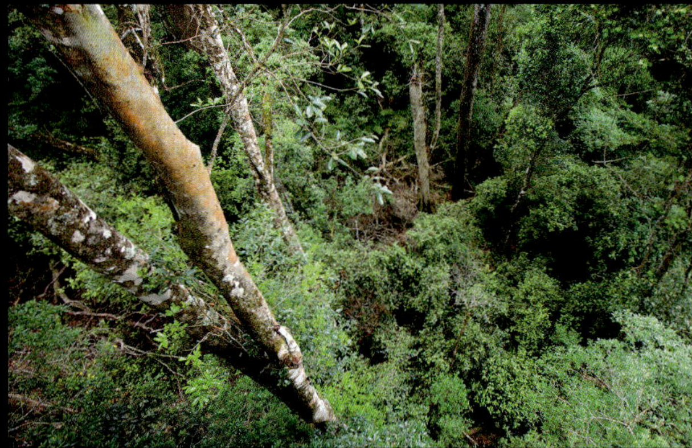

The vegetation strata of a forest

The different vegetation strata can clearly be recognised when looking in on the forest from above. Even in this small forest clearing, light penetrates far enough down for the shrubs of the middle strata to get enough light for healthy growth.

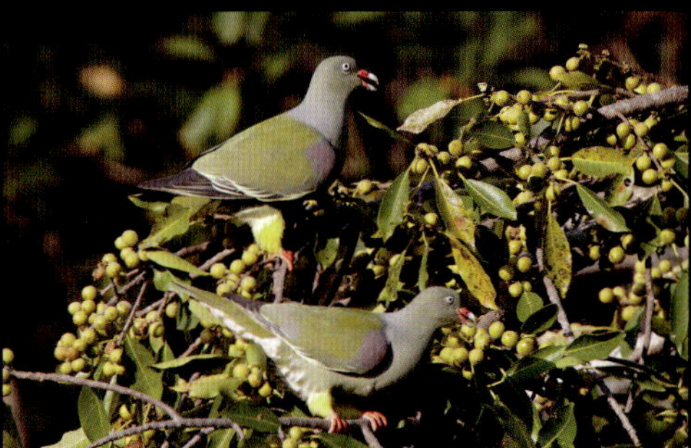

Green Pigeons

In closed evergreen woodlands of South Africa, the fairly common Green Pigeon is a familiar sight on berry-bearing trees when the fruit is ripe. It favours wild fig trees but others like the jackal-berry, tree-fuschia, milkwood, yellowwoods and others are also visited in the fruiting season. Foraging in small groups, they clamber parrot-like on the slender fruit-bearing branches in the canopy, never feeding on the ground. Like all forest bird species, they have a loud voice. Interestingly, they are among the comparatively small group of monogamous bird species, forming lifelong pair bonds.

Sand Forest with its deciduous canopy is quite rare in South Africa and occurs only in the far northern regions of KwaZulu-Natal (Mkuze, Phinda, Ndumu and Thembe Game Reserves) and in southern Mozambique, growing on ancient sand dunes. In South Africa these forests are known simply as Sand Forest, while in Mozambique they are known as Licuati Forest. Sand forests are thought to be remnants of coastal dune forests which have been separated from the ocean for more than a million years as the shoreline has shifted slowly eastwards over the millennia.

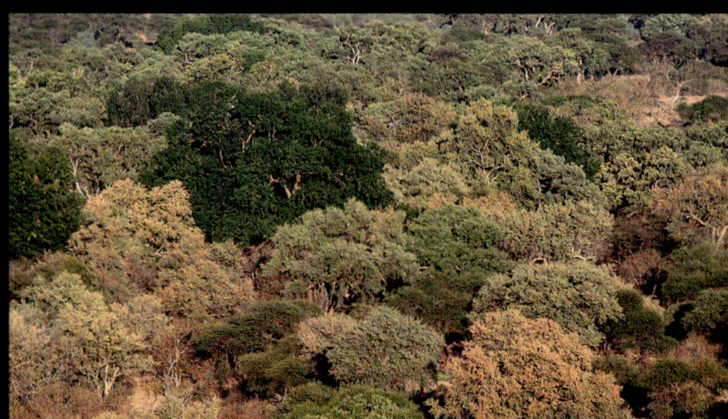

Sand forest

The Purple-crested Turaco is associated with woodland where it integrates with forest or coastal forest, but it is less of a forest species than its relative, the Knysna Turaco. It forages in the canopy of fruiting trees, perching on the end of branches to pick the fruit directly with its beak. When the turaco flies, it briefly exposes its breathtaking crimson primaries, which are not visible when the wings are folded.

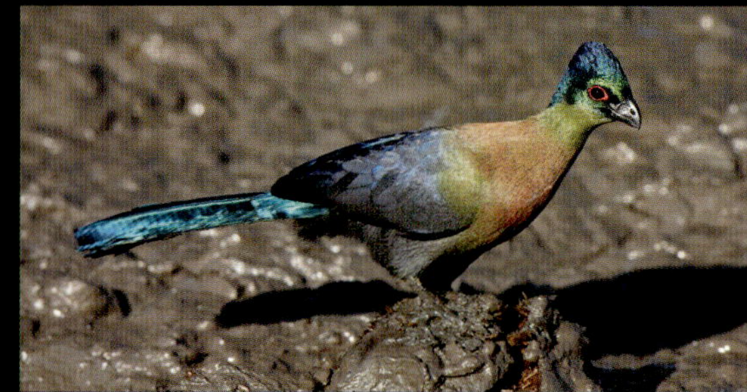

Purple-crested Turaco

Another typical forest bird is the shy Crested Guinea Fowl. It lives in small groups and forages on the forest floor where it is perfectly camouflaged in the dappled shade.

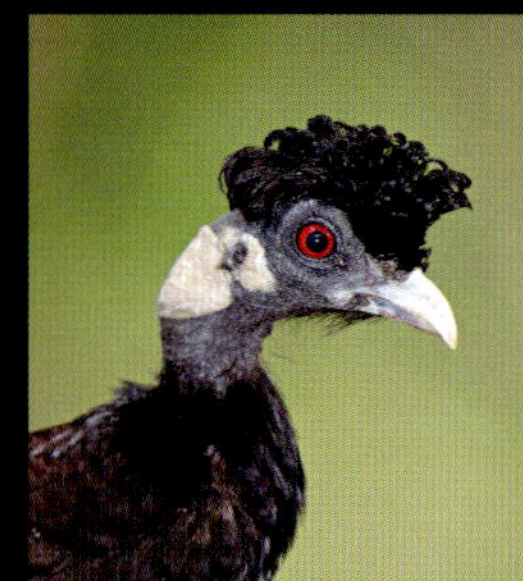

Crested Guinea Fowl

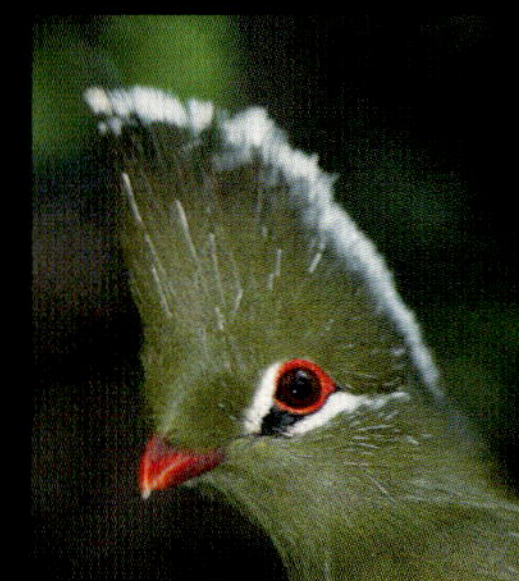

Knysna Turaco

Due to the high rainfall in forested areas, streams abound. In the diffused light, the conditions are perfect for the growth of mosses, both on rocks and on the humus-enriched ground. Mosses are small soft plants forming close green carpet-like growths on the forest floor, commonly growing close together in clumps or mats in damp shady areas. They do not flower and therefore do not produce seeds, but primitive spores, and have only simple leaves. Liverworts are closely related and also form dense mats on the forest floor. Their leaf-like structures are flat and ground-hugging.

Where there are marshy patches or trickling streams and enough light penetrates to the forest floor, many different kinds of ferns find a perfect habitat. Bracken ferns have a profuse growth form and often occur at forest margins, serving as a fire barrier and protecting the forest.

Although not strictly confined to the deep forest, the *Scadoxis puniceus* or blood lily, also called snake lily, is often found at forest margins where enough light penetrates to the

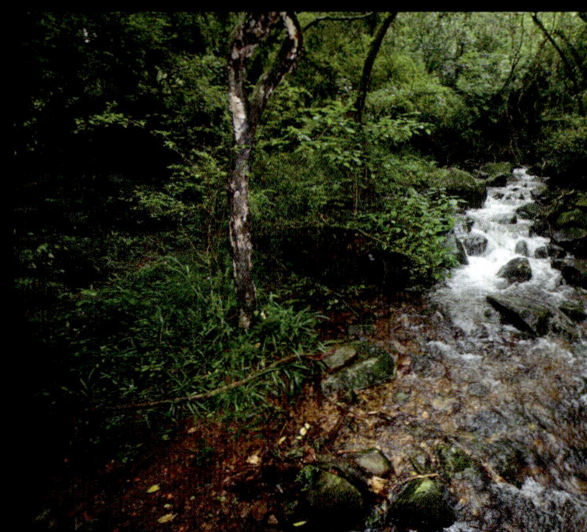

Moss covered forest floor

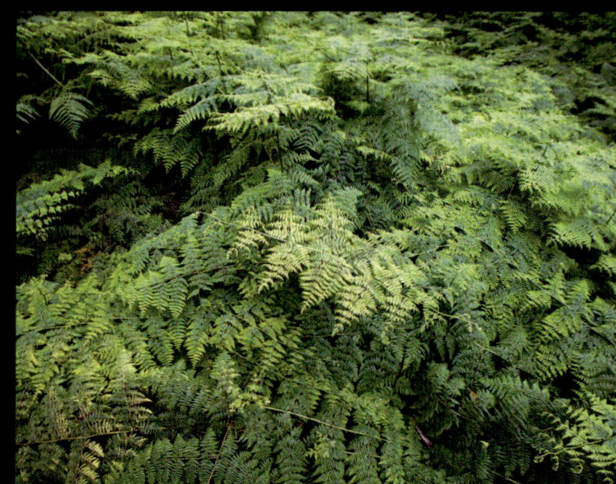

Bracken ferns

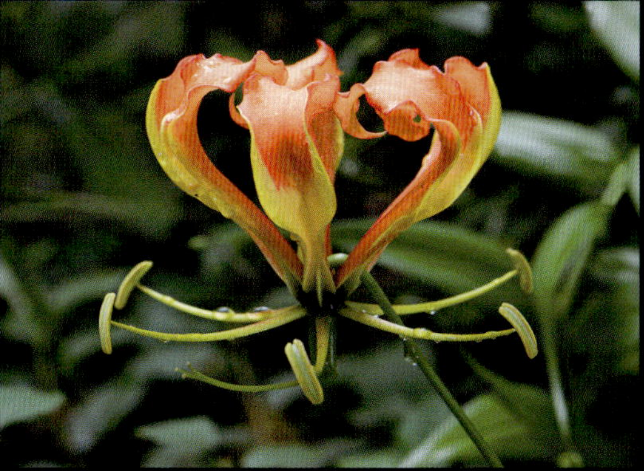

Gloriosa superba or flame lily

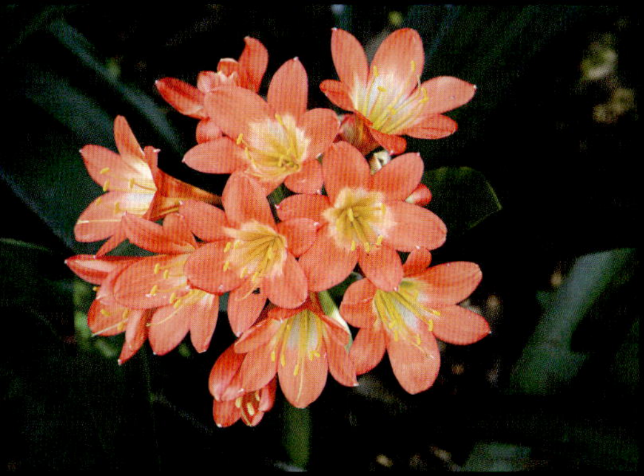

Clivia miniata or bush lily

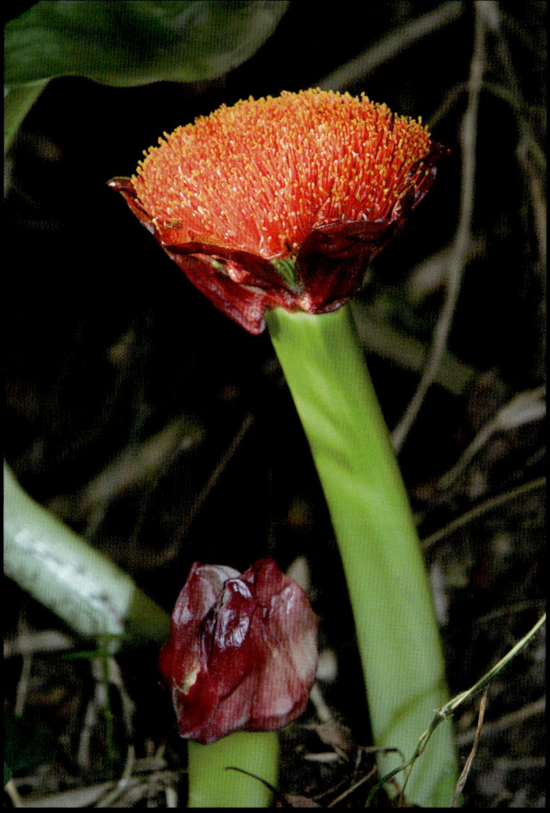

Scadoxis puniceus or blood lily

forest floor. The scrambler *Gloriosa superba* or flame lily also prefers forest margins but is mainly confined to coastal dune forest, thicket and the edges of sub-tropical forests. The *Clivia miniata* is a well-known and loved plant of the coastal bush and temperate forest.

Closely confined to forest habitat, the gregarious Sykes monkey (also called samango) feasts on ripe fruit when it is available, otherwise it looks for flowering plants or simply feeds on leaves. Like other primates, it lives in troops varying from four to 40 members with a specific dominance hierarchy. Being diurnal, they sleep in trees and start their day by sunning themselves before they start foraging.

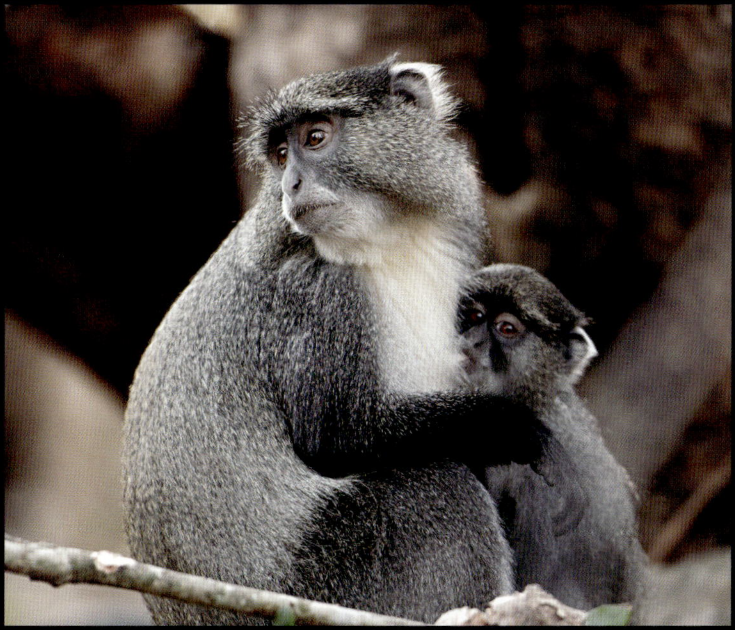

Sykes monkey

Shy and secretive, the red duiker keeps to itself most of the time. It feeds on freshly fallen leaves, flowers and wild fruit on the forest floor, but it may also browse on soft leaves of low shrubs. A typical forest antelope, it occurs in riverine forest or dense coastal bush. It is often seen in the St Lucia area of the iSimangaliso Wetland Park.

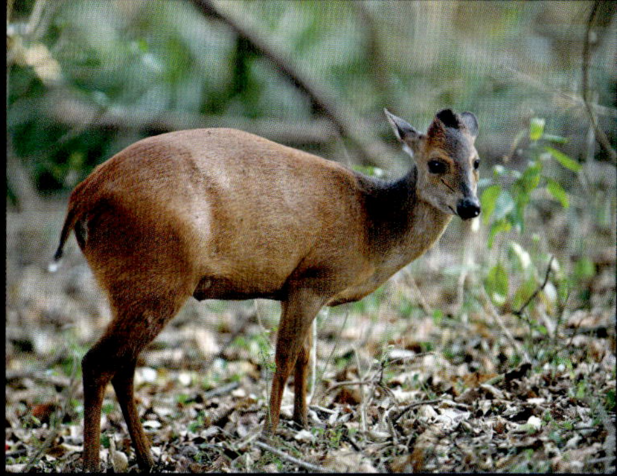

Red duiker

Epiphytes are quite common in forests. The term epiphyte describes a plant which, like a parasite, grows on a host, but unlike a parasite, it takes no nutrients from the tree itself, relying on nutrients from the air, rainwater and the compost that lies on tree branches. Their epiphytic way of life gives these plants advantages in the forest, allowing them access to more direct sunlight, a greater number of canopy animal pollinators and the possibility of dispersing their seeds via wind. Orchids are excellent examples of epiphytes. They have roots with a large surface area for rapid absorption of nutrients and water. Their secondary stems can store water so the plant can withstand very dry periods. They produce tiny seeds (measured in microns) that number in the hundreds of thousands. The balloon-like seed coat coupled with the small seed size enables orchid seeds to be dispersed over great areas by wind currents.

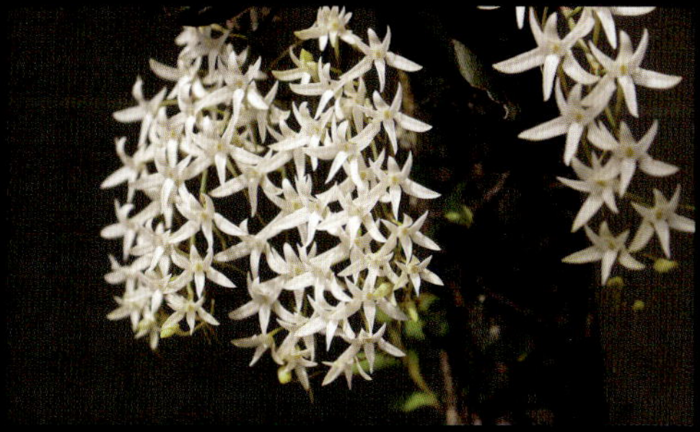

Epiphyte

South Africa's National Tree is the real yellowwood, *Podocarpus latifolius*. In forests, they can grow up to 40m in height with the base of the trunk sometimes up to 3m in diameter. One of the oldest trees left untouched by early woodcutters can be seen in the Tsitsikamma Forest near Plettenberg Bay. It is estimated to be between 600 and 800 years old, is 36.6m high and has a trunk circumference of 9m. A boardwalk path carries visitors through the dappled light of the lush, cool forest.

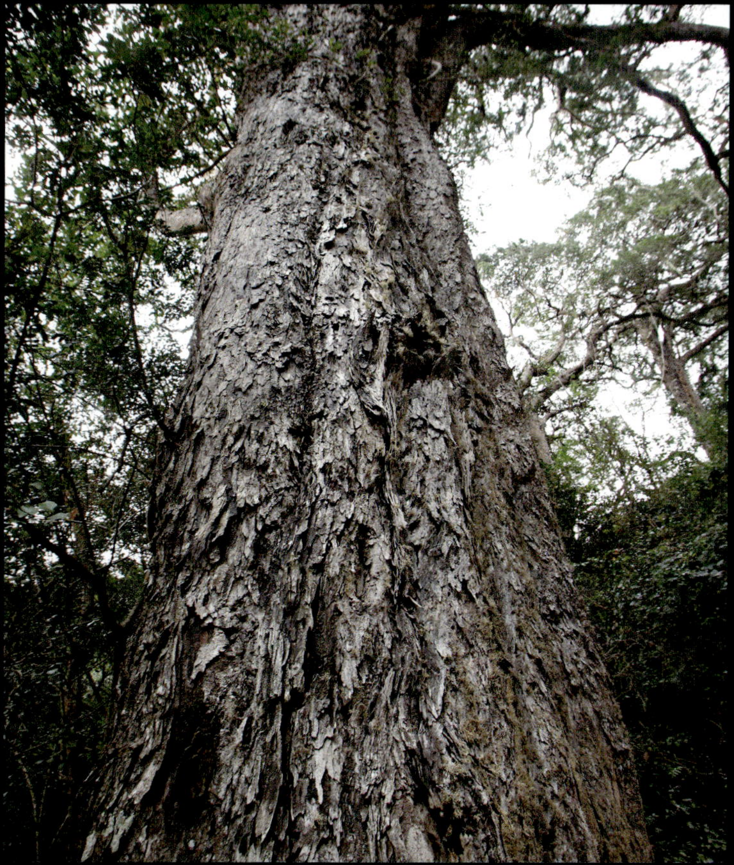

Real yellowwood or Podocarpus latifolius

Rivers and wetlands

Not a biome but an intrinsic part of them all

The inland aquatic systems are wide-ranging and varied but can be grouped together as wetlands. These vary from mountain sponges and midland marshes to swamp forests and estuaries – linked by rivers and streams. The wetlands share common and important functions in river catchments by providing a regular water supply, filtering the water naturally, reducing the effects of floods and droughts, and providing a vital wildlife habitat.

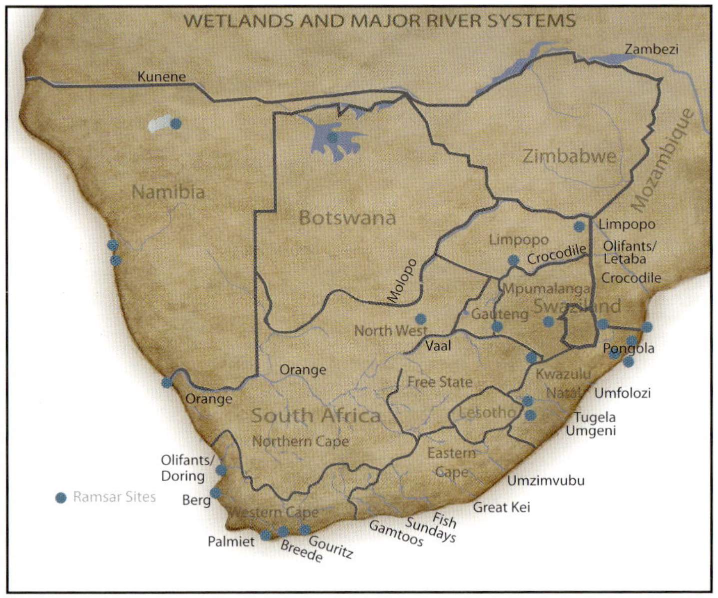

The form and abundance of wetlands vary considerably owing to regional differences in topography, climate, vegetation, soil, land use and hydrological conditions. Most wetlands have a high water table, water-carrying soil and water-loving plants, but in the semi-arid parts of southern Africa there are also numerous pans that support few if any plants and that may contain shallow water only once in five or more years.

Wetlands play an important role in maintaining biodiversity since they support an extraordinary variety of plant, bird and general wildlife. They also sustain livelihoods, as they provide access to an array of natural resources. In the past, the importance of wetlands was not understood and many were drained for agricultural and other development, or totally submerged in the construction of dams. Nowadays their value is increasingly understood and efforts are being made to rehabilitate former wetlands that have dried up.

Good places to experience wetlands

In southern Africa, there are various wetland sites that have been awarded Ramsar status. The Ramsar Convention is an intergovernmental treaty that provides the framework for national action and international cooperation for the conservation and wise use of wetlands and their resources.

South Africa presently has 20 sites designated as Wetlands of International Importance, with a surface area of 553 178 hectares. Namibia has four Ramsar sites, which are Walvisbay, Sandwich Harbour, Etosha Pan (Lake Oponono and Cuvelai Drainage) and the Orange River Mouth in conjunction with South Africa. Botswana presently has one huge site designated as a Wetland of International Importance, the Okavango Delta System with a surface area of 5 537 400 hectares. Mozambique presently has two sites designated as a Wetland of International Importance, with a surface area of 2 051 700 hectares, but one falls outside the limitations of southern Africa because the Zambezi River is regarded as the border.

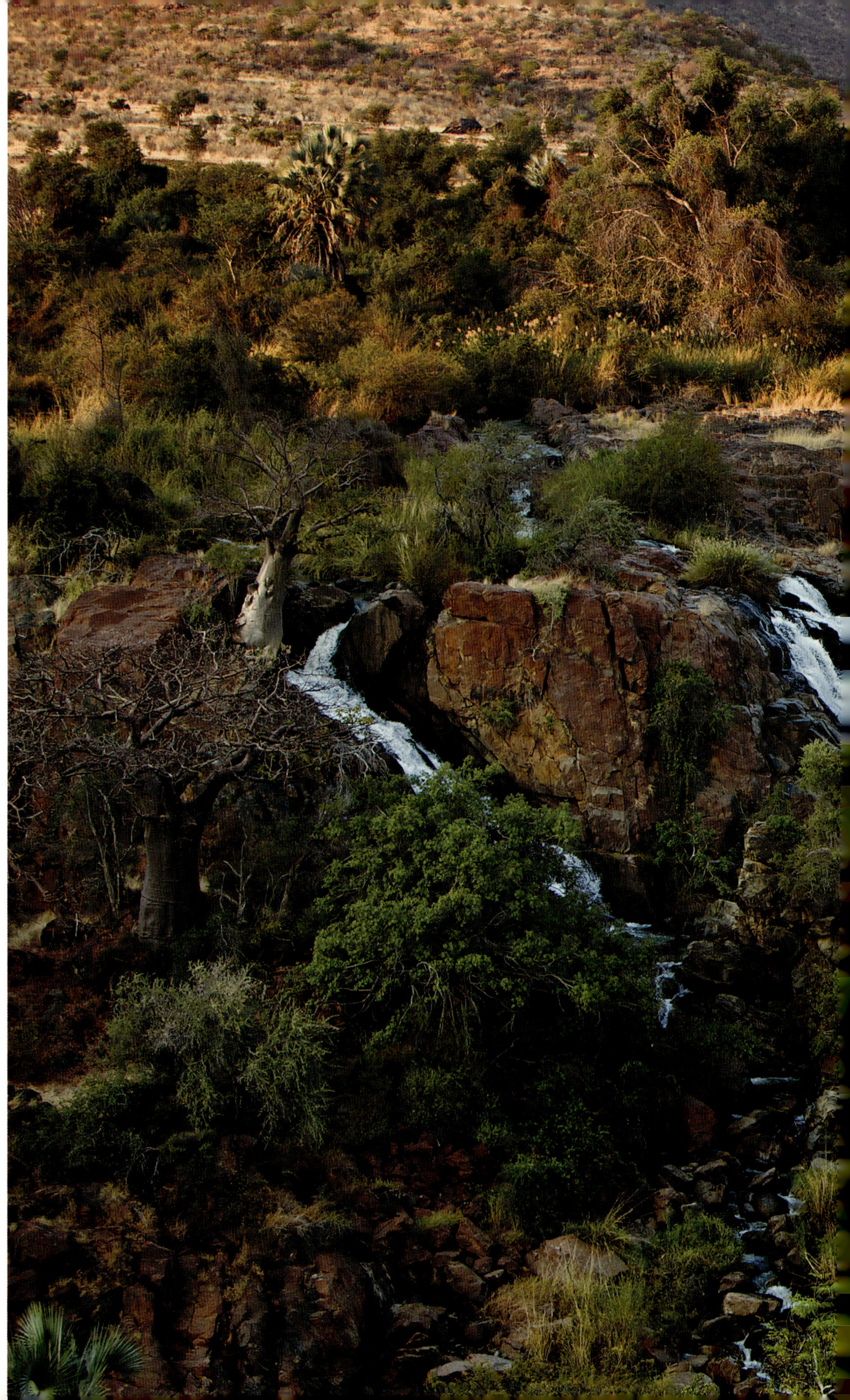

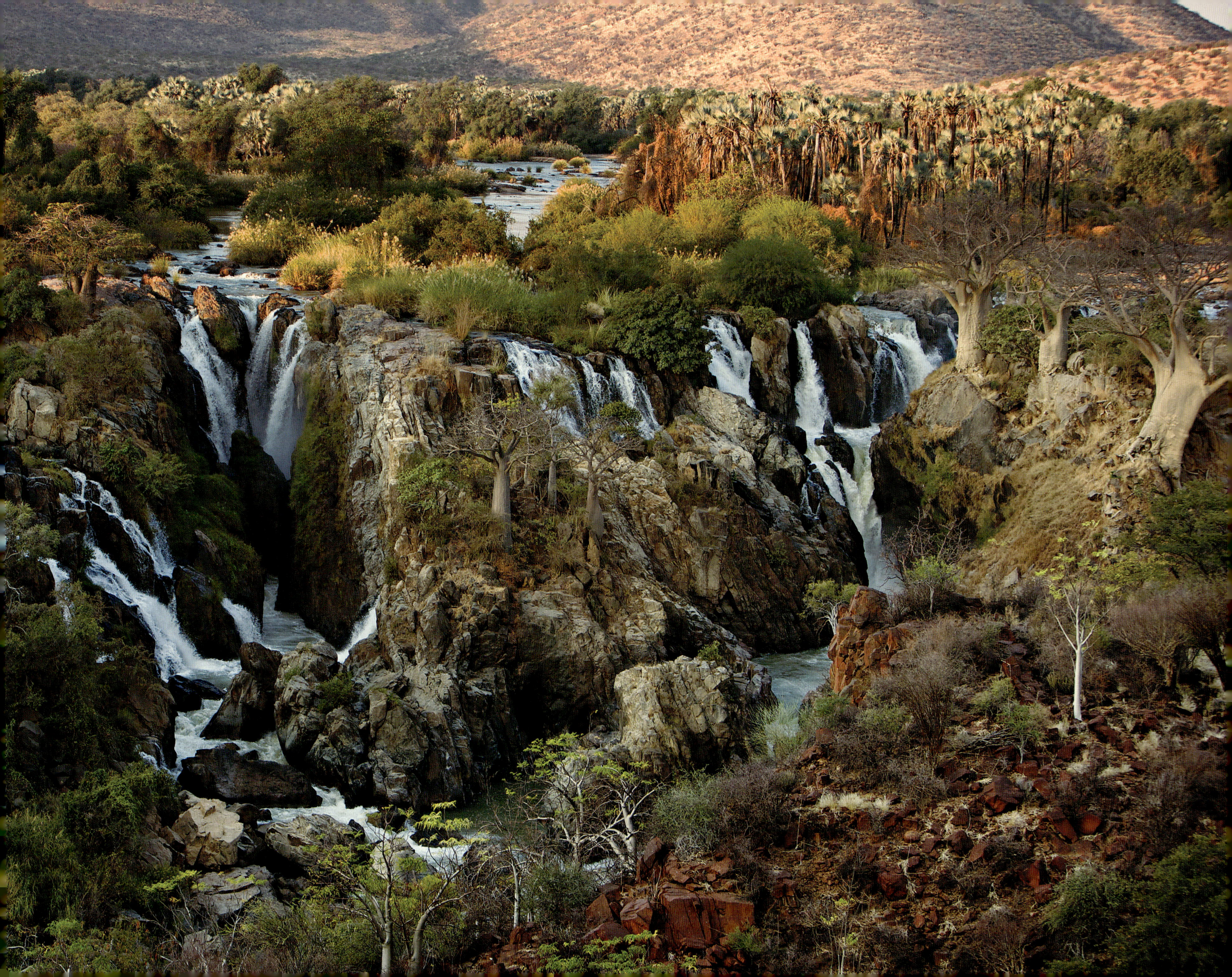

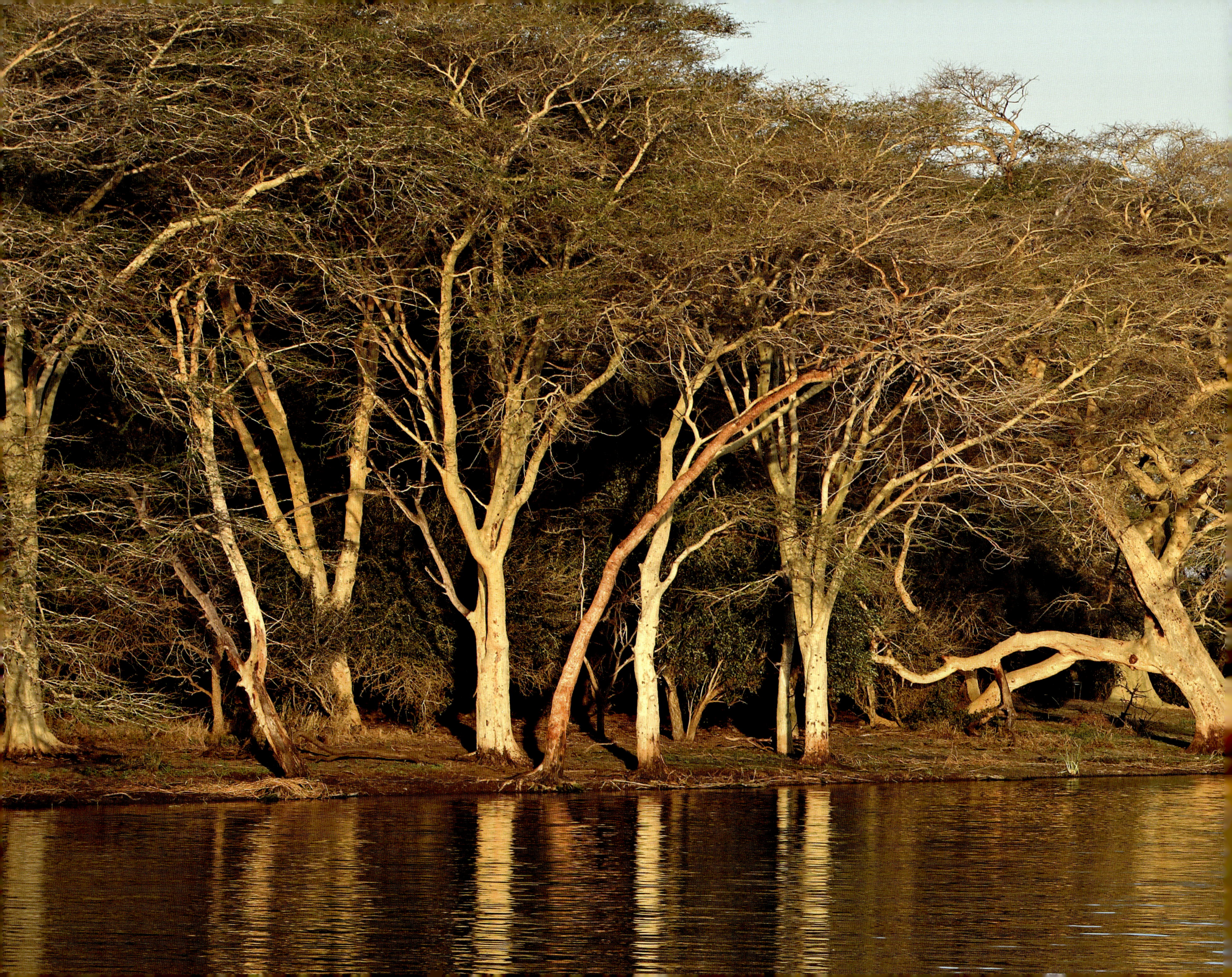

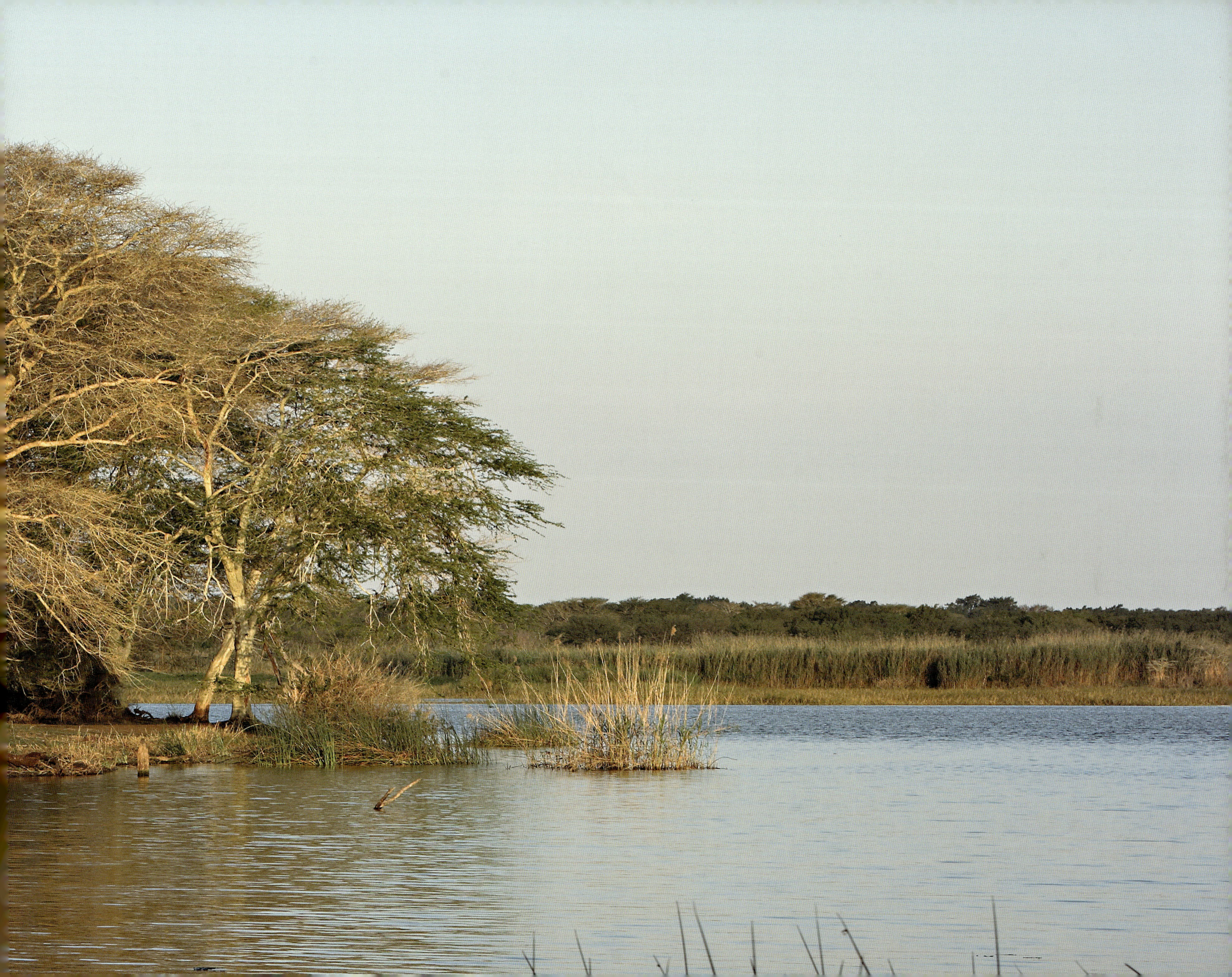

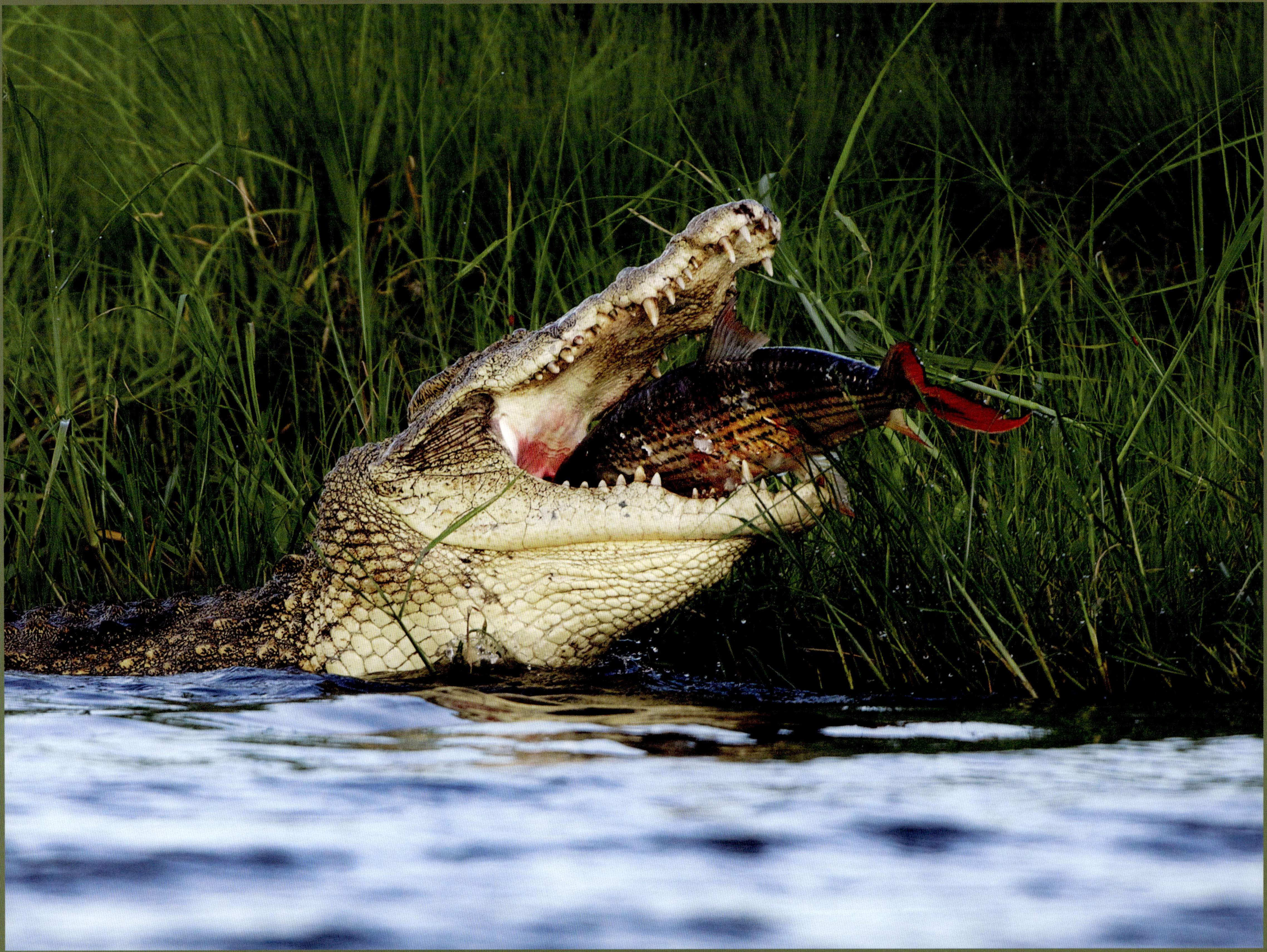

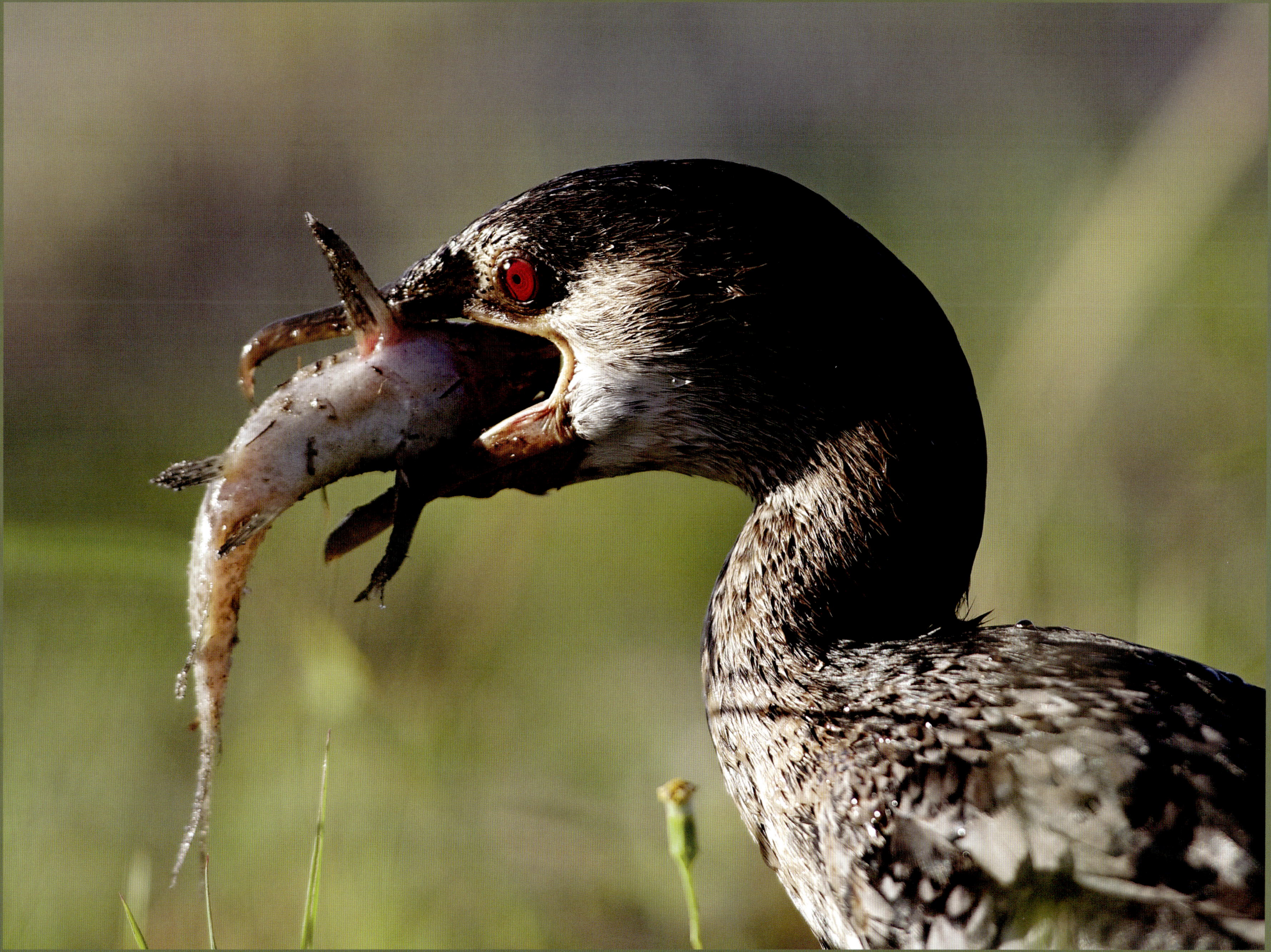

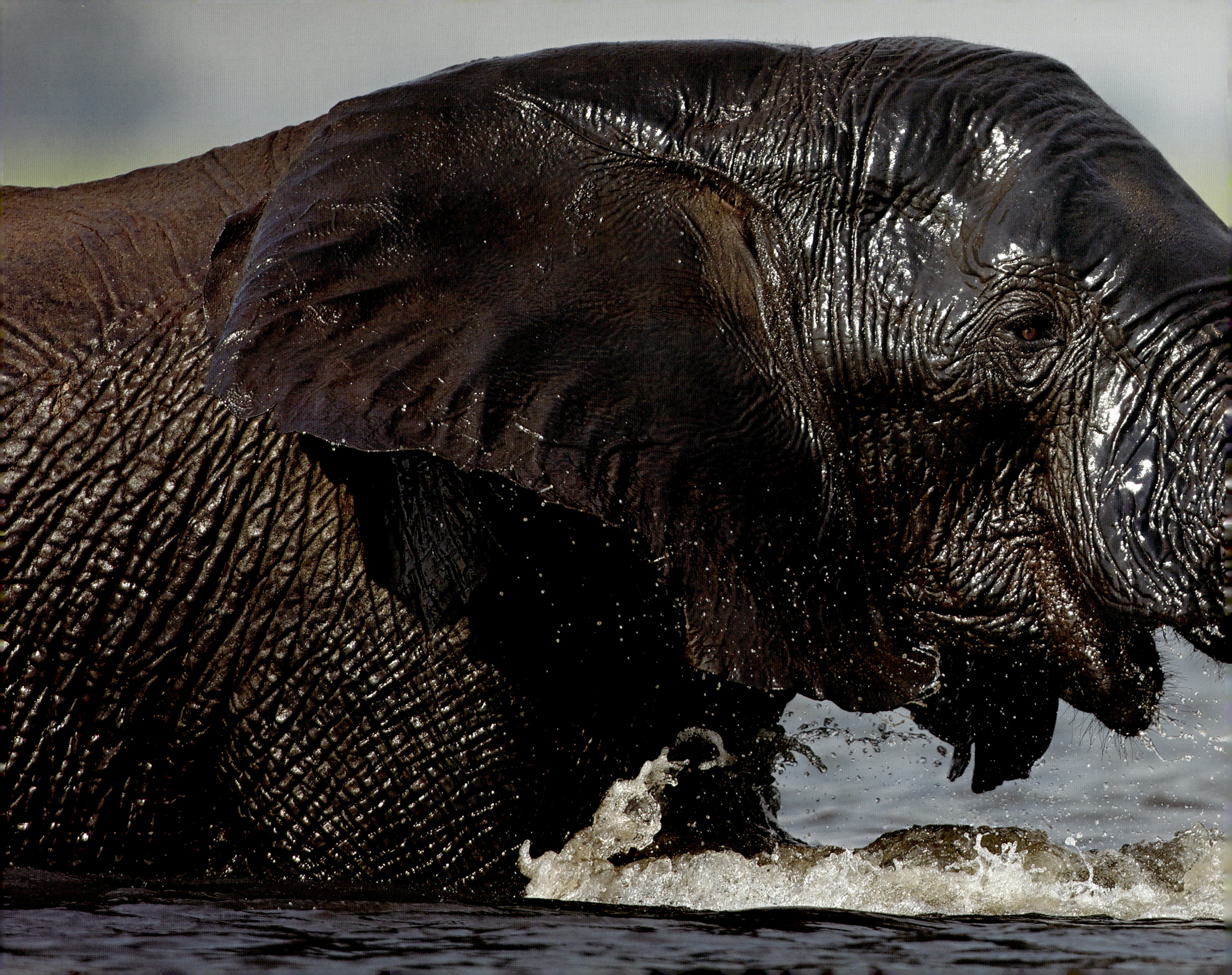

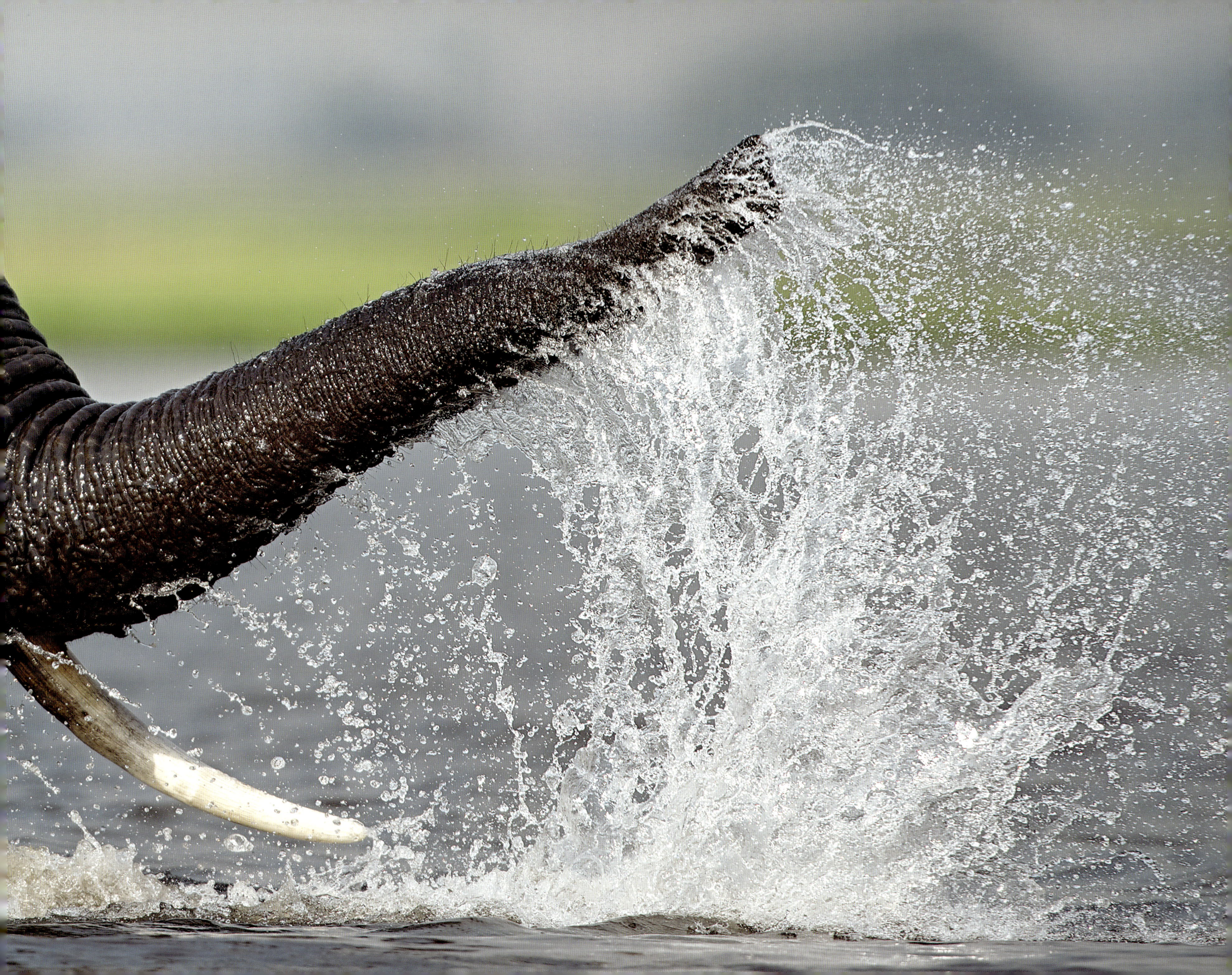

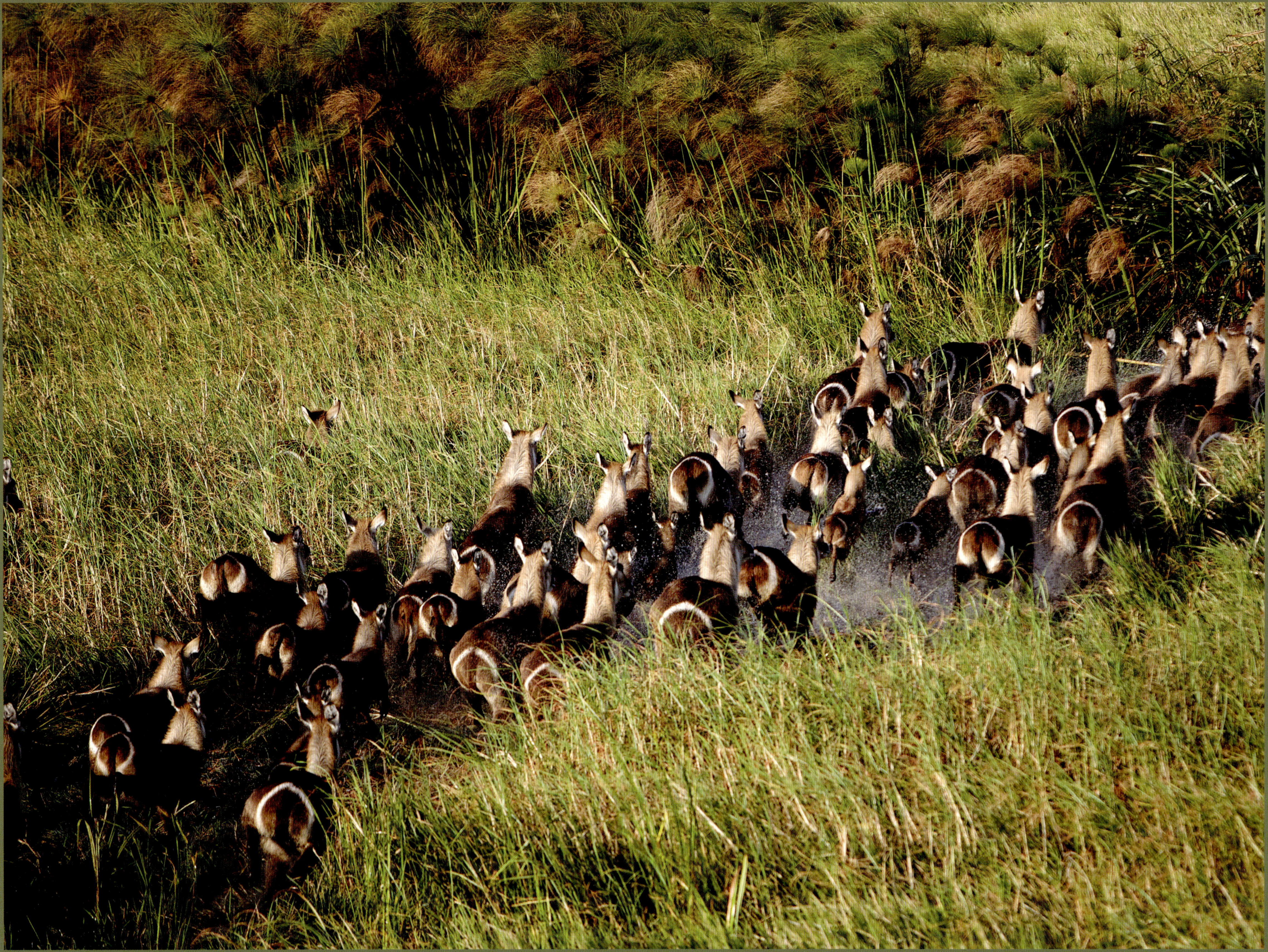

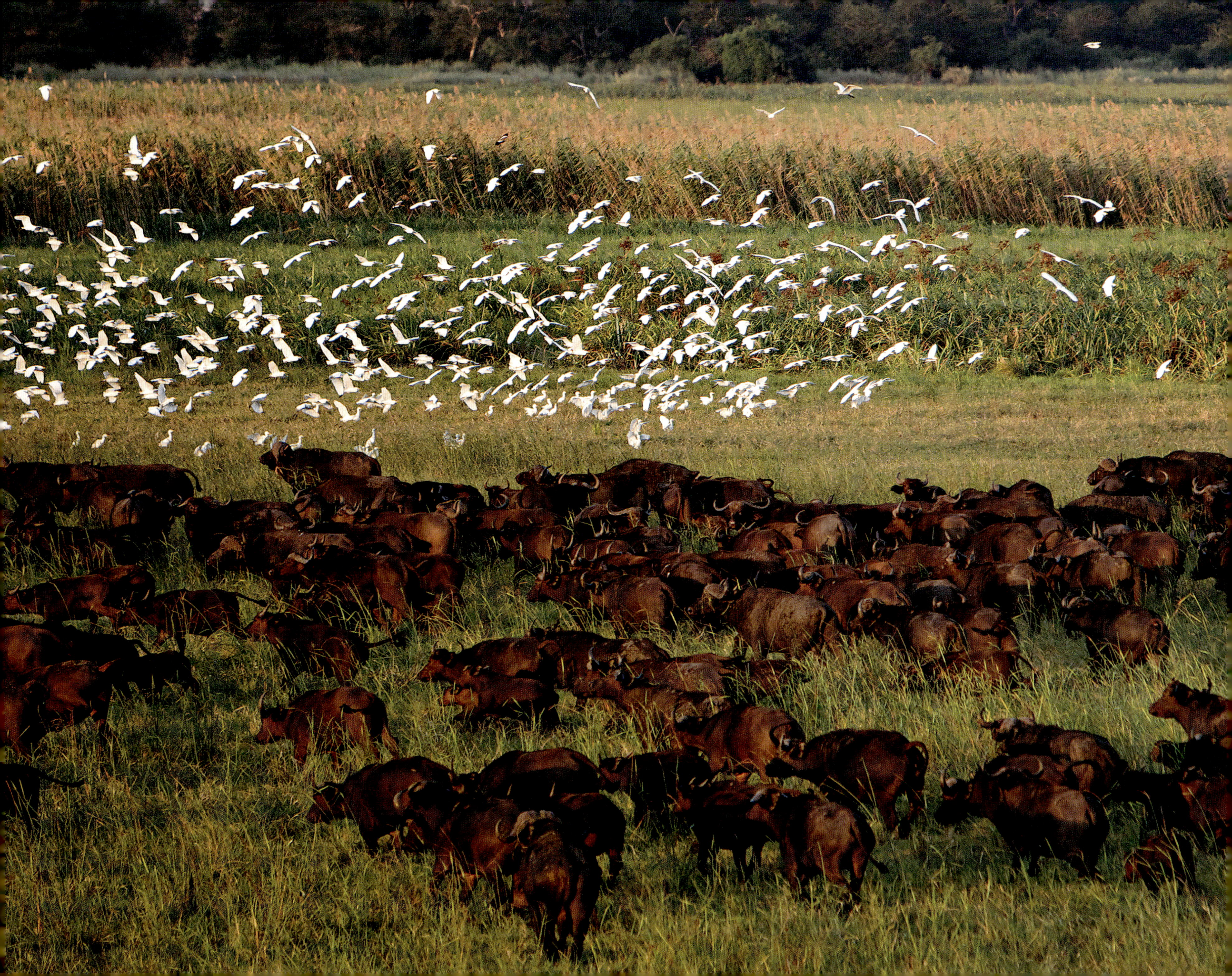

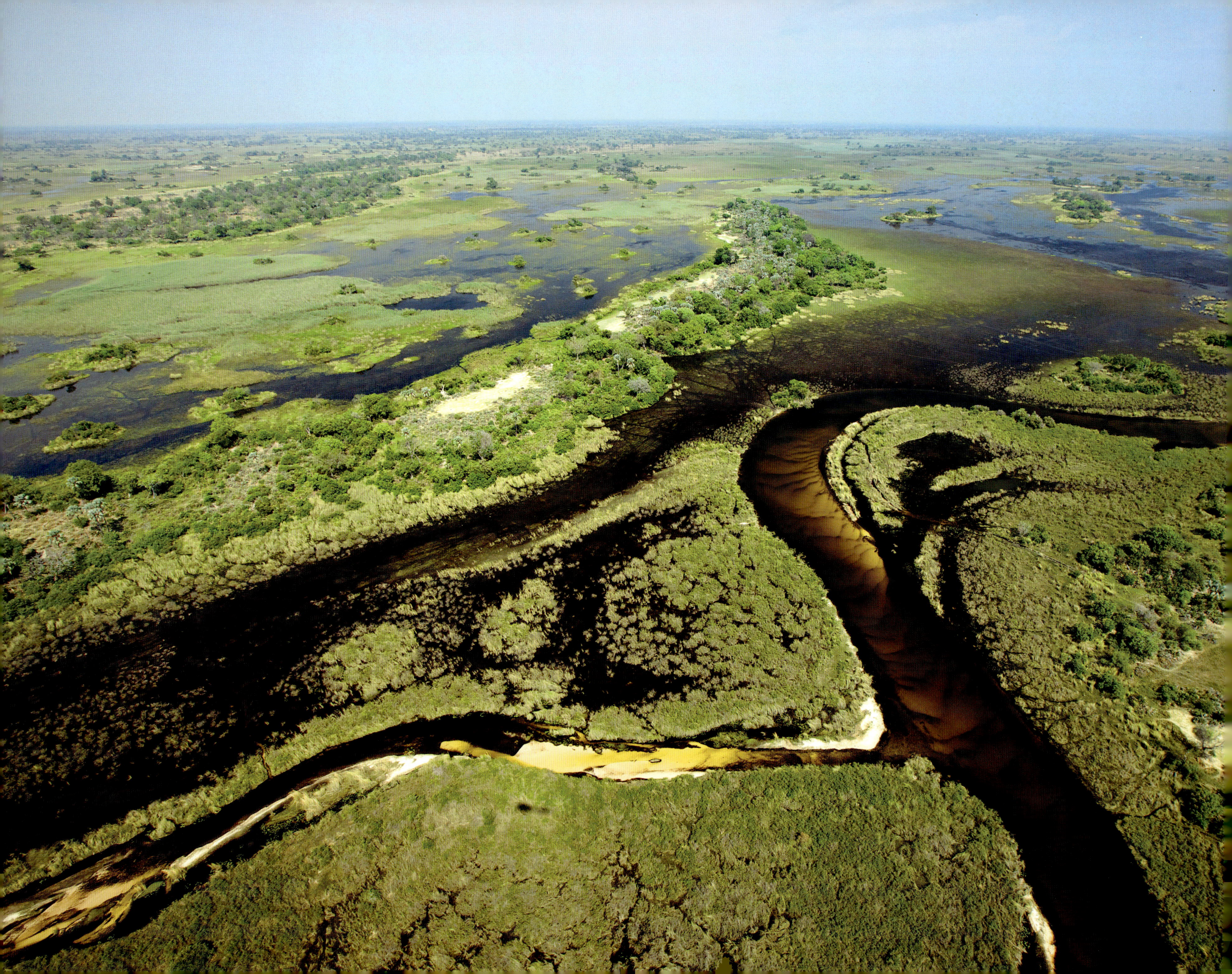

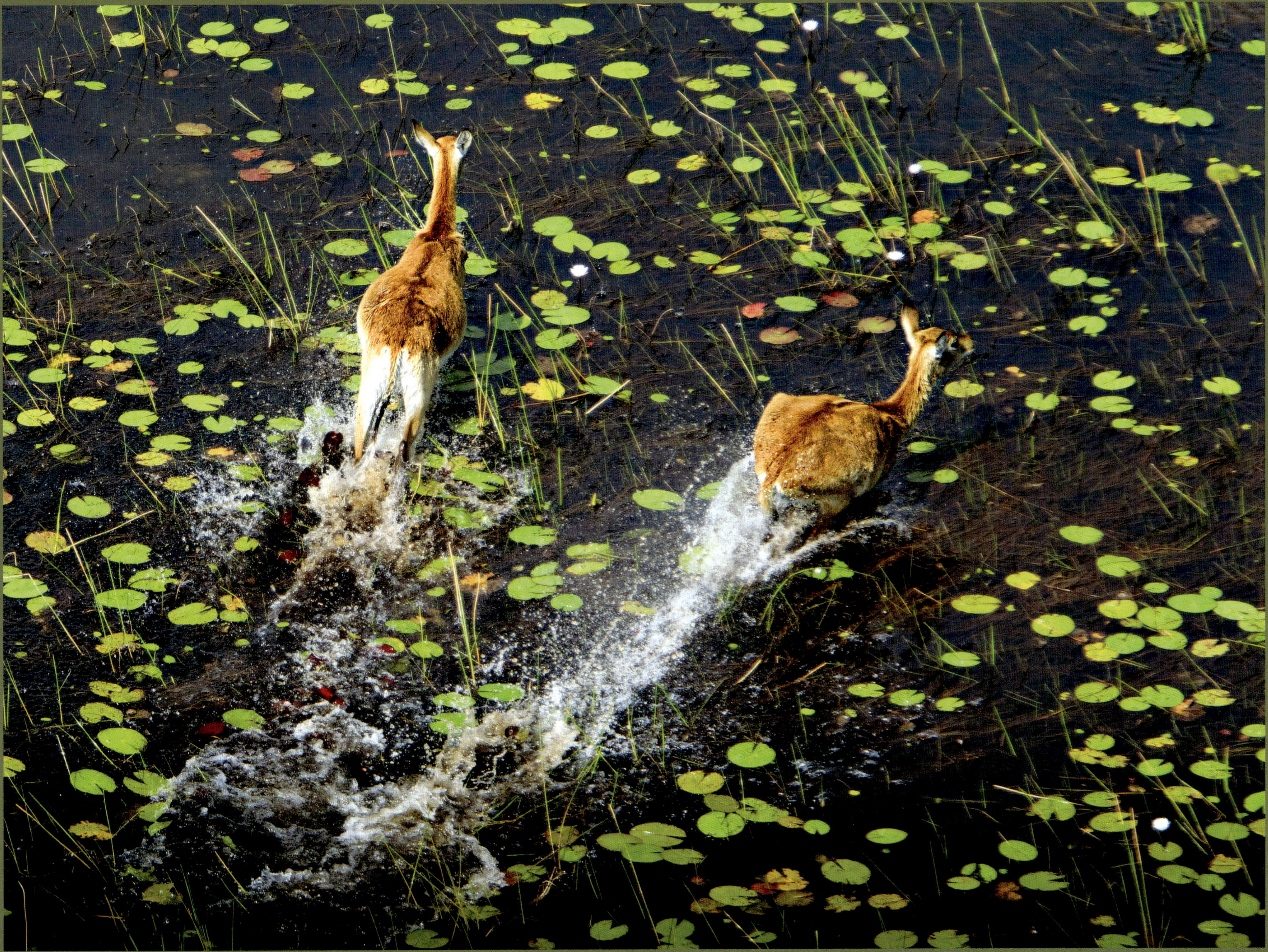

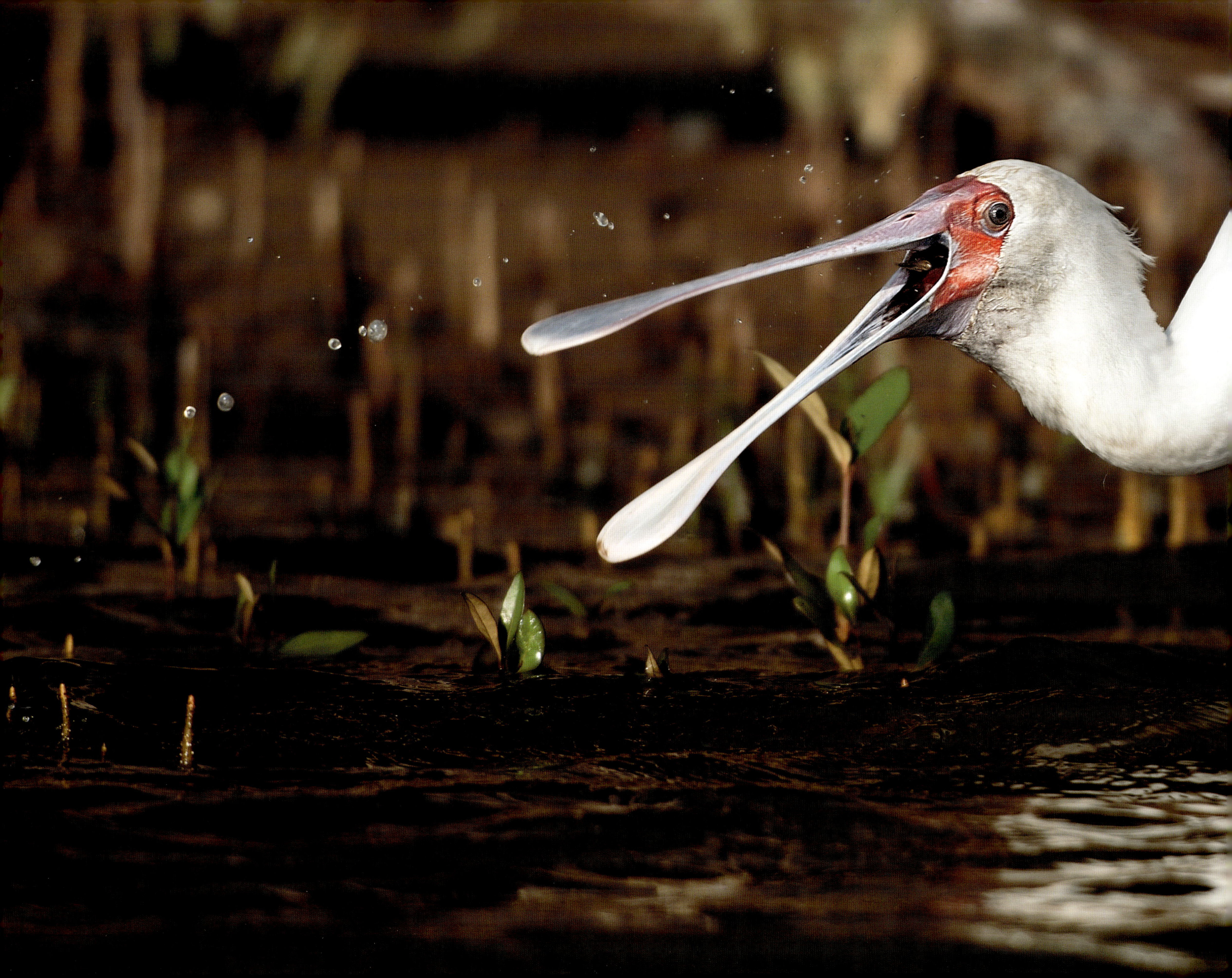

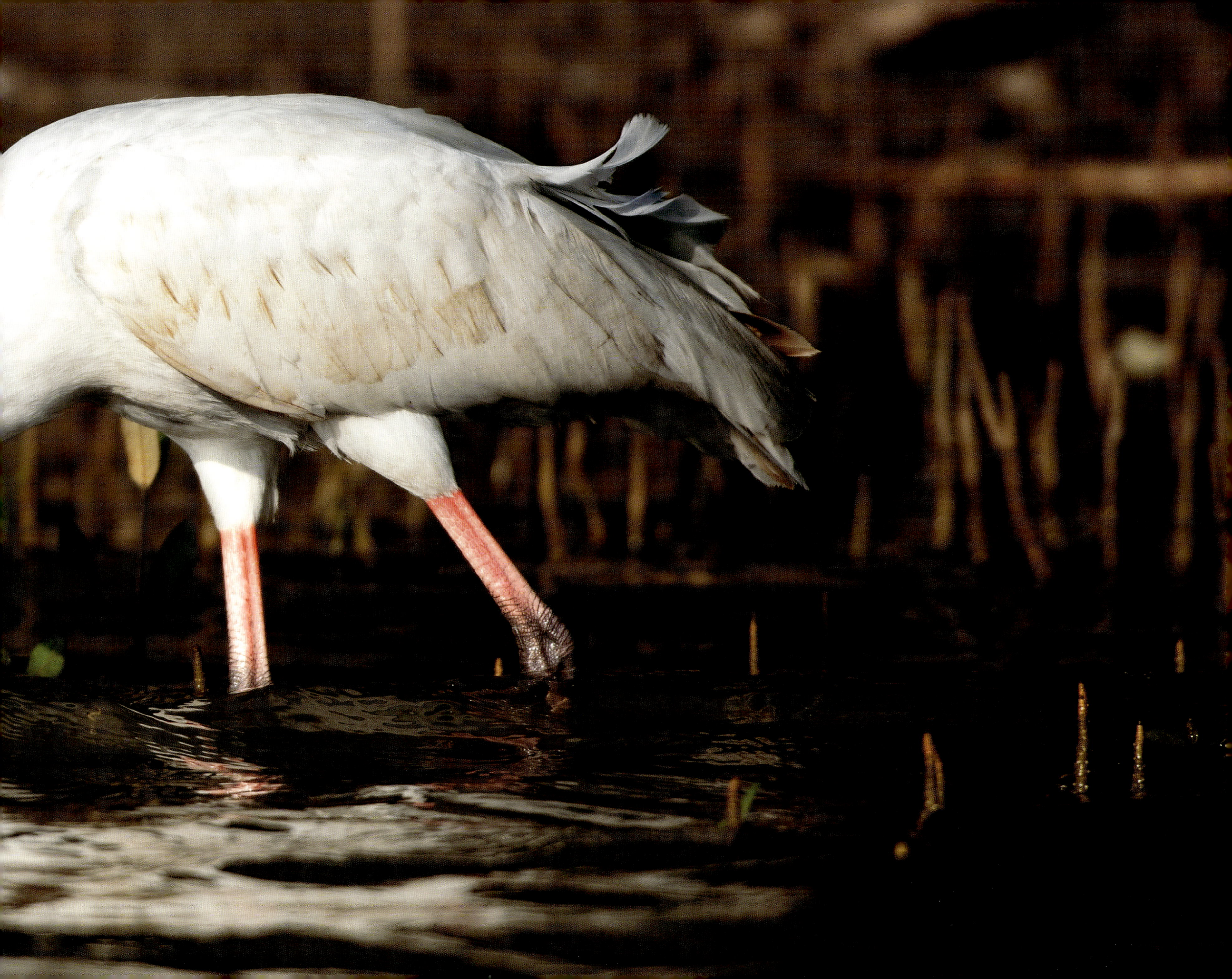

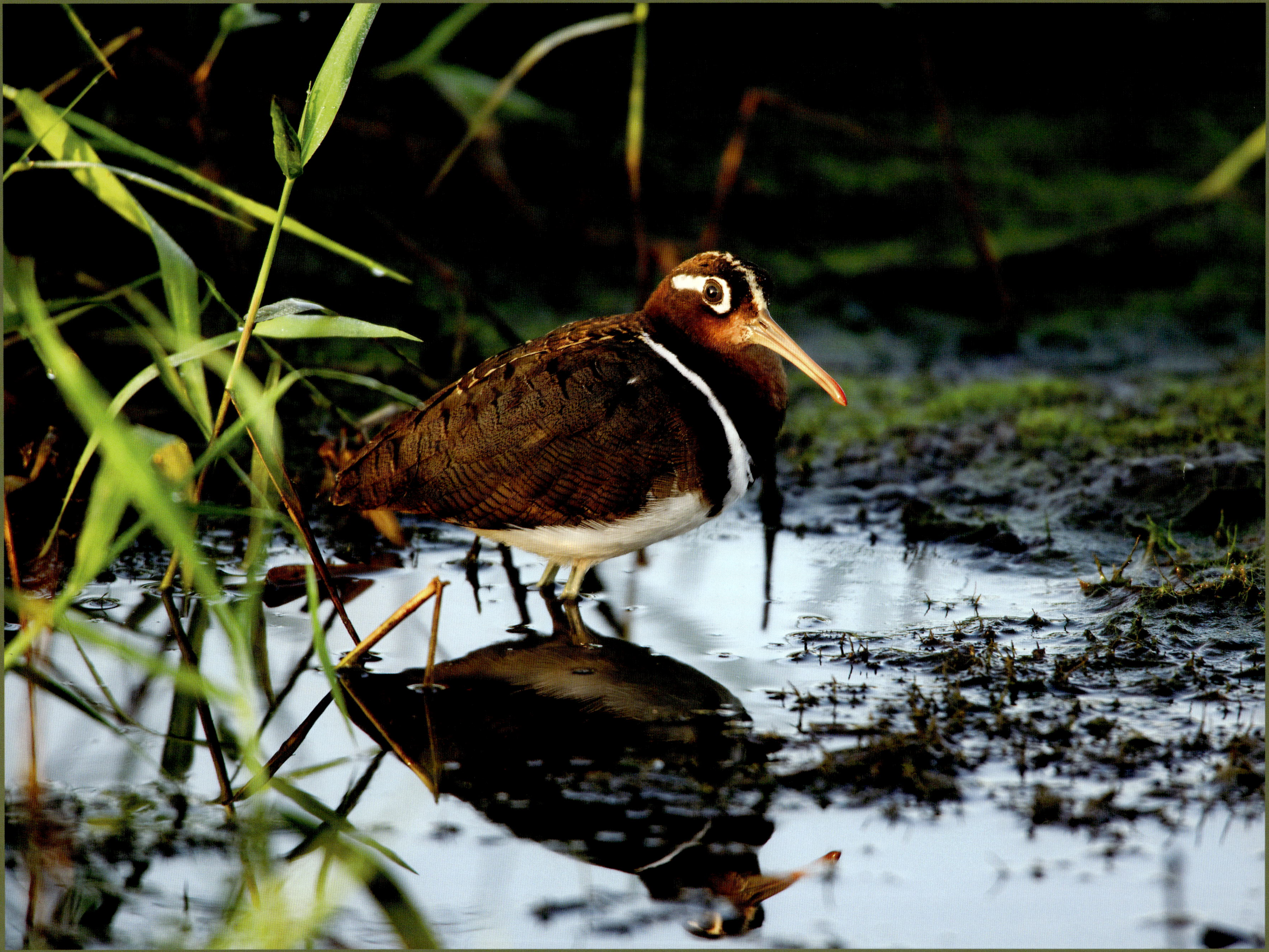

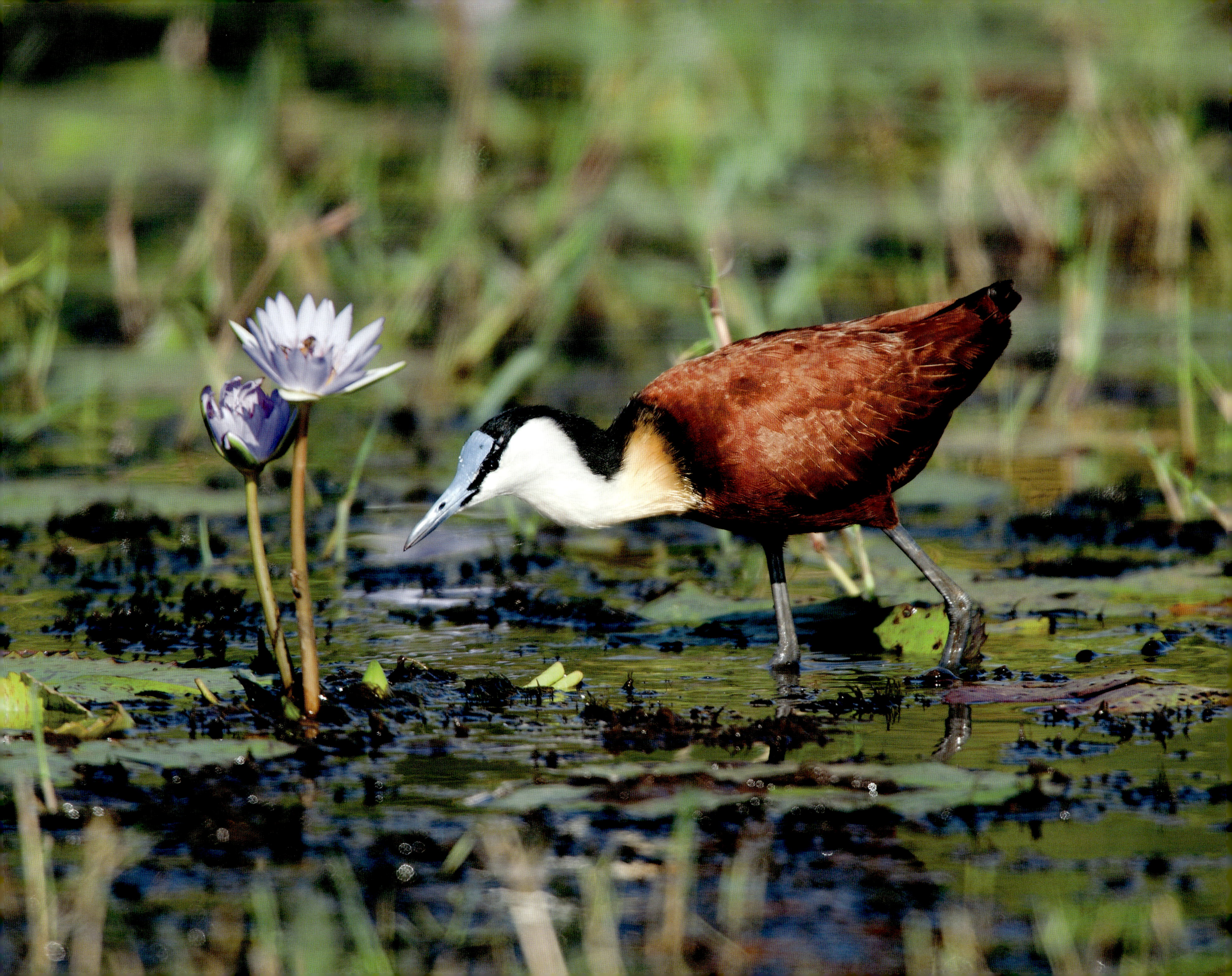

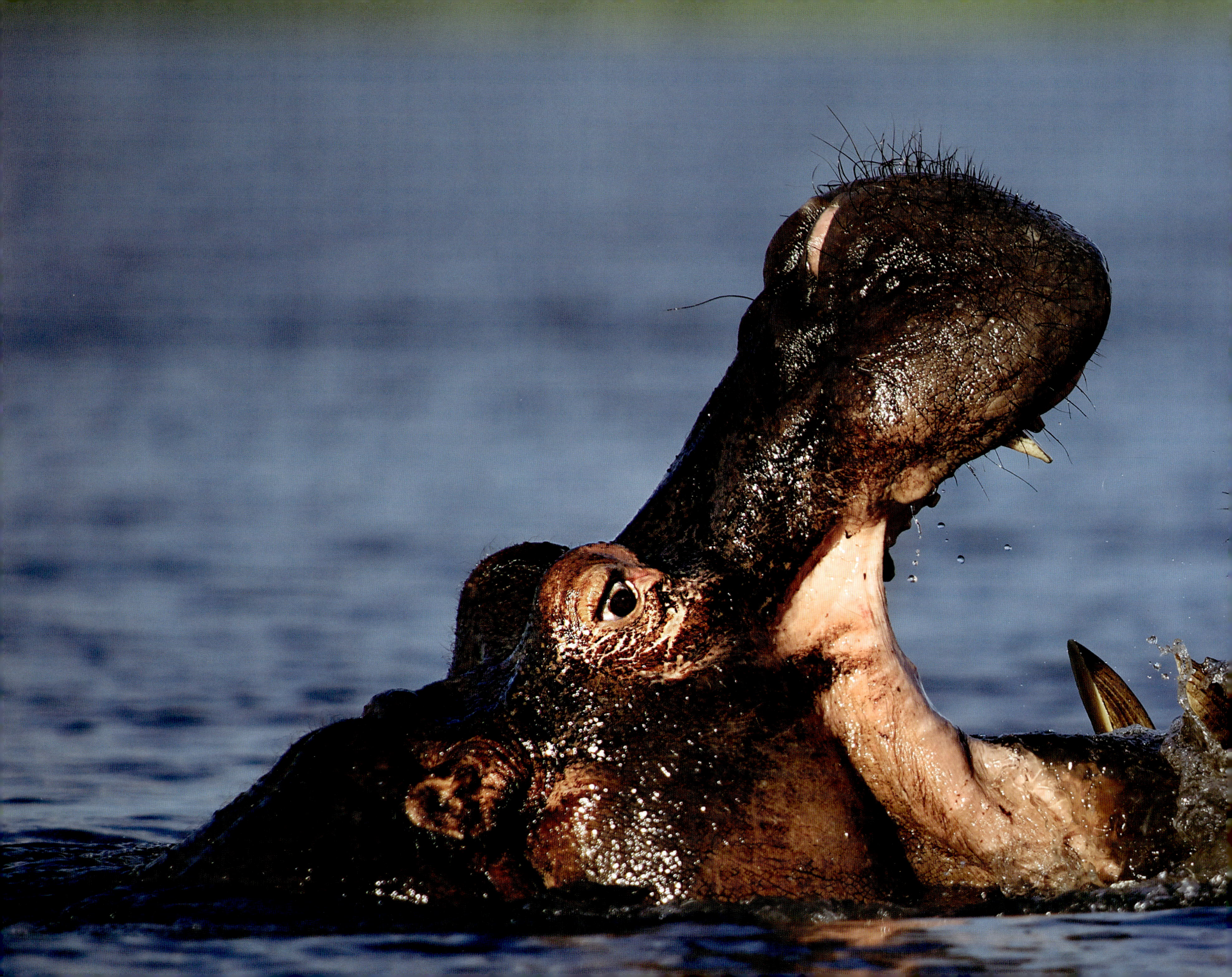

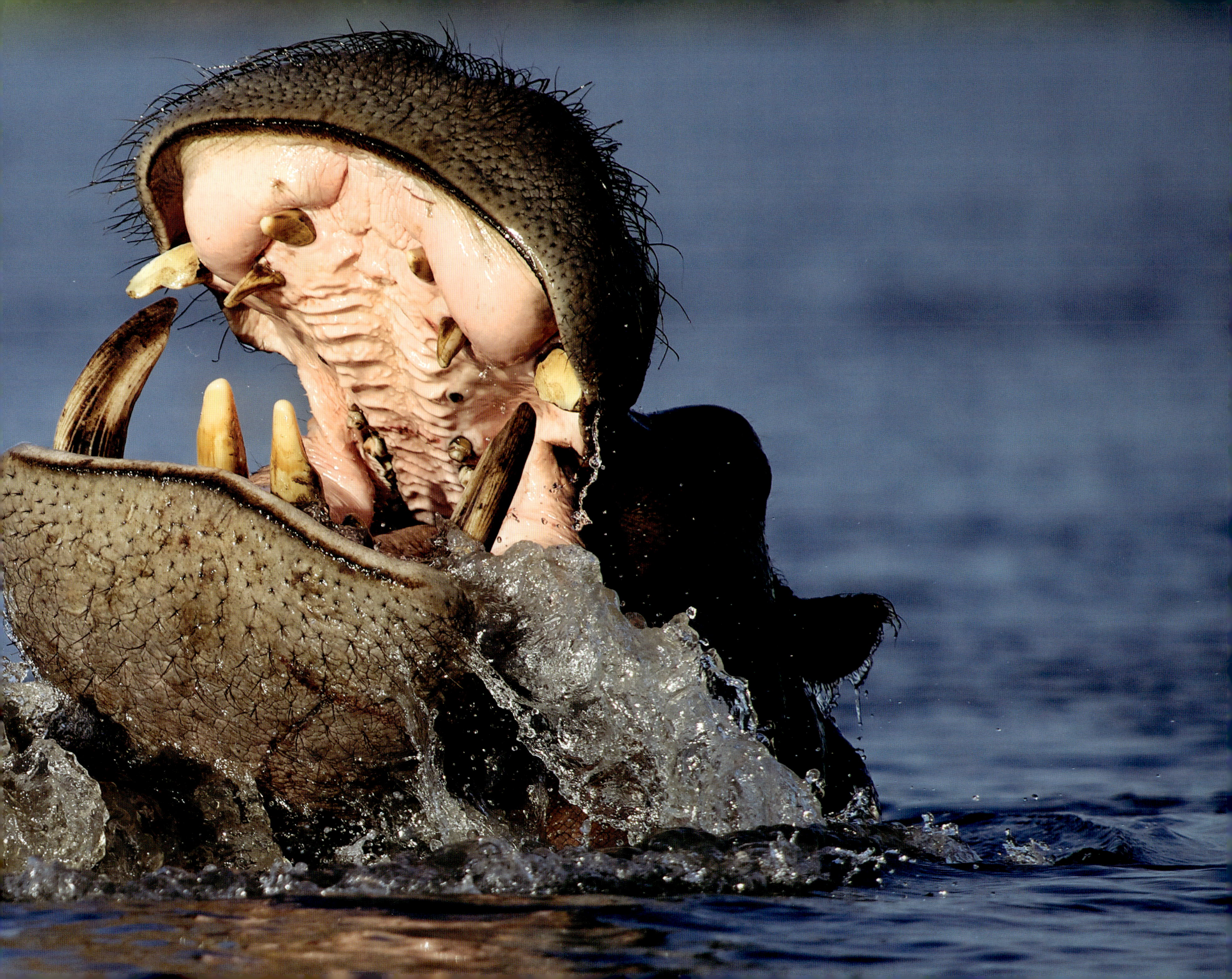

More about Rivers and Wetlands

Humankind relies on wetlands and so do animals big and small. On reaching water, elephant will satisfy their thirst, sometimes drinking up to 120 litres at a time, but once that is done, they often frolic around, splashing, rolling and submerging their huge bodies in the cooling liquid.

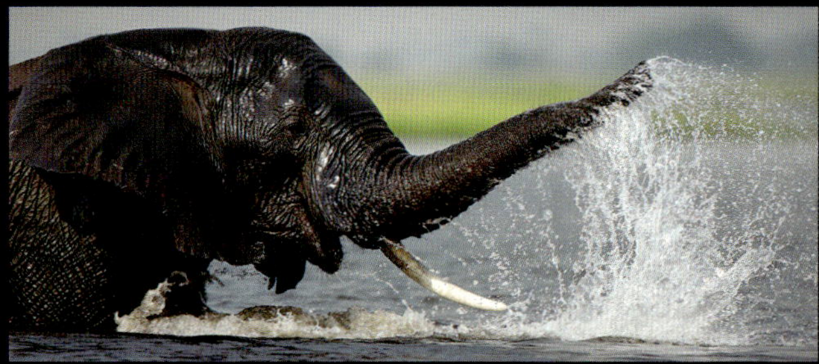
Elephant playing in the water

The Marromeu Complex, a Ramsar site in Mozambique, includes Sofala and Zambezia, the protected Marromeu Buffalo Reserve (Reserva especial de Marromeu) and four surrounding hunting concessions. This site comprises 688 000 hectares and includes a variety of wetland habitats such as the

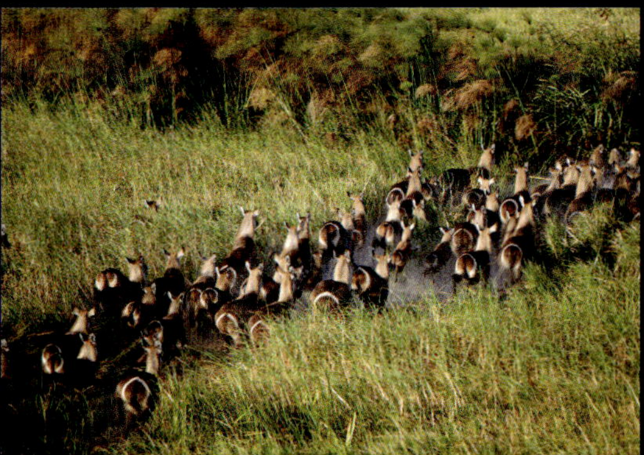
Waterbuck crossing the wetland at Sofala

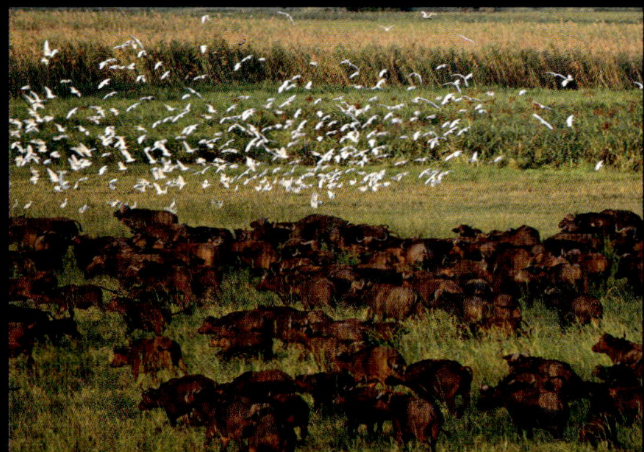
Buffalo are bulk grazers and favour grassy wetlands

Zambezian coastal flooded savannah freshwater swamps, mangroves and sea grass beds. Large herds of waterbuck and buffalo frequent these wetlands.

The Okavango Delta System is the largest inland delta in southern Africa – an area of 68 640km^2. Since it lies in a semi-arid area, most of the annual inflow is lost to evaporation, transpiration and seepage – only 3% of the water is discharged from the delta. The biodiversity within this Ramsar Wetland is amazing.

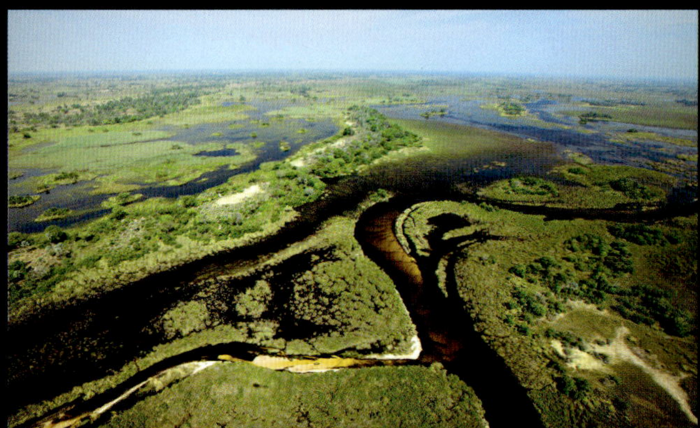
The Okavango Delta

The delta has a large population of sitatunga, *Tragelaphus spekei*, and red lechwe, *Kobus leche*, which do not occur elsewhere in the subcontinent. Many herbivores, including buffalo, zebra, elephant, hippo, blue wildebeest, waterbuck and common duiker, inhabit the place. In addition, predators such as lion, spotted hyena, wild dog, cheetah and leopard roam here too as they depend upon the high concentration of herbivores near permanent water bodies.

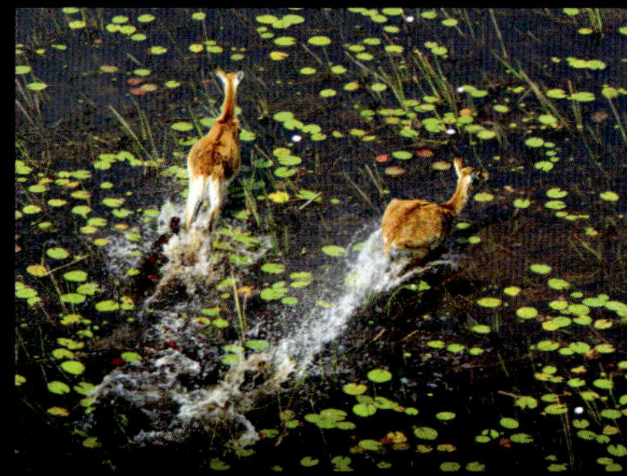
Red lechwe

Particularly prized as a game fish, the tigerfish, *Hydrocynus vittatus*, is commonly found in the Okavango Delta and the Zambezi River; the two biggest lakes along the Zambezi, Lake Kariba in Zimbabwe and Cahora Basa in Mozambique; and finally in the

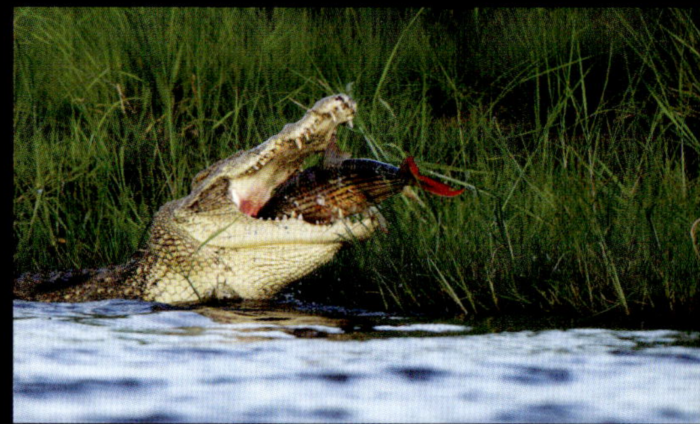
Crocodile swallowing tigerfish

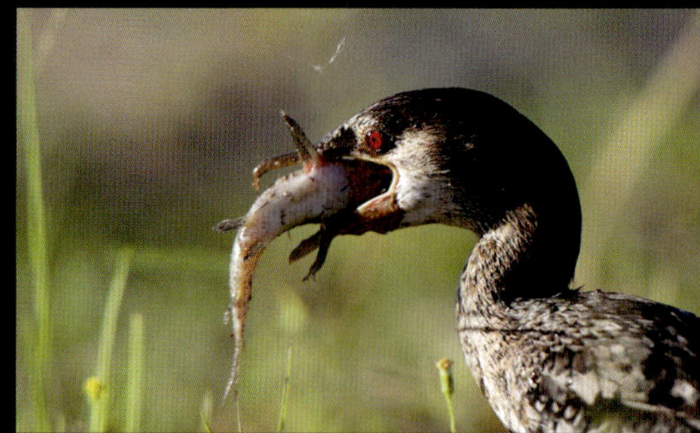
Reed Cormorant feeding on catfish

Jozini Dam in South Africa. These fish are fierce predators with distinctive, disproportionately large teeth, but they are swallowed easily and head first by the Nile crocodile that also shares their habitat.

The wetlands host an astounding variety of birds. The African Jacana is widespread and abundant. It favours areas with water lilies and walks on floating plants with a high-stepping gait, sometimes running to avoid sinking. The painted Snipe keeps to the reeds at the water's edge while the Reed Cormorant forages in deep water.

The most southernly distribution of the African Skimmer in southern Africa is the Okavango Delta and the Zambezi and Chobe rivers. They fly in flocks low over the water, wings held high and back, skimming the water for food, alternately flapping and gliding. The body

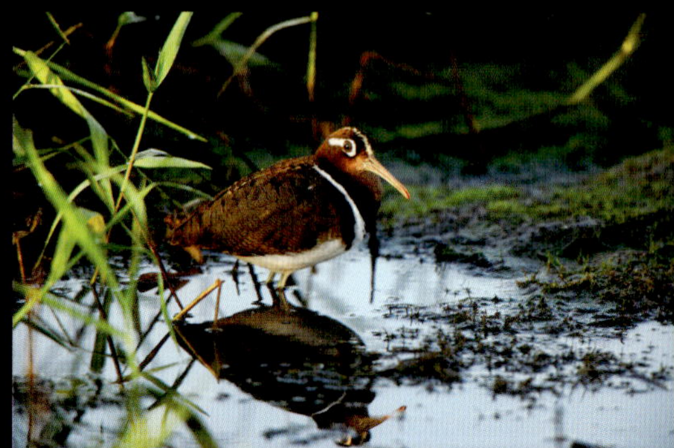
Painted Snipe

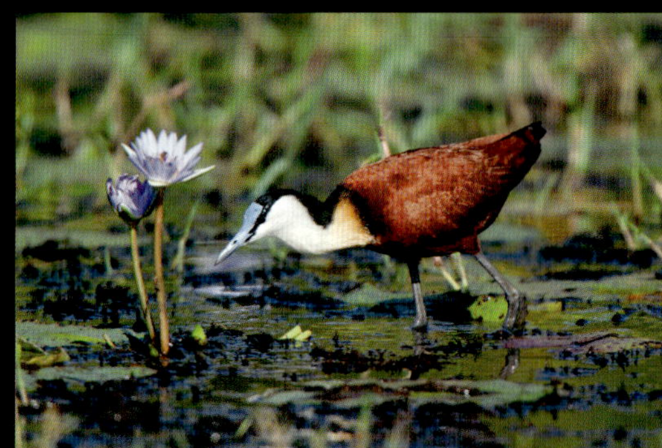
African Jacana

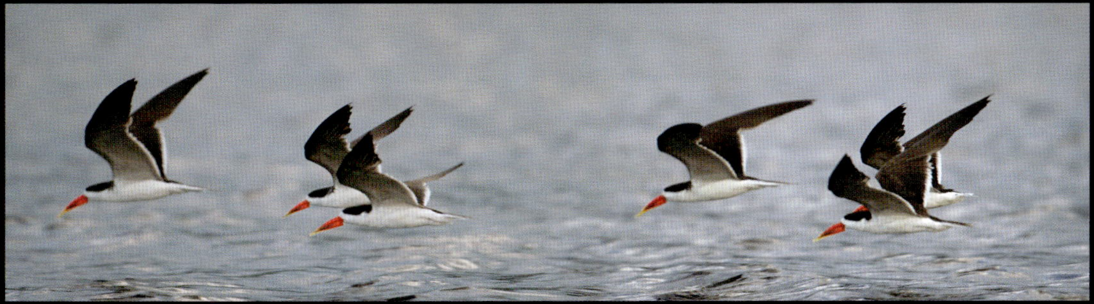

African Skimmers on the Zambezi

is tilted forward so that the lower mandible can slice through the water, picking up food and at the same time creating a characteristic wake.

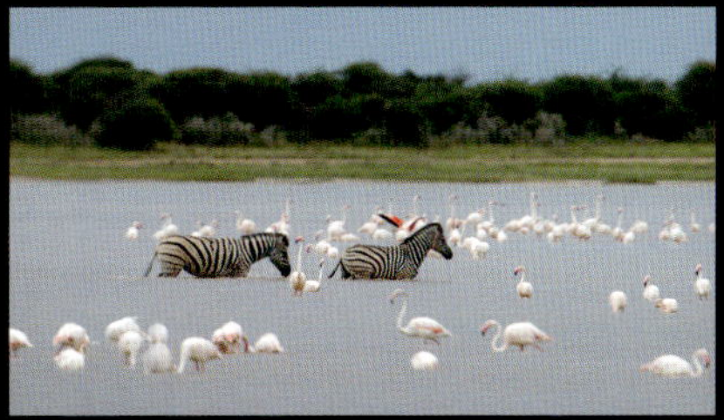

Fisher's Pan

The Etosha Pan is believed to have originated over 12 million years ago as a shallow lake fed by the Kunene River. Since then climatic and tectonic changes have lowered the water level so that the pan only holds water for a brief period each year. In December after good summer rains, the section called Fisher's Pan teems with flamingos and other water birds. Even zebra wade through the shallow lake, enjoying the short-lived abundance of water.

The African Spoonbill favours shallow aquatic habitats such as lake and river margins, seasonal pans, marshes, flood plains, mangrove swamps, and even sewage works and impoundments. It wades, foraging with its bill partly or wholly submerged, sweeping from side to side and snapping prey in its spoon-shaped bill.

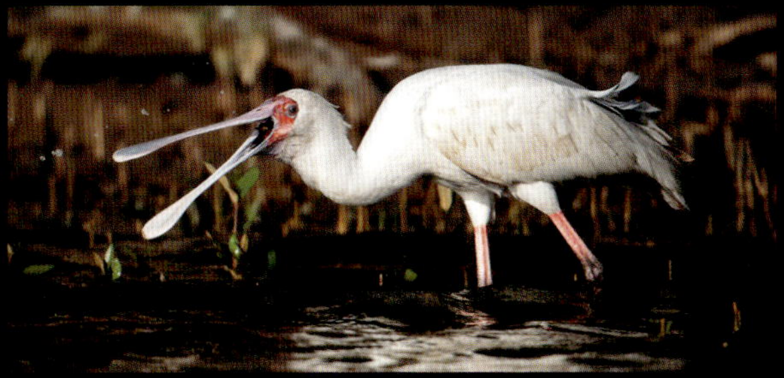

African Spoonbill

Hippo with their great size and barrel-shaped bodies occur widely in suitably deep water on the subcontinent. To maintain dominance, territorial bulls typically challenge others by opening their enormous mouths and displaying their canines.

Reaching heights of up to 25m, the very upright and tall Fever Tree, *Acacia xanthophloea*, with its distinctive yellow-green bark , graces riverbanks, pans and low-lying swampy areas. This preferred habitat gave cause for early European hunters and adventurers in Africa to give this tree its unwarranted and fearsome reputation and name. Those who camped near them often contracted high fevers which usually resulted in death. Although the connection was not made until many years later, the dreaded malaria-carrying *Anopheles* mosquito frequents the same habitat as the 'fever trees', which resulted in the unhappy association and name. Ironically, the bark of the fever tree is traditionally used by African tribes as a remedy to combat fever.

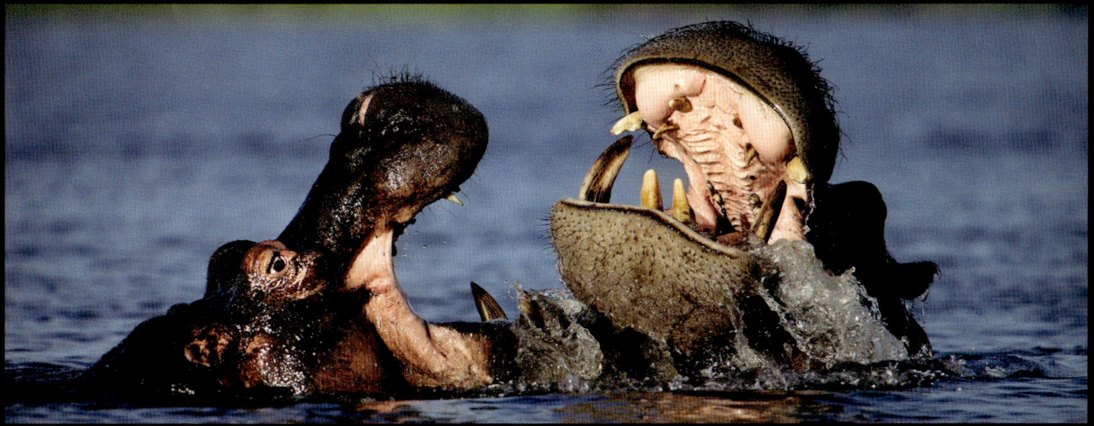

Hippo play-fighting

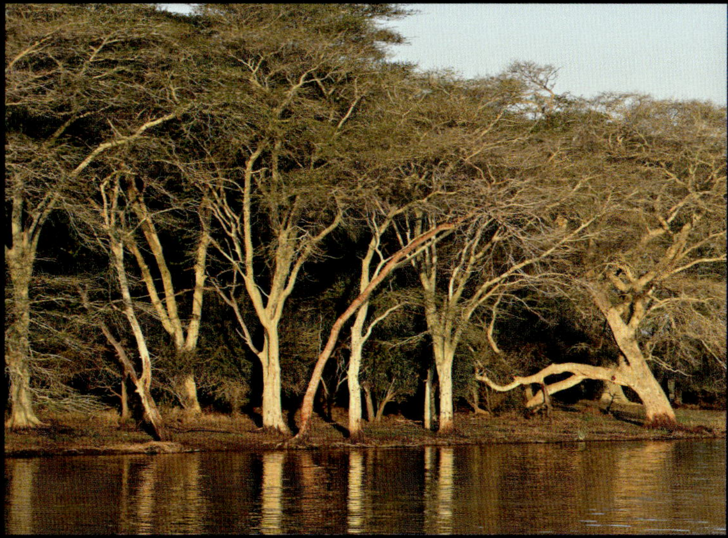

Fever Trees

The Kunene River is the most north-westerly border between the southern African subregion and Angola. At Epupa the river is half a kilometre wide and drops in a series of waterfalls over a distance of 1.5km. Although the greatest single drop is a mere 37m, the falls are a spectacular sight, especially when the river is in flood. 'Epupa' is the Herero word for foam, hence the name. The falls are a major visitor attraction in Namibia despite the difficulty in reaching the area, which is only accessible by four-wheel drive vehicles. It is a largely unspoiled environment with fig trees, baobabs, makalani palms, and coloured rock walls framing the falls.

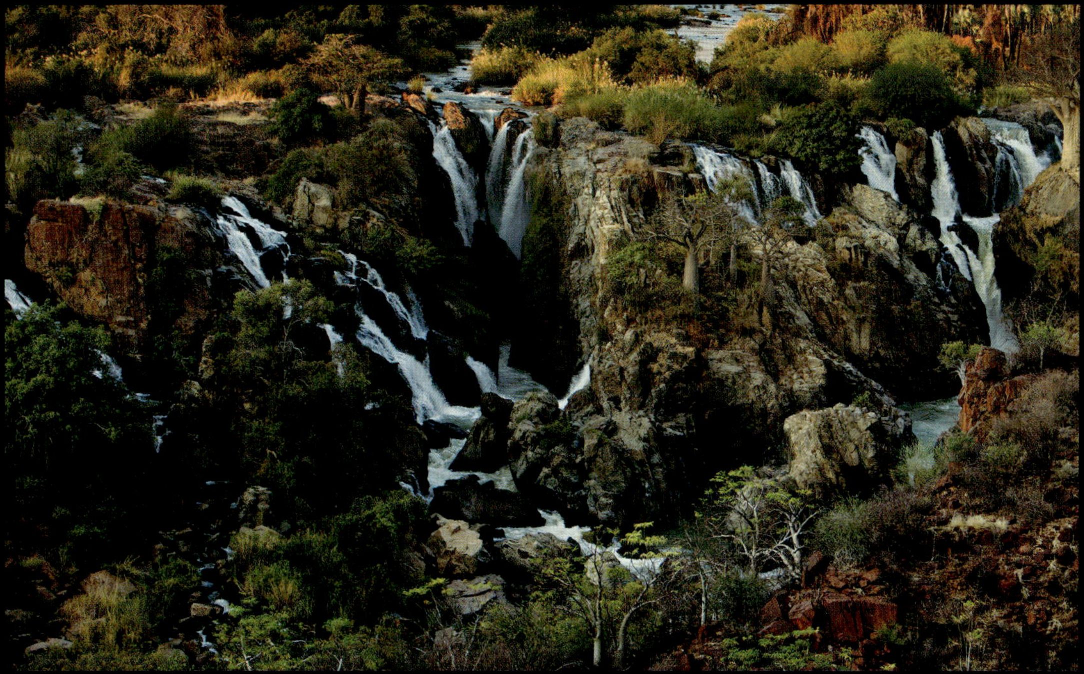

Epupa Falls in the Kunene River

National Parks and major Game Reserves in southern Africa

Transfrontier or Peace Parks in southern Africa

www.peaceparks.org

Established Peace Parks

|Ai-|Ais/Richtersveld; Kgalagadi; Greater Mapungubwe; Maloti-Drakensberg; Great Limpopo; Lubombo; Kavango/Zambezi; Lower-Zambezi-Mana Pools; Chimanimani

In process of development

- iSimangaliso Wetland Park to be included into the Greater Lubombo Transfrontier Conservation Area
- Nsubane Pongola Transfrontier Conservation Area between South Africa and Swaziland and later to be included into the Greater Lubombo Transfrontier Conservation Area
- Songimvelo-Malolotja Transfrontier Conservation Area also between Swaziland and South Africa and later to be included into the Greater Lubombo Transfrontier Conservation Area
- Usuthu-Tembe-Futi Transfrontier Conservation Area, incorporating Usuthu Gorge in Swaziland, a joint venture between Swaziland, Mozambique and South Africa and later to be included into the Greater Lubombo Transfrontier Conservation Area

National Parks

www.sanparks.org

Established National Parks

Addo Elephant; Agulhas; Augrabies; Bontebok; Camdeboo; Garden Route (Knysna, Tsitsikamma, Wilderness); Golden Gate Highlands; Karoo; Kgalagadi Transfrontier Park (shared with Botswana); Kruger; Mapungubwe; Marakele; Mokala; Mountain Zebra; Namaqua; Table Mountain; Tankwa; West Coast; |Ai-|Ais/Richtersveld Transfrontier Park (shared with Namibia)

Mega Parks in process of development

- Garden Route National Park (Knysna, Tsitsikamma, Wilderness)
- Greater Addo Elephant National Park (Addo, Woody Cape, Bird and St Croix Island)
- Baviaanskloof Mega Reserve (Baviaanskloof Reserve, Baviaanskloof Provincial Nature Reserve, Baviaanskloof Wilderness area, Baviaanskloof Conservation Area, Guerna Wilderness Nature Reserve, Kouga Wilderness Nature Reserve, Kouga Private Nature Reserve, Berg Plaatz Provincial Reserve, Sepree River Private Nature Reserve, Beakosneck Private Nature Reserve, Berg Plaatz Provincial Nature Reserve)

Provincial Parks and Private Game Reserves

Limpopo

www.limpopotourism.info
www.golimpopo.com

Mpumalanga

www.mpumalanga.com
www.mpumalangatourism.com

KwaZulu-Natal

www.kznwildlife.com
www.zulu.org.za

Eastern Cape

www.ecotourism.co.za
www.ecparks.co.za

Western Cape

www.capenature.org.za

Northern Cape

www.northerncape.org.za

North West

www.tourismnorthwest.co.za

Gauteng

www.sa-venues.com/gauteng_tourist_info.htm

Free State

www.freestatetourism.gov.za
www.explorefreestate.co.za

Zimbabwe

National Parks

www.zimbabwetourism.co.zw
www.wildlifetourism.net

Established National Parks

Chimanimani; Chizarira; Gonarezhou; Hwange; Lake Kariba; Mana Pools; Matobo; Matusadona; Nyanga; Victoria Falls; Zambezi

Game Reserves or Safari Areas

www.zimbabwetourism.co.zw
www.wildlifetourism.net

Established Safari Areas

Charara; Chewore; Chipinge; Chirisa; Dande; Deka; Doma; Hurungwe; Kazuma Pan; Malapati; Matetsi; Sapi; Tuli; Umfurudzi

Botswana

National Parks

www.botswanatourism.org
www.mewt.gov.bw/DWNP
Reservations: dwnp@gov.bw

Established National Parks
Chobe; Makgadikgadi/Nxai Pans; Kgalagadi Transfrontier Park (shared with South Africa)

Game Reserves
www.botswanatourism.org
www.mewt.gov.bw/DWNP
Reservations: dwnp@gov.bw

Established Game Reserves
Central Kalahari; Khutse; Moremi

MAJOR NATIONAL PARKS AND GAME RESERVES OF SOUTHERN AFRICA

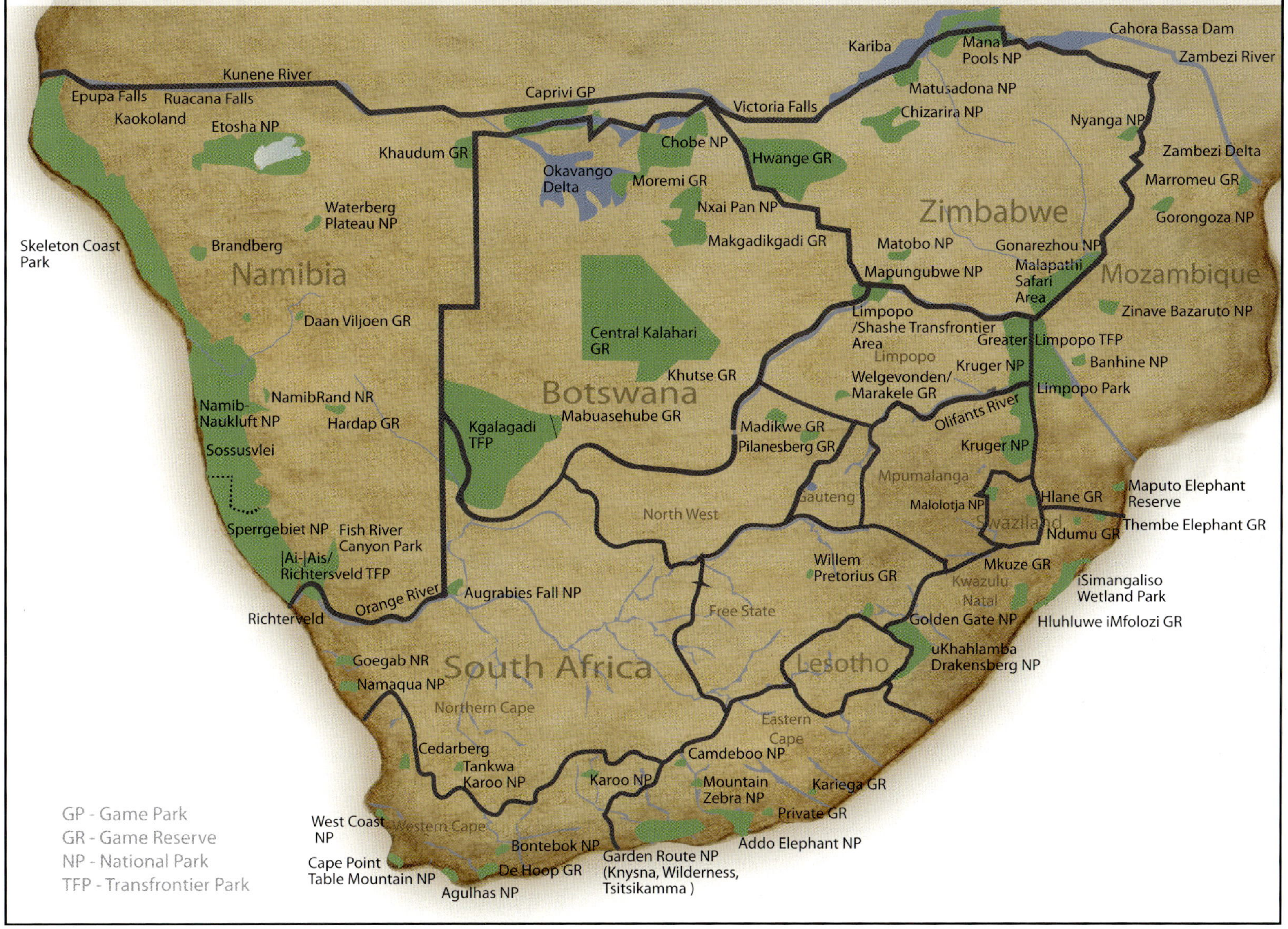

Namibia

National Parks
www.nmr.com.na
www.namibiatourism.com.na

Established National Parks
|Ai-|Ais/Richtersveld (shared with South Africa); Bwabwata; Etosha; Mudumu

Game Reserves, Parks and Private Game Reserves
www.nmr.com.na
www.namibiatourism.com.na

Established Reserves and Parks
Caprivi Game Park; Hardap Recreation Resort; Khaudum Game Park; Kongola; Naute Recreational Area; Namib Naukluft Park; Namib Rand Nature Reserve; National West Coast Tourist Recreational Area; Skeleton Coast Park; Tsao-bis Leopard Nature Park; Waterberg

Mozambique

National Parks
www.actf.gov.mz
www.gorongosa.net

Established National Parks
Banhine; Bazaruto; Gorongosa; Limpopo; Quirimbas; Zinave

Reserves
www.actf.gov.mz

Established Reserves and Wildlife Utilisation Areas
Gilè National Reserve; Maputo Special Reserve; Marromeu Buffalo Reserve; Niassa National Reserve; Pomene National Reserve; Chimanimani National Reserve; Futi Corridor; Maputo Protection Area; Coutada 4 – 15 in Manica and Sofala

Swaziland

National Parks
www.biggameparks.org

Established National Park
Hlane Royal National Park

Game Parks
www.biggameparks.org

Established Game Reserve
Mlilwane Wildlife Sanctuary; Mkaya Game Reserve

Lesotho

National Game Reserve and Conservation Area
www.lhda.org.ls
Bokong Nature Reserve; Maloti-Drakensberg Transfrontier Conservation Area

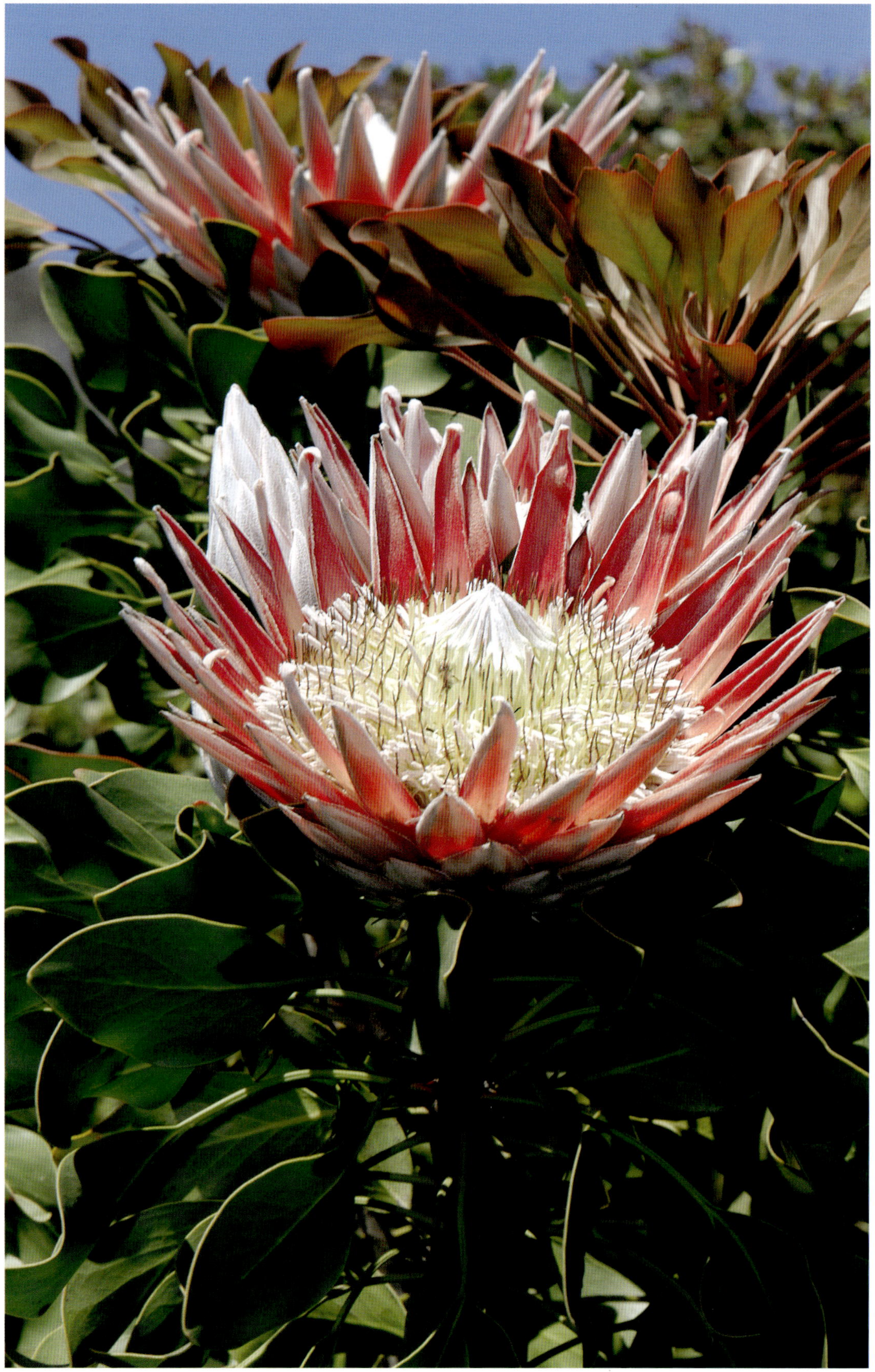

Selected bibliography

Cowling R, & Richardson D. 1995. *Fynbos: South Africa's Unique Floral Kingdom*. Cape Town, South Africa: Fernwood Press.

Hall-Martin, Anthony, & Carruthers, Jane (Eds). 2003. *South African National Parks: A Celebration.* Auckland Park, South Africa: Horst Klemm.

Germishuizen, Gerrit. 1997. *Wild Flowers of Northern South Africa*. Cape Town, South Africa: Fernwood Press.

Hockey PAR, Dean WRJ, & Ryan PG. 2005. *Roberts Birds of Southern Africa.* (Seventh Edition). Cape Town, South Africa: Trustees of the John Voelcker Bird Book Fund.

Lovegrove, Barry. 1993. *The Living Deserts of Southern Africa*. Vlaeberg, South Africa: Fernwood Press.

Manning, John. 2004. *Southern African Wild Flowers: Jewels of the Veld*. Cape Town, South Africa: Fernwood Press.

Manning, John. 2009. *Field Guide to Flowers of South Africa.* Cape Town, South Africa: Struik.

Nussey, Wilf. 1993. *The Crowded Desert: The Kalahari Gemsbok National Park*. Rivonia, South Africa: William Waterman.

Pooley, Elsa. 1998. *Field Guide to Wild Flowers KwaZulu Natal and the Eastern Region*. Cape Town, South Africa: Struik.

Skinner, John D, & Chimimba, Christian T. 2005. *The Mammals of the Southern African Subregion*. Pretoria, South Africa: University of Pretoria.

Van Wyk, Braam, & Malan, Sasa. 1988. *Field Guide to the Wild Flowers of the Witwatersrand and Pretoria Regions.* Cape Town, South Africa: Struik.

Van Wyk, Piet. 1984. *Veldgids tot die Bome van die Nasionale Krugerwildtuin*. Cape Town, South Africa: Struik.

ISBN: 978-0-620-53815-2
First edition
Published by **HPH Publishing**
PO Box 13244, Cascades, 3202
info@hphpublishing.co.za
www.hphpublishing.co.za
Tel: +27 86 171 0327
First published in 2012

Text by Ingrid van den Berg
Photography by Heinrich & Dana van den Berg, Philip & Ingrid van den Berg, Willem & Sandra von Willich
Edited by John Deane and Gillian Paizes
Proofread by Gillian Paizes
Design, typesetting and reproduction by Heinrich van den Berg
Printed and bound in Malaysia by Tien Wah Press (Pte.) Ltd